THE TRUTH OF
THE ORIGIN OF THE UNIVERSE
THE ORIGIN OF LIFE IN THE UNIVERSE - THE BLACK HOLE AND GOD

VOLUME-2

THE TRUTH OF
THE ORIGIN OF THE UNIVERSE
THE ORIGIN OF LIFE IN THE UNIVERSE - THE BLACK HOLE AND GOD

Dr. Sabrie Soloman
Ph.D., Sc.D., MBA, PE

KHANNA BOOK PUBLISHING CO. (P) LTD.
Publisher of Engineering and Computer Books
4C/4344, Ansari Road, Darya Ganj, New Delhi-110002
Phone : 011-23244447-48   **Mobile:** +91-99109 09320
E-mail : contact@khannabooks.com
Website : www.khannabooks.com

ISBN: 978-93-5538-296-2

**THE TRUTH OF THE ORIGIN
OF THE UNIVERSE (VOLUME-2)**
by **Dr. Sabrie Soloman**

First Edition: Dec 2024

Published by:
Khanna Book Publishing Co. (P) Ltd.
CIN: U22110DL1998PTC095547

Visit us at: www.khannabooks.com
Write us at: contact@khannabooks.com

To view complete list of books,
please scan the QR Code:

DEDICATIONS

William Thomson – The voice of his words of wisdom echoed through the core of the inner chambers of my heart, engraving channels of living knowledge, which generated floods of insight to illuminate my life`s journey.

William Thomson`s pure and innocent soul has reached the gates of heavens to bring me knowledge of Eternity. His living loving example compelled me to follow even his shadow to touch Eternity, while my feet are still on Earth!

His knowledge imparted on my mind inspiring me to write the "**The Origin of The Universe**" book – distinct from most other science books, which composed by worldwide prominent scientists highlighting unjustified and unproven hypotheses. Through **William Thomson** the factual truth of the Universe is recognized and logically revealed to my modest intellect to compose "**The Truth of The Origin of The Universe**," Thus, I humbly dedicate this book to him.

Joseph Rondinelli – While the size of the universe may not be comprehended, so also the goodness of the heart of my friend Joseph Rondinelli may not be grasped. Joseph permitted me to touch the limitless boundaries of the goodness of his loving nature. His friendship with me permitted me to observe a living loving example to follow.

Joseph Rondinelli is bountifully living loving those whom he cherished. He brought heavens highest values to my humble living earth. While I was temporarily incapacitated, he was my feet, hands, and even eyes sitting for hours by my side, debating our intellectual capacities to understand life under the domain of our **Creator, The Origin of Our Universe**.

When my mind acted as an interrupted volcano spouting novel thoughts of **The Origin of The Universe**, he came to my dwelling to rescue me out of the dense-smoke of my scattered knowledge --suffocating my mind. He instantly changed my environment to a more tranquil atmosphere enabling me once again recollect my thoughts, helping me to **Create Order out of Chaos**. For this and much more I am forever, indebted.

FOREWORD

Shortly before Frances R. Havergal, at age 36, wrote the hymn, "Lord, Speak to Me" she wrote in a letter, "… if I am to write any good, a great deal of living must go to a very little writing." Before putting pen to paper, a great deal of living has gone into what you are about to read.

When Dr. Soloman asked me to write a forward to his book on the origin of the universe, I was humbled and honored because this book is of unusual importance by an author with an exceptional life of service and greatly respected by all who know him and his work. I pray that those who read his 2,800 pages, 32 chapters will find its message clearly stated, giving clarity to a subject that has too long been delt with on a surface and non-scientific level.

Understanding the origin of the universe is key to understanding everything else and developing a truly Biblical worldview. A person's view on this subject is a window by which we view the world. Dr. Soloman gives us a verifiable frame to evaluate whether his findings are flawed or something that is understandably sensible to merit serious consideration.

When we tell others our view on evolution is based on scientific evidence rather than solely on the book of Genesis, we have a much stronger argument. Dr. Soloman builds the case that evolution, with his accompanying philosophy, is identified with a worldview at such a level that readers must sincerely struggle with how the theory could go against the evidence.

This book can train a whole new generation to confront and survive the worldview challenges they encounter all their life. Those of you who read this book in its entirety will be able to survive in a post-modern culture without being brainwashed by prejudice, faulty thinking, and preconceived assumptions.

Dr. Soloman has made exciting reading for the ordinary person, but it is thoroughly researched by scholarship to present a deep understanding of the subject. The situation we often find ourselves in is flawed thinking and lazy scholarship. To that end, I invite you to read Dr. Soloman's book which will prove to not only be enjoyable reading, but one of the most prolific and insightful tomes you will ever read on the origins of the universe.

Dr. Peter W. Teague,
President Emeritus
Lancaster Bible College | Capital Seminary & Graduate School

PREFACE

The Universe Had a Beginning

The scientific consent 100 years ago was that the universe was everlasting. This idea started to unravel with the inferences of Albert Einstein's Theory of Relativity back in 1916, where his equations pointed to an expanding universe. Yet he didn't like that conclusion and so supplemented a constant to his equation that invalidated the expansion. Later, he admitted it had been the major mathematical blunder in his life.

Then in 1929 astronomer Edwin Hubble acknowledged he observed galaxies expanding outward, which meant they had been much closer together in the past. Einstein, fascinated, wanted to see the indication for himself and in 1931 he went to the Mt. Wilson Observatory in Los Angeles, California. Einstein examined through the telescope, observed the evidence and then concluded, "I now see *the necessity* of a beginning." This began a change in the scientific boldness toward the cosmos.

Years after, in 1965, two U.S. scientists distinguished the leftovers of the original burst of energy of the creation occurrence characteristically so-called the "Big Bang." They both won a Nobel Prize in physics. One of them, Arno Penzias, later professed, "The best data we have [about the beginning of the universe] are exactly what I would have predicted had I nothing to go on but the first five books of Moses, the Psalms and the Bible as a whole" ("Clues to Universe Origin Expected," *The New York Times*, March 12, 1978, p. 1).

With the indication at hand, what was written in Genesis 1:1 truly surprised many scientists by its precision: "In the beginning, God created the heavens and the earth." Here it says the universe of matter and energy *appeared at a certain point in time* and was all created by an Ultimate Creator who existed before all of this occurred. It was an enormous proof of God's existence, with no real substitute enlightenments for a universe that, according to modern physics, *appeared out of nothing.*

The Universe is Fine-Tuned for Life

Almost 50 years ago, in 1973, cosmologist Brandon Carter instituted that the independent constants or laws in physics have one exceedingly rare characteristic in common—*they are precisely the values needed to establish and sustain a universe capable of producing life.* This is another vast and virtually accepted proof for a universe that has been prudently designed.

Scientists have instituted some 30 constants or laws of physics that rule the universe. All are dissimilar to each other and yet are finely tuned to unbelievable proportions to make life possible. The indication points to "Someone" spending much too much time tuning all of these laws so they would work in unison.

Amazingly, the Bible exposed this truth long before any scientist revealed these facts. As Jeremiah 33:25 states, "But I, the Lord, have a covenant with day and night, and *I have made the laws* that *control* earth and sky" (*Good News Translation*).

The Origin of Life – The Genetic Code

Opposing to what many have been led to trust, scientists have no genuine explanation for how life arose.

Even the well-known atheist and evolutionist Richard Dawkins acknowledged concerning the appearance of life, "*Nobody knows how it happened*" (*Climbing Mount Improbable,* 1996, p. 282). Moreover, one of the pioneers of the DNA code, the atheist Francis Crick, determined, "An honest man, armed with all the knowledge available to us now, could only state that in some sense, *the origin of life* appears at the moment to be *almost a miracle,* so many are the environments which would have had to have been gratified to get it going" (*Life Itself: Its Origin and Nature,* 1981, p. 88).

In the past 60 years, biologists have found that life started with a massive amount of accurate information *already embedded* in the cell. The human genome unaided is a molecule with almost 3 billion genetic letters, all precisely well-organized to give instructions to the cell. Furthermore, scientists have never instituted inorganic matter to generate a coded system of information and the machinery to interpret it. From the most primitive cells to human beings, all have the same elementary operating system of mind-boggling intricacy, with codes, transmitters and receivers all working together.

Furthermore, the origin of life aenigma has a "chicken-and-egg question"—which came first, the chicken or the egg? In this case, to get life to transpire, you need *both* the complete genetic code and the proteins—the machine parts—that read the code and build new proteins. Deprived of the code, you can't shape proteins. And without proteins, you can't process the code. So how could both have risen at the same time?

Biological life Exquisitely Programmed – "Robotic Machines"

In order to comprehend what is fashioning inside a cell, a good illustration is picturing a large city swarming with life and movement.

Biochemist Michael Denton defines the cell this way: "To clutch the reality of life as it has been discovered by molecular biology, we must amplify a cell a billion times until it is twenty kilometers in diameter and looks like a giant airship large enough to cover a great city like London or New York. What we would then see would be an entity of unmatched intricacy and adaptive design . . .

"We would see around us, in every direction we looked, all sorts of robot-like machines. We would notice that the simplest of the functional components of the cell, the protein molecules, were astonishingly complex pieces of molecular machinery, each one consisting of about three thousand atoms arranged in highly organized 3-D spatial conformation.

"We would wonder even more as we watched the strangely purposeful activities of these weird molecular machines, particularly when we realized that, despite all our accumulated knowledge of physics and chemistry, the task of designing one such molecular machine—that is one single functional protein molecule—would be completely beyond our capacity at present" (*Evolution: A Theory in Crisis,* 1986, p. 329).

This is the reason biochemists have a hard time have faith in explaining that blind evolution can build such machinery—and obtain all the parts to work together from the start. Furthermore, to keep the human body operational, biologists calculate that "about *330 billion cells* are substituted *daily,* equivalent to about 1 percent of all our cells" (Mark Fischetti, "Our Bodies Replace Billions of Cells Every Day," *Scientific American,* April 1, 2021).

"We take life for granted," adds Douglas Ell, "because it is everywhere. Our planet is overrun by *biological machines.* There are at least 10 million different types (species) of machines; some estimate that tens of millions of other types (species) have not yet been discovered . . .

"Coordinated systems permit blue whales to dive thousands of feet below sea level without being crumpled and sing complex songs that travel across oceans. Other systems guide bees to do a dance that tells other bees where to find the best sources of pollen. There are systems for hiding, systems for fighting, systems for reproducing, systems for obtaining food, systems for communicating, and so on" (*Counting to God,* p. 110).

Such findings show that everything about life is programmed to the last detail and that virtually nothing has been left to chance. Does this superb design point to evolution or to God? The answer is clear.

The Earliest Evidence of Life

Though Darwin titled his book *On the Origin of Species by Means of Natural Selection,* he was never able to substantiate that assumption. Many people adopt that the theory of evolution, with its countless mutations and natural selection as the means of change, can account for the origin and development of all the living things on this planet.

Yet this is sleight of hand, since evolution can account for *micro*evolution, or changes *within* the species (such as dogs of varying sizes, shapes and colors), but not *macro*evolution, or changes from one kind of creature to another. Natural selection can tell you something about the *survival* of the species, but nothing about the *arrival* of the species. It undoubtedly cannot trace the origin of the approximately *10 million species* on earth. These are classified into some 33 main body types or phyla, such as sponges, worms, insects and mammals.

Darwin foretold that as more of the fossil record was revealed, it would display types of species gradually appearing, beginning with one or a few, and then multiplying from simple to more complex life forms. He wrote, "If numerous species . . . have really *started into life at once,* the fact would be *fatal* to the theory of evolution through natural selection" (*Origin of Species,* 1859, p. 305). Yet that is precisely what has been found—major body types appearing at what's considered the *beginning* of the fossil record rather than in deposits laid down later.

Scientists call this "the Cambrian Explosion," referring to major types of plants and animals suddenly appearing *fully formed* in that fossil layer. This is the *opposite* of what Darwin and evolutionists had claimed would be found—and they have no real explanation or answers. Of the 33 main body types, 23 of them (or 70%) appear at the recognized *beginning* stage of the fossil record.

The issue in question here, by analogy, would be as discovering together such diverse inventions as a washing machine, a refrigerator, a bicycle, a car and an airplane. While they do have some shared features, they have very discrete purposes and resolutions. Likewise, the major types of creatures found in the Cambrian layer, such as sponges, worms, trilobites and jawless fish, are very diverse, complex and appear suddenly, with no evidence of these main body types evolving from other creatures.

As paleontologist Niles Eldredge admitted: "If life had evolved into the wondrous profusion of creatures little by little, then there should be some fossiliferous record of those changes . . . But no one has found any evidence of such in-between creatures . . . All of the fossil evidence to date has failed to turn up any such missing links" (George Alexander, "Alternate Theory of Evolution Considered," *Los Angeles Times,* Nov. 19, 1978). Indeed, Darwin has been let down by the fossil record!

Earth Has "Just Right" Conditions to Sustain Life

In 1966, Carl Sagan presented the famous TV documentary series *Cosmos.* He believed in order to have life you just needed two conditions—a right kind of star and a planet at the right distance. This supposition showed to be totally baseless.

Now, more than 50 years later, scientists have come to the comprehension that *more than 200 conditions* have to be "just right" for life to exist and thrive. The probabilities are about 1 in $10^{2,685,000}$. The probability of that happening comes out at about **1 in $10^{2,685,000}$**, or 10 followed by **2,685,000** zeros. For comparison, the Universe only has 10^{80} atoms. The infographic finishes by letting you know that the probability of you existing as you is pretty much zero.

As author Eric Metaxas explains: "Today there are more than 200 known parameters necessary for a planet to support life—*every single one of which must be perfectly met, or the whole thing falls apart.* Without a massive planet like Jupiter nearby, whose gravity will draw away asteroids, a thousand times as many would hit Earth's surface. The odds against life in the universe are simply astonishing" ("Science Increasingly Makes the Case for God," *The Wall Street Journal,* Dec. 25, 2014).

The Bible tells us: "The Lord is God. He made the skies and the earth. He put the earth in its place. He did not want the earth to be empty when he made it. He created it *to be lived on.* I am the Lord. There is no other God" (*Isaiah 45:18, Easy-to-Read Version*).

The Universe Mathematically Designed

Amazingly, the universe has been discovered to be mathematically designed. It follows orderly laws that can be described in mathematical terms. Sir James Jeans, one of the great astronomers of the 20th century, remarked: "From the intrinsic evidence of his creation, the Great Architect of the Universe *now begins to appear as a pure mathematician . . .* The universe begins to look more like a great thought than like a great machine" (*The Mysterious Universe,* 1930, pp. 134, 137).

A substantial problem for evolutionists and atheists is this: *Evolution can't do math,* since it is founded on random variations and mutations, and math necessitates an intelligent agent who can first formulate a mathematical blueprint of laws before creating things so they will be logical. This is why the present cosmos can be traced back to mathematical rules.

As Einstein distinguished, "The most incomprehensible thing about the universe is that it is comprehensible." He meant that it could be understood in *mathematical* terms but that an explanation for that was outside math.

As far back as the early 1900s, scientists were discovering the laws that govern the subatomic realm, the tiny microcosm designated by quantum mechanics. It has very diverse rules than our macro world and appears to make room for such things as free will to arise.

Many scientists came to comprehend that *not all is determined by matter and energy.* Experiments illustrate that an observer can change a particle through means of observing it. The consequences are that we can determine the outcome of our lives by the choices we make.

It brings to mind what God said: "*Today I have given you a choice between life and death, success and disaster. I command you today to love the Lord your God. I command you to follow him and to obey his commands, laws, and rules. Then you will live . . .*" (*Deuteronomy 30:15-16, ERV*).

Science Points to The Existence of God

Cutting-edge science is continually showing more intricacy and deeper design, not only in the cosmos, but also in all living things.

The biblical patriarch Job once challenged skeptics to look at the design of the creatures around them and notice they witness to a Supreme Designer and Creator. He stated: "Even birds and animals have much they could *teach* you; ask the creatures of earth and sea for their wisdom. All of them know that *the Lord's hand made them*" (*Job 12:7-9, GNT*).

Therefore, by exploring all the evidence and grasping where it leads, we hope you will believe in God, the Creator and the "Origin of the Universe." K*eep* believing in Him, and earnestly seek His will for your life!

Sabrie Soloman

ACKNOWLEDGMENTS

The Author is extremely grateful for the opportunity to express his heartfelt acknowledgments to the individuals who have played a vital role in the successful publication of "The Truth of The Origin of The Universe" book by Khanna Book Publishing Company.

First and foremost, the author extends his deepest gratitude to the Director and Editor-in-Chief, Mr. Mukul Seth, for his unwavering dedication and expertise in shaping the book's narrative. His keen eye for detail and impeccable editing skills have significantly enriched the content and ensured its quality.

Also, the author expresses his sincere appreciation to the Editorial Team, comprising of Mr. Mukesh Sharma and Mr. Yogesh Kumar. Their valuable feedback, insightful suggestions, and meticulous attention to detail have greatly enhanced the overall coherence and readability of the manuscript.

A special mention goes out to the Marketing Team, led by Mr. Rakshit Khanna for his tireless efforts in promoting the book and reaching a wider audience. Their innovative marketing strategies and relentless pursuit of excellence have played a crucial role in generating interest and enthusiasm among readers.

Furthermore, I am truly grateful for the dedication and hard work of the Sales Team, consisting of Mr. Nand Lal, Mr. Surinder Kumar, and Mr. Roshan Lal. Their exceptional sales acumen and customer-centric approach have boosted sales and fostered positive relationships with clients and partners.

The author would also like to extend his heartfelt thanks to Creation Ministries International for its invaluable contributions to the research and development of the book. Their pioneering work in the field of creation science has laid the foundation for a deeper understanding of the origins of the universe.

Sabrie Soloman

ABOUT THE AUTHOR

Dr. Sabrie Soloman, Ph.D., Sc.D., MBA, PE – He is the Chairman & CEO of American SensoRx, Inc., USA; Founder of Advanced Manufacturing Technology Post Graduate Studies at Columbia University, NY, USA; Professor of Advanced Technology at Columbia University. Dr. Soloman authored numbers of technical books published and translated worldwide: Sensors Handbook (2 editions), Sensors and Control Systems in Manufacturing (2 editions); Affordable Automation; Introduction to Electromechanical Engineering; Modern Welding Technology; 3D Printing Technology; 3D Bioprinting Technology & Design to name a few. Dr. Soloman holds numerous Patents, Technical Awards, and several US Product Registrations. Dr. Soloman is considered an international authority on advanced manufacturing technology, robotics, biomedical engineering, pharmaceuticals, and automation in the microelectronic, automotive, beef, pork, poultry industries. He has been and continues to be instrumental in developing and implementing several industrial and modernization programs through the United Nations to Europe, Asia, and African Countries. He is the first to introduce and implement unmanned flexible synchronous/asynchronous manufacturing systems in the microelectronic and the meat industries, and the first to incorporate advanced vision technology in wide array of robot/micro-robot manipulators. Dr. Soloman was selected to deliver the US presidential closing address "Innovative Remote Sensors Technology," at the Universal Design Conference," New York, USA. Dr. Soloman was the President of the International Christian Union at New Castle-Upon-Tyne University. He debated the "Origin of the Universe" before numerous attendees presenting his case against intellectuals and proponents of Big Bang's Singularity in Great Britain.

CONTENTS

CHAPTER 8: MOLECULAR INTELLIGENT DESIGN   57–100

CHAPTER 9: THE BLACKHOLE   101–142

CHAPTER 10: OUR MILKY WAY GALAXY — 143–192

CHAPTER 11: THE UNIVERSE IS FINELY TUNED FOR LIFE — 193246

CHAPTER 12 : DIVINE DESIGN   247-296

INTRODUCTION

A SOPHISTICATED GUIDE TO UNDERSTANDING INTELLIGENT DESIGN

In the realm of cosmic understanding, the concept of Intelligent Design stands as a testament to the exquisite complexity of our universe. This theory, intricate as the cosmic ballet it describes, proffers a perspective on the world that goes beyond the limits of what we can observe. This guide dives deep into the theories of Intelligent Design and Molecular Intelligent Design, presenting an enlightening exploration of the universe's origins.

Delving into the domain of cosmic comprehension, the theory of Intelligent Design exists as an homage to the exceptional intricacy that permeates our universe. This notion, as labyrinthine as the celestial choreography it details, provides a worldview that extends far beyond the boundaries of observable matter. Through the lens of this guide, we explore not only the concept of Intelligent Design but also delve into the equally complex idea of Molecular Intelligent Design, thereby unfolding an enlightening and exhaustive analysis of the cosmic origins that birthed our universe

An Overview of Intelligent Design

The paradigm of Intelligent Design is an intriguing one, suggesting that behind the intricate tapestry of our universe lies the careful hand of a purposeful architect. This theory, in its essence, seeks to offer an alternative to the randomness of evolution, asserting that the multifaceted intricacies observed in the cosmos and within life forms are beyond the scope of mere chance or the mechanics of natural selection. Instead, it proffers the hypothesis that the finesse and strategic nuances observed in our universe are suggestive of a meticulously orchestrated design.

The conceptual framework of Intelligent Design presents a captivating proposition. It conjectures that our universe, with its detailed and complex patterns, isn't simply a result of accidental factors but rather an intentionally organized project of a conscious designer. This perspective essentially contests the element of randomness linked to evolutionary processes. It emphasizes that the variety of complexity and details present in our cosmic setting and various life forms surpass the capacity of sheer luck or natural selection. The theory advances an alternative view - it proposes that the subtlety, strategic elements, and minute intricacies apparent in our universe imply a delicately crafted, meticulously planned structure, carefully designed with strategic intent, thus insinuating a purposeful architect at play.

Intelligent Design vs. Evolution

The dialogue between the philosophies of Intelligent Design and Evolution has captured the imaginations of many, fostering rich and heated discourse. Evolution, a concept widely accepted in the scientific atheist community, postulates that the universe and life within it were birthed and molded over vast expanses of time through the arbitrary process of natural selection. On the contrary, the school of Intelligent Design advocates for an alternative and realistic perspective, asserting the universe and its inhabitants are the intentional work of an Eternal master designer. The crux of their argument? The universe and its living entities exude such meticulous precision and profound complexity that attributing their origins to pure chance seems inadequate, if not improbable. Instead, they assert, it appears far more plausible to credit their existence to an Eternal intelligent cause, a celestial architect if you will. Nonetheless, this ongoing debate, while intellectually stimulating, invites us to probe deeper into the nature of our existence and the origin of the universe.

The exchange of ideas between the philosophical tenets of Intelligent Design and the scientific theories of Evolution has been a captivating point of debate, generating stimulating discussions brimming with fervor. Evolution, which is largely embraced by scholars within the atheist scientific realm, puts forward the hypothesis that the birth and evolution of our universe, and the life within, spanned across immense durations of time, facilitated by the non-deterministic mechanisms of natural selection. Meanwhile, the proponents of Intelligent Design bring a contrasting viewpoint to the table. They argue fervently that the universe and all the lifeforms that inhabit it, aren't merely a product of random chance, but are the deliberate creations of a superior entity. Why do they believe this? The complexity, meticulousness, and precision are visible in the universe and its living organisms appear too sophisticated and profound to be simply credited to a haphazard game of chance. Instead, they contend, a far more believable theory is to attribute the origins of the universe and life to the actions of an intelligent agency - an extraordinary cosmic engineer, so to speak. However, this fascinating exchange of ideas isn't just an intellectual exercise, but also a trigger for deeper reflections about the nature of our existence and the universe we inhabit.

The Concept of Molecular Intelligent Design

Venturing into the microcosm of life, Molecular Intelligent Design offers a fascinating lens through which to observe and understand the world and the origin of life. This facet of the Intelligent Design theory illuminates the intricacies that lie at the heart of all living entities, suggesting a grand design behind even the tiniest of biological mechanisms. The central idea revolves around the proposition that even the minute constituents of organisms exude an intricacy that seems too sophisticated to be a product of random evolution.

Molecular Intelligent Design seeks to uncover evidence of a master designer within the very building blocks of life. By examining the complex mechanisms and structures at a molecular level, it advocates that these are the products of intentional design rather than haphazard genetic mutation and selection. The exactness of molecular structures and processes, from the precision of protein folding to the efficiency of cellular function, is put forth as testimony of a grand, orchestrated blueprint.

This microscopic exploration emphasizes the profound complexity nestled within life forms, asserting that such extraordinary details suggest the guiding hand of an intelligent designer. Yet, it's important to understand that this theory is not without its detractors. Critics argue that attributing these complexities to a divine influence rather than the natural processes of evolution undermines the scientific method.

Nevertheless, the concept of Molecular Intelligent Design continues to intrigue and challenge our understanding of life's origins. As we delve deeper into the world of biomolecular mechanisms, we are left to ponder whether the precision we observe is merely a product of natural selection, or if it could indeed hint towards a cosmic architect's master plan. As such, Molecular Intelligent Design invites us to question, explore, and marvel at the complexity of life on a minute scale, while simultaneously considering its implications on a universal level.

Taking a step into the microscopic world of life, the field of Molecular Intelligent Design offers a mesmerizing view into the intricacies of biological organisms. This sub-discipline of the Intelligent Design theory emphasizes the marvel of complexity found even in the smallest life forms, pointing towards the existence of an intricate grand design that seems to be governing the mechanism of these biological structures.

The crux of Molecular Intelligent Design theory lies in the contention that the minutest elements of organisms - their very cells and molecules - radiate an eloquence and complexity in their structures and functions that seems implausible to have resulted from a sequence of random evolutionary events. It asserts that the exquisite and meticulous architecture of these molecular structures could hardly be the product of chance, but rather indicative of an intelligent architect.

Advocates of Molecular Intelligent Design

As advocates of Molecular Intelligent Design inspect the fundamental units of life, they unearth indications of a master architect hidden in these microcosmic structures. They put forth the idea that the remarkable cellular structures and mechanisms, such as the process of protein folding, cellular metabolism, genetic transcription, and molecular signaling, are likely the result of intentional, meticulous design instead of haphazard genetic mutations and selections guided by natural selection.

Darwinian Model of Evolution and Natural Selection

Their focus is on the exactitude and precision of molecular mechanisms - from the systematic orchestration of cellular processes to the incredible efficiency of various metabolic pathways - as tangible proof of a grand, orchestrated design. By illuminating these miniature yet elaborate systems of life, they provide a platform to question the currently prevailing Darwinian model of evolution and natural selection.

However, despite providing an awe-inspiring perspective on life, this theory also attracts substantial criticism. Critics assert that attributing the remarkable complexity of life to a divine or superior intellect as opposed to the natural processes of evolution is an unscientific approach that undercuts the methodical inquiry characteristic of the scientific method.

Yet, Molecular Intelligent Design persistently inspires and instigates us to reconsider our understanding of life and its origins. By examining the deep mysteries hidden within the biomolecular processes, we are compelled to reflect on the precision observed in these microcosms of life. It prompts us to consider whether this precise execution is simply the result of unguided evolution or whether it indicates a grander plan laid out by an intelligent cosmic architect. Therefore, the discipline of Molecular Intelligent Design encourages a deeper inquiry and appreciation for life's complexities at a microscopic level, while concurrently contemplating its wider implications at a universal scale.

The Impact of Intelligent Design on Faith

Theories like Intelligent Design can wield an immense influence on personal faith, serving as a framework that resonates deeply with spiritual seekers. It stands as an affirmation of a universe born out of purpose and intentionality, carefully crafted by an entity of divine proportions. This convergence of faith and science can be a comforting harbor for those wrestling with existential questions, seamlessly bridging the gap between spiritual beliefs and the complex realities of the cosmos.

Intelligent Design could also act as a beacon, illuminating a path for spiritual seekers looking for signs of the divine in the physical world. It provides a narrative that validates faith, translating the abstract concept of a divine creator into a tangible framework that aligns with the observable intricacies of the universe. Such validation could potentially strengthen the bond between believers and their faith, fostering a deeper understanding of their spiritual journey.

Importantly, Intelligent Design doesn't claim to solve all the mysteries inherent in faith and spirituality. However, it provides a unique prism through which the enigmatic elements of existence can be viewed, harmonizing spiritual beliefs with a scientific perspective. This alignment serves as a reminder of the timeless connection between faith and the quest for knowledge, reinforcing the significance of spiritual perspectives in our journey to understand the universe and our place within it.

Intelligent Design Resonates with Spiritual Seekers

While faith is, by nature, a deeply personal and individual journey, theories like Intelligent Design can serve as guiding stars, offering frameworks that resonate with spiritual seekers. By providing a scientifically-aligned narrative that complements spiritual beliefs, Intelligent Design can potentially enhance one's understanding and appreciation of both the physical and metaphysical dimensions of existence.

Seeking Expert Guidance on Intelligent Design

Navigating the complexities of theories such as Intelligent Design may initially appear daunting. Yet, it's not a journey one has to undertake alone. The academic world is rich with scholars and scientists who have dedicated their intellectual pursuits to unraveling these cosmic concepts. Their collective wisdom serves as a beacon of knowledge, illuminating the path for those with a curiosity to explore these sophisticated perspectives on the origins of the universe.

Intelligent Design, with its intricate propositions and profound implications, certainly presents a fascinating intellectual pursuit. However, understanding its nuances requires a patient and open mind, often best fostered under the guidance of seasoned experts. Their insights can provide you with a firm foundation of understanding, allowing you to confidently engage with the theories and debates that surround this intriguing perspective.

It's important to remember that although Intelligent Design offers a compelling viewpoint, it doesn't purport to solve all of life's enigmas. It presents an alternative perspective, one that invites contemplation, dialogue, and even skepticism. However, the beauty of this exploration lies not in the pursuit of definitive answers, but in the stimulation of thought, the provocation of questions, and the expansion of one's intellectual horizon.

Whether you're on a spiritual quest seeking validation or merely intrigued by the alternative explanations of our universe's intricate design, immersing yourself in the study of Intelligent Design can offer fresh and enlightening perspectives. As you delve deeper into this cosmic ballet, guided by the expertise of seasoned scholars, you might just find that this intellectual journey enriches your understanding of existence, and perhaps even brings you a step closer to the elusive harmony between faith and science.

Sabrie Soloman

INTELLIGENT DESIGN IN EVERYDAY REASONING

INTELLIGENT DESIGN

Intelligent Design (ID), argument planned to reveal that living organisms were created in more or less their current shape by an "intelligent designer."

Intelligent design was articulated in the 1990s, mainly in the United States, as a clear repudiation of the theory of biological evolution progressed by Charles Darwin (1809–82). Constructing on a version of the argument from design for the existence of God initiated by the Anglican clergyman William Paley (1743–1805), followers of intelligent design envisaged that the functional parts and systems of living organisms are "irreducibly complex," in the logic that none of their constituent parts can be detached deprive of causing the entire system to terminate functioning. From this premise, they conditioned that no such system could have originated about through the incremental alteration of functioning precursor systems by method of random mutation and natural selection, as the typical evolutionary account upholds; as a substitute, living organisms must have been created all at once by an intelligent designer. In Darwin's Black Box: The Biochemical Challenge to Evolution (1996), the American molecular biologist Michael Behe, the principal scientific representative for intelligent design, presented three main illustrations of irreducibly complex systems that supposedly cannot be explicated by natural means:

1. the bacterial flagellum, employed for locomotion, Figure 7.1,

2. the cascade of molecular reactions that occur in blood clotting, Figure 7.2, or coagulation, and

3. the immune system.

Fig.7.1: *Flagellum Protein Motor*

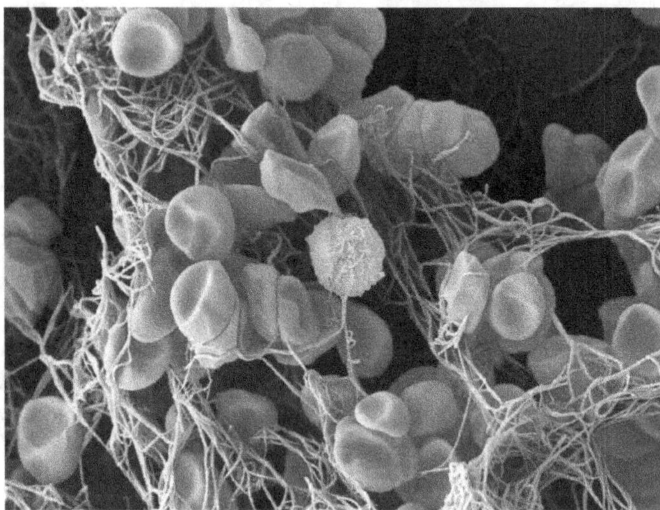

Fig.7.2: *Functions of Blood: Clotting*

Intelligent design was extensively professed as being associated with scientific *creationism*, the idea that scientific realities can be presented in sustenance of the *divine* creation of the countless forms of life. Followers of intelligent design upheld, nevertheless, that they acquired no position on creation and were indifferent with biblical literalism. Subsequently, they did not challenge the predominant scientific interpretation on the age of Earth, nor did they counteract the incidence of small evolutionary changes, which are sufficiently observed and superficially work by natural selection. Comparable to prior advocates of creationism, they inscribed decrees or originated lawsuits designed to allow the teaching of their interpretation as an *alternative* to evolution in American public schools, where teaching in any form of *religion* is constitutionally prohibited. In the main case on the issue, *Kitzmiller* v. *Dover Area School District* (2005), regarding a school district in Dover, Pennsylvania, a federal court ruled that intelligent design was not plainly separate from creationism and consequently, should be omitted from the curriculum on the foundation of earlier decisions, notably *McLean* v. *Arkansas* (1982).

Adversaries of intelligent design contended that it rests on an essential misinterpretation of natural selection and that it disregards the existence of ancestral systems in the evolutionary history of plentiful organisms. Some distinguished that the disagreement had been contested by Darwin himself in direct response to Paley. Beginning in the 1990s, conceptual progressiveness in *molecular biology* shed additional light on how irreducible complexity can be accomplished by natural means. Evolutionary biologists projected numerous methods to clarify Behe's three examples of complexity, including:

1. the self-organizing nature of biochemical systems,
2. the built-in redundancy of complex organic structures (if one crucial step is absent, other processes can achieve the same result), and
3. the role of versatile exploratory processes that, in the course of their normal physiological functioning, can help give rise to useful new structures of the body.

Meanwhile, intelligent design look as if powerless of creating a scientific research program, which unavoidably widened the gap between it and the recognized norms of science.

CREATIONISM

Fig.7.3: Adam and Eve

Creationism, the faith that the universe and the numerous forms of life were created by God out of nothing (*ex nihilo*). Even if the idea of God as creator is as ancient as *religion* itself, contemporary creationism is basically an answer to evolutionary theory, which can explicate the *diversity* of life without alternative to the doctrine of God or any other divine power. It may also reject the big-bang model of the appearance of the universe. Mainstream scientists commonly discard creationism. Creationism is principally accompanying with *conservative* branches of Protestant Christianity, nevertheless, there are proponents of Jewish and Islamic creationism as well as supporters in other branches of Christianity.

Biblical, or young-Earth, creationists have faith in the story told in Genesis of God's six-day creation of all things is exactly correct and that Earth is only a several thousand years old, as *extrapolated* from the biblical genealogies that start with Adam, the first *man*.

Others, such as old-Earth creationists, have faith in a creator made all that exists, but they may not hold that the Genesis story is an exact history of that creation. These creationists frequently accept *fossils* and other geological evidence for the age of Earth as factual and may or may not hold that God employed the big bang in the creation of the universe. Both types of creationists, though, have faith in changes in organisms may encompass changes within a species (often understood as the "kind" mentioned in Genesis ١:٢٤) or descending changes such as negative mutations, but they do not have faith in any of these changes can lead to the evolution of an inferior or simpler species into an advanced or more-complex species. Therefore, the theory of biological evolution is disputed by all creationists, Figure 7.3.

Fig.7.4: Charles Darwin cartoon - Man Is but a Worm Anti-Evolution Book Sale

Creationism turns out to be the object of attention between conservative religious groups subsequent the publication in 1859 of *On the Origin of Species* by Charles Darwin (1782–1809), the foremost organized statement of evolutionary theory, Figure 7.4. Within two decades most of the scientific community had recognized some form of evolution, and majority of churches in the end followed suit. In the early 20th century, some state administrations in the United States barred the teaching of evolution on the grounds that it denies the biblical creation story, which they well-thought-out a revealed truth. The consequence was the well-known Scopes Trial (the so-called "Monkey Trial") of 1925, in which a high-school teacher, John T. Scopes, was found guilty of unlawfully teaching the theory of evolution (he was later acquitted on a technicality). Creationism has basically been promulgated by conservative Protestant Christians.

Fig.7.5: William Paley, (1743–1805)

In 1950 Pope Pius XII released an encyclical letter confirmatory that there is no intrinsic struggle between the theory of evolution and the teachings of the Roman Catholic Church, on condition that Catholics still have faith in that humans are endowed with a soul created by God. In 1996 Pope John Paul II lengthened and reiterated the church's position, upholding evolution as "more than a hypothesis."

Brian Greene asks Richard Dawkins: Does God exist?

Commencement in the late 20th century, many creationists supported a view known as intelligent design. This view, which appealed to draw from modern science, was a modern explanation of the argument from design for the existence of God as set forth by the Anglican clergyman William Paley (1805–1743), Figure 7.5. Intelligent design is not acknowledged by all creationists, nevertheless, since many of its advocates leave open the distinctiveness and attributes of the "intelligent designer" of the universe, rather than equating it with the God of the Hebrew Bible and the New Testament. Today most creationists in the United States favor the elimination of evolution from the public-school curriculum or at least the teaching of creationism side by side evolution as what they consider an equally legitimate scientific theory.

SHIFTED POSITION

Various people today—both Christians and non-Christians—undertake that "creation" and "evolution" are opposing beliefs: That a person either have faith in a creator God, or they have faith in a natural process like evolution. I understand this perspective as I been acquainted with many colleagues adopting these two views. Some of them still believe that God is the creator of all things, but their perspective on evolution has shifted. they now see evolution as merely a way of understanding God's creation, through the lens of science.

How did they change their mind about evolution? They claim that first step came across faithful, trustworthy Christians who didn't fit into the boxes of "creationist" or "evolutionist." These Christians allegedly showed them a deeper and better manner to resolve their faith with the discoveries of science. They helped them to comprehend the scientific indication behind evolution, and allegedly to more faithfully understand the biblical account of creation. The list of these helping hand Christians were also influenced by Christians includes Francis Collins, Tim Keller, John Walton, and N.T. Wright, among many others.

Unfortunately, any one believes that the All-Power Omnipotent Creator needs assistance from a created being to form the universe and sustain life is in fact reducing the capability of God in performing the act of creation. Man's scientific intelligence has reduced the Infinitely intelligent God equating Him as a being needs to experimentally perfect His creation to Man, Mammals and Plants:

- In the beginning God created the heavens and the earth
- Then God said, "Let there be light," and there was light. And God saw that the light was good. Then he separated the light from the darkness. God called the light "day" and the darkness "night." ...]
- *And evening passed and morning came, marking the* **first day**. ...]
- Then God said, "Let the land sprout with vegetation—every sort of seed-bearing plant, and trees that grow seed-bearing fruit. These seeds will then produce the kinds of plants and trees from which they came." And that is what happened. The land produced vegetation—all sorts of seed-bearing plants, and trees with seed-bearing fruit. Their seeds produced plants and trees of the ***same kind***. And God saw that it was good.
- *And evening passed and morning came, marking the* **third day**. ...]
- Then God said, "Let the earth produce every sort of animal, each producing offspring of the ***same kind***— livestock, small animals that scurry along the ground, and wild animals." And that is what happened. God made all sorts of wild animals, livestock, and small animals, each able to produce offspring of the ***same kind***. And God saw that it was good. ...]
- Then God said, "Let us make human beings in our image, to be like us. They will reign over the fish in the sea, the birds in the sky, the livestock, all the wild animals on the earth, and the small animals that scurry along the ground."
- So, God created human beings in his own image. In the image of God, he created them; male and female he created them.
- Then God looked over all he had made, and he saw that it was very good!
- And evening passed and morning came, marking **the sixth day**.

God created man, animals, marine life, and plants all according to its kind. Every type of life stayed in its own kind of specie. God did not have an open door for animals to advance to human beings. Many species may have adapted to suit their environment according to the province of God programming their DNA to be activated, when necessary, Figure 7.6. Nonetheless, not any specie ever crossed over its designated channel to become another specie.

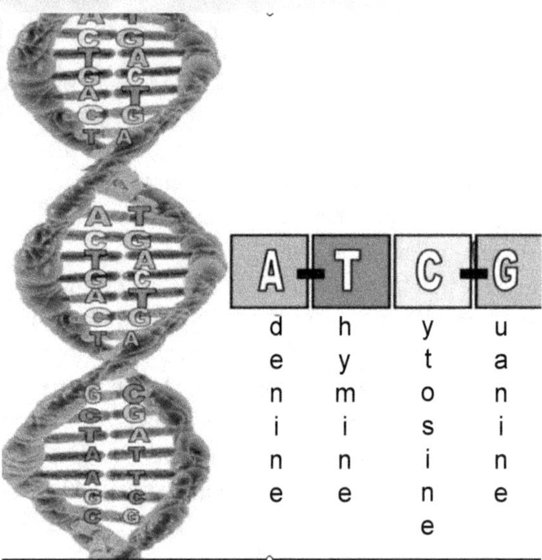

Fig.7.6: DNA - Each cell in your body contains 46 chromosomes with your complete set of genetic information on each strand. If you were able to completely uncoil a single strand of human DNA, it would be over 6 feet tall. But just how much genetic information is stored on one single strand? Let's find out. One strand of human DNA consists of approximately 3.1467 billion base pairs (a number just slightly less than the number of seconds in 100 years: 3.15576 billion seconds).

Man is the only creation that took the same image of God, however, in a much-limited form of body, soul and spirit. No other creature had such a privilege to be formed in the image of God, Figure 7.7. Even angels or archangels in heavens have not been given such privilege.

Fig.7.7: The Lord Formed the Man from the Dust of the Ground Genesis 2:7

Combining creation and evolution is like lighting a feeble candle to assist the scorching sun in midday! It is useless. God's creative power is infinite and *could* not be expressed through an evolutionary process. Scientific Evolutionary power is vastly limited, and vastly insignificant to assist the infinity in any conceivable way.

It is similar to a vast deep boundless ocean asking for a molecule of water to quench its thirst. It is illogical and contradictory in its intention, as the ocean cannot become thirty and the molecule of water, in any case is not needed. Creation is in no need to Evolution-science. God's Sacred Word and Wisdom do not need man's scientific-evolutionary thinking.

B.B. Warfield (1851-1921)

Theologian, key defender of the doctrine of Biblical inerrancy

Quotes from *B.B. Warfield, Figure 7.8*: 'I do not think that there is any general statement in the Bible or any part of the account of creation, either as given in Genesis 1 and 2 or elsewhere alluded to, that need be opposed to evolution. [...] There is no *necessary* antagonism of Christianity to evolution, *provided that* we do not hold to too extreme a form of evolution.'

As a faithful believer and a Professor of Science and Technology, this type of doctrine defender did not understand the Darwinian Evolution. I would have advised him to reconsider educating himself first before holding such title as a doctrine defender.

Fig.7.8: B.B. Warfield (1851-1921)

Karl Barth (1886-1968)

Theologian, prominent member of "Confessing Church" that opposed Hitler and Nazism

Quotes from Kar; Barth: 'The creation story deals only with the becoming of all things, and therefore with the revelation of God, which is inaccessible to science as such. The theory of evolution deals with that which has become, as it appears to human observation and research and as it invites human interpretation. Thus, one's attitude to the creation story and the theory of evolution can take the form of an either/or only if one shuts oneself off completely either from faith in God's revelation or from the mind (or opportunity) for scientific understanding, Figure 7.9.

Fig.7.9: Karl Barth (1886-1968)

Once again, this theologian understood little about the Theory of Evolution. Also, he did not fully grasp the true meaning of Colossians 1:16-17 "for through Him (Jesus Christ) God created everything in the heavenly realms and on earth. He made the things we cannot see—as such as thrones, kingdoms, rulers and authority in the unseen world. Everything was created through Him and for Him. He existed before anything else, and **holds** all creation together.

Indeed, God holds all creation together. He **holds** the 72.2 trillion cells in our adult body. He holds 2-3 trillion Galaxies in the realm of heaven. He sustains all species together, each according to its kind. This is Contrary to the Darwinian Evolutionary Theory.

Billy Graham (1918-2018)

Evangelist, pastor, and author

Fig.7.10: *Billy Graham (1918-2018)*

Quoted from Billy Graham: 'I don't think that there's any conflict at all between science today and the Scriptures. I think that we have misinterpreted the Scriptures many times and we've tried to make the Scriptures say things they weren't meant to say. I think that we have made a mistake by thinking the Bible is a scientific book. The Bible is not a book of science. The Bible is a book of Redemption, and of course I accept the Creation story, Figure 7.10. I believe that God did create the universe. I believe that God created man, and whether it came by an evolutionary process and at a certain point He took this person or being and made him a living soul or not, does not change the fact that God did create man. [...] whichever way God did it makes no difference as to what man is and man's relationship to God.'

I am totally in harmony with Billy Graham's statement that the Bible's main message is to redeem the falling man and honor the resurrected Jesus our savior and redeemer. However, Billy Graham statement seems a highly diplomatic response, however, naive. He appeared to refrain from entering into the arena of science and evolution to offer view or to offer a faithful review of the perception of an ungodly evolution of a beast to become a man in the Image of God. Nonetheless, The Bible is not an ignorant book of science. Actually, if science does not yield to the Word of God, science becomes degenerate science. If we ignored such matter, we may end up repeating the tragedy of Galil Galileo, (*Galileo* di Vincenzo Bonaiuti de' Galilei (15 February 1564 – 8 January 1642), who was an Italian astronomer, physicist and engineer).

If the Clergymen at the time of Galileo examined the Bible --"Isaiah 40:22" *'God sits above the* **circle of the earth**. *The people below seem like grasshoppers to him! He spreads out the heavens like a curtain and makes his tent from them,'*-- they would have discovered the Earth is a sphere, contrary to their belief. Similarly, the science of the "Theory of Evolution" is an excellent example of a "degenerate science."

C.S. Lewis (1898-1963)

Author, scholar, and apologist

Quoted from C.S. Lewis: 'We must sharply distinguish between Evolution as a biological theorem and popular Evolutionism or Developmentalism which is certainly a Myth. [...] To the biologist Evolution [...] covers more of the facts than any other hypothesis at present on the market and is therefore to be accepted unless, or until, some new supposal can be shown to cover, still more facts with even fewer assumptions, Figure 7.11.

Fig.7.11: *C.S. Lewis (1898-1963)*

For long centuries God perfected the animal form which was to become the vehicle of humanity and the image of Himself. He gave it hands whose thumb could be applied to each of the fingers, and jaws and teeth and throat capable of articulation, and a brain sufficiently complex to execute all the material motions whereby rational thought is incarnated. The creature may have existed for ages in this state before it became man: it may even have been clever enough to make things which a modern archaeologist would accept as proof of its humanity. But it was only an animal because all its physical and psychical processes were directed to purely material and natural ends. Then, in the fullness of time, God caused to descend upon this organism, both on its psychology and physiology, a new kind of consciousness which could say "I" and "me," which could look upon itself as an object, which knew God, which could make judgments of truth, beauty, and goodness, and which was so far above time that it could perceive time flowing past.'

C.S. Lewis was partially correct to make a distinction between, "Evolution as a biological theorem and popular Evolutionism or Developmentalism which is certainly a Myth." Nonetheless, he had elevated "Evolution … biological theorem" as an unproven theory. Therefore, it should not be considered "theory" as it still an unproven hypothesis. A theory such as Gravitational Theory and the Three Laws of Sir Isaac Newton, as well as Einstein Theory of Relativity, they are true tested and validated theories, unlike The Darwinian Evolution Theory or rather (Hypothesis).

Nonetheless, C.S. Lewis also said the following statement: *'For long centuries God perfected the animal form which was to become the vehicle of humanity and the image of Himself. He gave it hands whose thumb could be applied to each of the fingers, and jaws and teeth and throat capable of articulation, and a brain sufficiently complex to execute all the material motions whereby rational thought is incarnated. The creature may have existed for ages in this state before it became man'*

The above statement is shocking to say the least, in particularly when uttered by one of my favorite 20th century philosophers, C.S. Lewis. He said: *'For **long centuries** God **perfected** the animal form which was to become the vehicle of **humanity** and the image of Himself.'* –This statement shows that C.S. Lewis perceived the creator was engaged in exploring the possibility of developing a **human** form from an **animal** form. The Creator toiled for **centuries** to perfect a lower specie animal to become a higher specie human to take even the image of the creator Himself.

Again, C.S. Lewis did not grasp one of the vital attributes of God, which is the **absolute perfection**. God does not need time to perfect any of His creation. God formed Man from the dust of the Earth, not in centuries, not in years, not in months or weeks, but in a single day and by the His breath God breathed on Man instant life. Additionally, God is **outside time** as He is from everlasting to everlasting. He created time, especially for His creation, who are residing below the threshold of Heavens. God knows the end from the beginning, as He is omniscient. It is clearly stated in Genesis 1:26-28, 31

- Then God said, "Let us make human beings in our image, to be like us. They will reign over the fish in the sea, the birds in the sky, the livestock, all the wild animals on the earth, and the small animals that scurry along the ground."

- So, God created human beings in his own image. In the image of God, he created them; male and female he created them.

- Then God looked over all he had made, and he saw that *it was very good*!

- And evening passed and morning came, marking **the sixth day**

John Stott (1921–2011)

Evangelical leader, spokesperson, author, and apologist

'It is most unfortunate that some who debate this issue of evolution begin by assuming that the words "creation" and "evolution" are mutually exclusive. If everything has come into existence through evolution, they say, then biblical creation has been disproved, whereas if God has created all things, then evolution must be false. It is, rather, this naïve alternative which is false, Figure 7.12.'

In spite of John Stott's fame and eloquent speeches, in addition to his intellectual overwhelming power when he stood high behind his pulpit, where I have attended his sermons several times in Newcastle upon Tyne, England

Fig.7.12: John Stott (1921–2011)

– sadly, he is belittling the true understanding of "Creation," while he is giving credence to the hypothesis of "Evolution."

"Creation," is not a hypothesis or a theory. It is a fact of existence to all created lives on Earth and in the Universe. Before the act of creation, the Lord God had distinctively seen all His creation including all the necessary, and most likely, the essential changes (mutations). Thus, he made provisions that each specie to stay in line, according to its kind, as He intended it to be.

The potential of any change that may occur in any specie has already been pre-planned even before the creation of the universe, and before the creation of any form of life. Nonetheless, the change in any specie may take place only to suit the demand of the new environment. The internal potential power of the specie to adapt to the new environment is resident in the DNA, which secures that potential to change. Therefore, the Creator has embedded His unique program in the core of the DNA triggered only as demanded by the new environment. Evolution of the specie to jump its kind has no place in God's perfect plan.

Indeed, it is naïve of Stott to consider Evolution as an Act to be adopted as Creation!

Pope Benedict XVI (1927–)

Pope Emeritus of the Roman Catholic Church

We cannot say: creation or evolution, inasmuch as these two things respond to two different realities. The story of the dust of the earth and the breath of God [...] does not in fact explain how human persons come to be but rather what they are. It explains their inmost origin and casts light on the project that they are. And, vice versa, the theory of evolution seeks to understand and describe biological developments. But in so doing it cannot explain where the 'project' of human persons comes from, nor their inner origin, nor their particular nature. To that extent we are faced here with two complementary—rather than mutually exclusive—realities.

Fig.7.13: Pope Benedict XVI (1927–)

With all due respect to Pope Benedict, "Darwinian Evolution" is not another reality, while "Creation is the very core of reality." Equating Evolution as another reality puts the infinite Creator in the same footage as a human scientist attempting to find alternative to creation and even to the Creator Himself. If Pope Benedict had the time to study the depth of the Evolution hypothesis, he would have realized that it is simply collective observations combined with humanistic assumptions to form an imaginative conclusion aimed to substitute the infinite Creator with a disputed idea of a mere mortal. I humbly suggest Pope Benedict to examine Genesis one more time to discover that all created life forms came to existence in mere 6 days. The life forms were fully functioning, plants, animals, and marine life. While Man was created in the image of God Himself, mature adult with outstanding reasoning capabilities and unsurpassed memory.

I do not deny that you, as well as many people may be faced with "*Evolution*" and "*Creation*" as "*complementary*" entities—"*rather than mutually exclusive—realities.*" This sort of submissiveness, is not founded on any Biblical substance.

Isaiah 55:8-9: "For my thoughts are not your thoughts, neither are your ways my ways, saith the LORD. For as the heavens are higher than the earth, so are my ways higher than your ways, and my thoughts than your thoughts."

Genesis 1:30-31: And to every beast of the earth and every bird of the air and every creature that crawls upon the earth—everything that has the breath of life in it—I have given every green plant for food." And it was so. And God looked upon all that He had made, and indeed, it was very good. And there was evening, and there was morning— the sixth day.

Pope Francis (1936–)

Current Pope of the Roman Catholic Church

[God] had created beings and allowed them to develop according to the internal laws that he gave to each one, so that they were able to develop and to arrive at their fullness of being. He gave autonomy to the beings of the universe at the same time at which he assured them of his continuous presence, giving being to every reality. And so, creation continued for centuries and centuries, millennia and millennia, until it became what we know today, precisely because God is not a **demiurge or a magician**, but the creator who gives being to all things. […] The Big Bang, which nowadays is posited as the origin of the world, ***does not contradict the divine act of creating, but rather requires it. The evolution of nature does not contrast with the notion of creation, as evolution presupposes the creation of beings that evolve***, Figure 7.14.

Fig.7.14: *Pope Francis (1936–)*

It evident that Pope Francis' believe in "Evolution," "Big Bang" which is inconsistent with God's Word. "When we read in Genesis the account of creation, we run the risk of imagining that God was a magician with a magic wand able to do everything. But that is not so."

These words might sound as if they came from some controversial preacher. But they did not. They were spoken by Pope Francis to an audience from the Pontifical Academy of Science Oct. 27, according to Newsweek. In that speech, he officially affirmed that the Catholic Church believes in ***both evolution and the Big Bang theory***, although with the addition of God into the equation.

"*The Big Bang that today is considered to be the origin of the world does not contradict the creative intervention of God,*" Francis said to his audience. "*On the contrary, it requires it.*"

Merriam-Webster defines the Big Bang as, "the cosmic explosion that marked the beginning to the universe according to the big bang theory." Based on that definition, I am inclined to give that particular point to the pope. Genesis 1:1 (NIV) says, "God created the heavens and the earth." There is no explanation given of exactly how he did that. There is not a remote possibility, that God chose to create our universe through the rapid expansion of highly condensed matter. He spoke and it became fully matured.

Pope Francis then went into further detail about exactly how the universe was created. He said creation was not one single event, but that it "***went forward for centuries and centuries, millennia and millennia until it became what we know today.***"

That was the part where Francis violated Biblical teaching. His statement brings up two assertions, both of which I totally disagree with. The biggest issue, of course, is that of evolution. But hidden within that is another assertion — that creation itself took millenniums and millenniums of years instead of 6 days of 24 hours each.

The first chapter of Genesis: The Bible clearly states creation as taking six days. In the NIV, each set of verses describing a day ends with, "And there was evening, and there was morning — (First day .. Second day .. Third day .. etc.)." According to God and Science, a website dedicated to showing evidence for God, there is some debate among scholars as to the meaning of the particular Hebrew word used to mean "day" because it can occasionally mean a long period of time. However, the fact that each day is described as having an evening and a morning seems to me to indicate pretty strongly that they were, in fact, literal days.

There is another point that convinces me that indeed, each day was literal 24 hours. According to Genesis 1:11, plants were created on the third day. However, the sun, whose light plants need in order to survive, did not come into existence until day four. As the Institute for Creation Research points out, that would be fine so long as the days were actually literal. But if each day represented thousands of years, the plants would have died long before the sun came into being, let alone any other forms of life.

Putting all of that aside, the biggest issue I had with Pope Francis' speech was his assertion that evolution actually happened.

"Evolution in nature is not in contrast with the notion of divine creation because evolution requires the creation of the beings that evolve," Francis said.

He went on to say that God has indeed used evolution to create the forms of life we know today.

"(God) created beings and left them to develop according to the internal laws that he gave each one, so that they would reach their fulfillment," Pope Francis said.

This is an argument I have heard quite often in attempts to reconcile evolution and the Bible. Is it not possible that God simply created a handful of life forms and then guided evolution to do the rest? I believe the answer to be a resounding "No."

When Genesis 1 talks about the creation of the plants, the sea life, the birds and the land animals, the Bible uses a variation of the phrase "God created them … according to their kind" each time. That sounds to me as if each type of animal was created specifically. Rather than making a few birds, a few fish and a few land animals, I believe God carefully created each type of animal.

The pope said God is not a magician, and I agree with him. God is far more powerful than a simple charlatan performing parlor tricks on the sidewalk. God created the entire universe. Accordingly, why would a God with that much power create a prototype and then sit back and watch for thousands of years as it slowly turned into what he meant for it to be? Would he not simply jump straight to the final model? Certainly, he would.

There is one final factor that leads me to believe God did not use evolution to perfect life. Near the end of each set of creation verses, the Bible says, "And God saw that it was good." To me, that sounds like he was satisfied with his work. This is further evidenced by his resting on the seventh day. He was happy with his creation.

For any creative type individual, it is obvious how critical creatives are of their own work. Often, they slave over their work until they are satisfied.

God is the master creator. Would he not be the same way? Why would he paint the canvas of our universe, call it good and then sit back and watch the canvas morph into what he really wanted it to be? As the very standard of perfection, would he not pour his heart into his work until he was completely satisfied? True, this is no scientific argument. It is only conjecture from a human mind whose love of creating is dwarfed into nothingness compared to that of the creator. Nonetheless, somehow, one just cannot envision the being who the infinite creator Himself as being satisfied with anything less than exactly what He set out to make.

In the end, the debate is not really about science versus religion or creationism versus evolution. At the end, I believe it all comes down to this — **do we accept the Bible, word for word, as the true and inerrant word of God, regardless of what the world tells us?**

Fig.7.17: *Sacred Geometry*

That may be the most important question we ever ask ourselves. Do we believe it? There can be no middle ground, no riding of the fence. Either we believe or we do not. It is as simple as that.

What or Who is The Sacred - Manifestations

Sacred, the power, being, or realm understood by religious persons to be at the core of existence and to have a transformative effect on their lives and destinies. Other terms, such as *holy, divine, transcendent, ultimate being* (or *ultimate reality*), *mystery,* and *perfection* (or *purity*) have been used for this domain, Figure 7.17. "*Sacred*" is also an important technical term in the scholarly study and interpretation of religions.

For a discussion of dogmatic interpretations of the divine as a being or force .

THE CONCEPT OF SACRED

It was during the first quarter of the 20th century that the concept of the sacred became dominant in the comparative study of religions. _Nathan Söderblom (1866 - 1931), an eminent Swedish churchman and historian of religions, asserted in 1913 that the central notion of religion was "holiness" and that the distinction between sacred and profane was basic to all "real" religious life, Figure 7.18.

In 1917 Rudolf Otto's *Heilige* (Eng. trans., *The Idea of the Holy*, 1923) appeared and exercised a great influence on the study of religion through its description of *religious* man's experience of the "numinous" (a mysterious, majestic presence inspiring dread and fascination), which Otto, a German theologian and historian of religions, claimed, could not be derived from anything other than an a priori sacred reality. Other scholars who used the notion of sacred as an important interpretive term during this period included the sociologist Émile Durkheim in France, and the psychologist-philosopher Max Scheler in Germany. For Durkheim, sacredness referred to those things in society that were forbidden or set apart; and since these sacred things were set apart by society, the sacred force, he concluded, was society itself. In contrast to this understanding of the nature of the sacred, Scheler argued that the sacred (or infinite) was not limited to the experience of a finite object. While Scheler did not agree with Otto's claim that the holy is experienced

Fig.7.18: *Nathan Söderblom (1866 - 1931)*

through a radically different kind of awareness, he did agree with Otto that the awareness of the sacred is not simply the result of conditioning social and psychological forces. Though he criticized Friedrich Schleiermacher, Figure 7.19, an early 19th-century Protestant theologian, for being too subjective in his definition of religion as "the *consciousness* of being absolutely dependent on God," Otto was indebted to him in working out the idea of the holy. Söderblom recorded his dependence on the scholarship of the history of religions (*Religionswissenschaft*), which had been a growing discipline in European universities for about half a century; Durkheim had access to two decades of scholarship on nonliterate peoples, some of which was an account of actual fieldwork. Scheler combined the interests of an empirical scientist with a philosophical effort that followed in the tradition of 19th-century attempts to relate human experiences to the concept of a reality (essence) that underlies human thoughts and activities.

Since the first quarter of the 20th century many historians of religions have accepted the notion of the sacred and of sacred events, places, people, and acts as being central in religious life if not indeed the essential reality in religious life. For example, phenomenologists of religion such as Gerardus van der Leeuw and W. Brede Kristensen have considered the sacred (holy) as central and have organized the material in their systematic works around the (transcendent) object and (human) subject of sacred (cultic) activity, together with a consideration of the forms and symbols of the sacred. Such historians of religions as Friedrich Heiler and Gustav Mensching organized their material

Fig.7.19: *German Theologian and Philosopher (1768-1834) - Germany - Friedrich Schleiermacher*

according to the nature of the sacred, its forms and structural types. Significant contributions to the analysis and elaboration of the sacred have been made by Roger Caillois, a sociologist, and by Mircea Eliade, an eminent historian of religions.

BASIC CHARACTERISTICS OF THE SACRED

The term sacred has been used from a wide variety of perspectives and given varying descriptive and evaluative connotations by scholars seeking to interpret the materials provided by anthropology and the history of religions. In these different interpretations, however, common characteristics were recognized in the sacred, as it is understood by participant individuals and groups: it is separated from the common (profane) world; it expresses the ultimate total value and meaning of life; and it is the eternal reality, which is recognized to have been before it was known and to be known in a way different from that through which common things are known.

The term sacred comes from Latin *sacer* ("set off, restricted"). A person or thing was designated as sacred when it was unique or extraordinary. Closely related to *sacer* is *numen* ("mysterious power, god"). The term numinous is used at present as a description of the sacred to indicate its power, before which man trembles. Various terms from different traditions have been recognized as correlates of *sacer*: Greek *hagios*, Hebrew *qadosh*, Polynesian *tapu*, Arabic *ḥaram* (not to be confused with *ḥarām*, "forbidden"); correlates of *numen* include the Melanesian *mana*, the Sioux *wakanda*, old German *haminja* (luck), and Sanskrit *Brahman*.

Besides the dichotomy of sacred–profane the sacred includes basic *dichotomies* of pure–unpure and pollutant–"free." In ancient Rome the word *sacer* could mean that which would pollute someone or something that came into contact with it, as well as that which was restricted for divine use. Similarly, the Polynesian *tapu* ("*tabu*") designated something as not "free" for common use. It might be someone or something specially blessed because it was full of power, or it might be something accursed, as a corpse. Whatever was tabu had special restrictions around it, for it was full of extraordinary energy that could destroy anyone unprotected with special power himself. In this case the sacred is whatever is uncommon and may include both generating and polluting forces. On the other hand, there is the pure–impure *dichotomy*, in which the sacred is identified with the pure and the profane is identified with the impure. The pure state is that which produces health, vigor, luck, fortune, and long life. The impure state is that characterized by weakness, illness, misfortune, and death. To acquire purity means to enter the sacred realm, which could be done through purification rituals or through the fasting, continence, and meditation of *ascetic* life. When a person became pure, he entered the realm of the divine and left the profane, impure, decaying world. Such a transition was often marked by a ritual act of rebirth.

AMBIVALENCE IN MAN'S RESPONSE TO THE SACRED

Because the sacred contains notions both of a positive, creative power and a danger that requires stringent prohibitions, the common human reaction is both fear and fascination. Otto elaborated his understanding of the holy from this basic *ambiguity*. Only the sacred can fulfill man's deepest needs and hopes; thus, the reverence that man shows to the sacred is composed both of trust and terror. On the one hand, the sacred is the limit of human effort both in the sense of that which meets human frailty and that which prohibits human activity; on the other hand, it is the unlimited possibility that draws mankind beyond the limiting temporal–special structures that are constituents of human existence.

Not only is there an ambivalence in the individual's reaction to the numinous quality of the sacred but the restrictions, the tabus, can be expressive of the creative power of the sacred. Caillois has described at length the social mechanism of nonliterate societies, in which the group is divided into two complementary subgroups (moieties), and has interpreted the tabus and the necessary interrelationship of the moieties as expressions of sacredness. Whatever is sacred and restricted for one group is "free" for the other group. In a number of respects—*e.g.*, in supplying certain goods, food, and wives—each group is dependent on the other for elemental needs. Here the sacred is seen to be manifested in the order of the social–physical universe, in which these tribal members live. To disrupt this order, this natural harmony, would be sacrilege, and the culprit would be severely punished. In this understanding of the sacred, a person is, by nature, one of a pair; he is never complete as a single unit. Reality is experienced as one of prescribed relationships, some of these being vertical, hierarchical relationships and others being horizontal, corresponding relationships.

Another significant ambiguity is that the sacred *manifests* itself in concrete forms that are also profane. The transcendent mystery is recognized in a specific concrete symbol, act, idea, image, person, or community. The unconditioned reality is manifested in conditioned form. Eliade has elucidated this "dialectic of the sacred," in which the sacred may be seen in virtually any sort of form in religious history: a stone, an animal, or the sea. The ambiguity of the sacred taking on profane forms also means that even though every system of sacred thought and action differentiates between those things it regards as sacred or as profane, not all people find the sacred manifested in the same form; and what is profane for some is sacred for others.

The sacred appears in *myths*, sounds, ritual activity, people, and natural objects. Through retelling the *myth* the divine action that was done "in the beginning" is repeated. The repetition of the sacred action symbolically duplicates the structure and power that established the world originally. Thus, it is important to know and preserve the eternal structure through which man has life, for it is the model and source of power in the present.

The recognition of sacred power in the *myth* is related to the notion that sound itself has creative power—in particular special, *sacred sounds*. Sometimes these sounds are words, such as the name of god, divine myth, a prayer, or hymn; but sometimes the most sacred sounds are those that do not have a common meaning, for example, the Hindu *om*, the Buddhist *oṃ maṇi padme hūṃ*, or the Jewish and Christian "Hallelujah."

Closely connected with verbal expressions of sacred power are activities done in *worship*, in sacraments, sacrifices, and festivals. Part of the importance of religious ritual is that in the realm of the sacred all things have their place. In order for human existence to *prosper* (or even continue) it must correspond as closely as possible to the divine pattern (destiny, or will). Different religious traditions have different theological and philosophical formulations of the meaning of sacraments. In Roman Catholic Christianity, a sacrament is "an outward and visible sign of an inward and invisible grace, Figure 7.20." In Brahmanic Hinduism a *saṃskāra* (sacrament) is a

Fig.7.20: Roman Catholic Christianity, a Sacrament

sacred act that perfects a person and that culminates at the end of a series of *saṃskāra*s in a spiritual rebirth, a symbolic "second birth." In both of these cases, the sacred action establishes the relation between the divine and human worlds.

Other sacred activity includes initiation, sacrifice, and *festival. Initiation rites* among nonliterate societies both expose and establish the world view of the participants. The initiate learns the eternal order of life as proclaimed in the myth. Life is viewed essentially as the work of supernatural beings, and the initiate in this ritual is taught this secret of life and how to gain access to divine benefits. The initiate learns the tabus and is often given a sacred mark—*e.g.,* circumcision, tattoo, or incisions—to express physically that he is part of the sacred (original) community. In other religions, such as Christianity, Buddhism, and Hinduism, an initiate to a special holy (often monastic) community within the larger religious community is designated by a change in name and wearing apparel, denoting his special relation to the sacred.

In *festivals* and sacrifices two religious functions are often combined:

1. to provide new power (energy, life) for the world, and
2. to purify the corrupted, defiled existence.

Religious festivals are a return to *sacred time*, that time prior to the structured existence that most people commonly experience (profane time). Sacred calendars provide the opportunity for the profane time to be rejuvenated periodically in the festivals. These occasions symbolically repeat the primordial chaos before the beginning of the world; and just as the world was created "in the beginning," so in the repetition of that time the present world is regenerated. The use of masks and the suspension of normal tabus express the unstructured, unconditioned nature of the sacred. Dancing, running, singing, and processions are all techniques for re-creation, for stimulating the original power of life. Ritual activity moves power in two directions:

1. it concentrates it in one place, time, and occasion, and
2. it releases power into the everyday stream of events through its self-abundance—the primal vibration reverberates throughout existence.

The new energy dispels the old, depleted, polluted energy; it cleanses the constricted, clogged, hardened channels of life.

One of the most important forms in which man has access to the sacred is in the sacrifice. The central procedure in all sacrifices is the use of a victim or substitute to serve as a mediator between the sacred and profane worlds. The sacrifice (Latin *sacri-ficium,* "making sacred") is a consecration of an offering through which the profane world has access to the sacred without being destroyed by the sacred. Instead, the sacrificial object (victim) is destroyed in serving as a unique, extraordinary channel between these two realms. In sacrificial rites it is important to duplicate the original (divine) act; and because creation is variously conceived in different religious traditions, different forms are preserved: the burning or crushing of the "corn mother," the crushing of the *soma* stalks, the slaughter of the lamb without blemish, the blood spilling of a sacred person, such as the firstborn.

Sacredness is manifested in sacred officials, such as priests and kings; in specially designated sacred places, such as temples and images; and in natural objects, such as rivers, the sun, mountains, or trees. The priest is a special agent in the religious cult, his ritual actions represent the divine action. Similarly, the king or emperor is a special mediator between heaven and earth and has been called by such names as the "son of heaven," or an "arm of god."

Just as certain persons are *consecrated*, so specific *places* are designated as the "gate of heaven." Temples and shrines are recognized by devotees as places where special attitudes and restrictions prevail because they are the *abode* of the sacred. Likewise, certain *images* of God (and sacred books) are held to be uniquely powerful and true (pure) expressions of divine reality. The image and the temple are, in traditional societies, not simply productions by individual artists and architects; they are reflections of the sacred essence of life, and their measurements and forms are specified through sacred communication from the divine sphere. In this same *context*, natural objects can be imbued with sacred power. The sun, for example, is the embodiment of the power of life, the source of all human consciousness, the central pivot for the eternal rhythm and order of existence. Or, a river, such as the Nile for the ancient Egyptians and the Ganges for the Hindu, gave witness to the power of life incarnated in geography. Sacred mountains (*e.g.,* Sinai for Jews, Kailāsa for Hindus, Fujiyama for Japanese) were particular loci of divine power, law, and truth.

DIMENSIONS OF THE SACRED

The sacred, by definition, permeates all dimensions of life. Within the kind of religious uneasiness that is expressed in sacred *myth* and ritual, though, there is a distinct focus on time, space (cosmos), and active mediators (heroes, ancestors, divinities). When existence is seen in terms of the contradiction of sacred and profane—which assumes that the sacred is wholly other than, yet necessary for, everyday existence—it is significant to know and to get in contact with the sacred. In determined festivals men celebrate sacred time; a sacred agenda marks off the intervals of man's life, and these sacred festivals offer the pattern for productive and joyous living.

Seasonal sacred calendars are particularly important in predominantly farming societies. In the very order of nature, people observe that diverse seasons have their distinct values. These diversities are celebrated with spring festivals (when the world is re-created through ritual expressions of generation) and harvest festivals (of thanksgiving and of protecting the life force in seeds for the next spring). Here time is viewed as cyclical, and one's life is marked by those rituals in which one recurrently returns to the divine source.

Likewise, the mythologies and rituals mark off the world (universe) into places that have distinct sacred meaning. The territory in which one lives is real to that extent as it is in interaction with the divine reality. Within this territory is life; outside it is disorder, danger, and demons. Throughout most of history the "sacred universe" was corresponding with a certain territory, and one could speak literally of Christian lands, the Jewish homeland, the Islamic world, the place of the noble people (*Āryāvarta*, Hindu), or the central kingdom (China). Sanctifying one's possession of land with certain rituals was equal to founding an order with divine sanction. In Vedic ritual, for example, the erection of a fire altar (in which the god Agni—fire—was present) was the founding of a universe on a microcosmic scale. Once a cosmos is recognized, there are certain spaces that are especially sacred. Certain rivers, mountains, groves of trees, caves, or human constructions such as temples, shrines, or cities provide the "gate," "ladder," "navel," or "pole" between heaven and earth. This sacred place is that which both permits the sacred power to stream into existence and provides order and steadiness to life.

Another dimension of the sacred is divine or heroic movement: the pivotal action done by creative or caring agents. One's spiritual ancestors require not be biologically distinct ancestors; they may not even be human. They are the vital forces on which survival hinges on and can be personified in animal skills (longevity, rebirth, magical skills), in the "habits of the ancients," or through a special hero who has offered present existence with material and spiritual aids. If the notion of sacred appearance is extended to include the social relationships (especially anathemas) in a community, then communal relations can be regarded as a dimension through which the sacred is appeared. Here human values are sacralized by social restraints that prescribe—*e.g.,* with whom one can eat or whom one can marry or kill. The founding of a community necessitates establishing certain relationships; and these relationships are sacred when they bear the power of decisive, eternal, cosmic force. For example, the consecration of a king or emperor in outdated agricultural societies was the formation of a system of loyalty and order for society.

By extending the notion of "sacralization" to include human restructuring of experience within the framework of any absolute norm, the sacred can be seen in such dimensions of life as history, self-consciousness, *aesthetics*, and philosophical image (conceptualization). Each of these methods of human experience can develop the creative force whereby some people have "become real" and grew the most profound understanding of themselves.

CRITICAL PROBLEMS

Phenomenologists of religion who use the concept "sacred" as a universal term for the basis of religion vary in their assessment of the nature of the sacred manifestation. "Otto and van der Leeuw" hold (in different formulations) that the sacred is a reality that *transcends* the uneasiness of the sacred in symbols or rituals. The forms (ideograms) through which the sacred is articulated are secondary and are simply reactions to the "wholly other." "Kristensen and Eliade," on the other hand, regard the sacred reality to be obtainable through the specific symbols or conducts of capturing the sacred. Thus, "Kristensen" seats importance on how the sacred is apprehended, and Eliade describes different modalities of the sacred, while "Otto" looks outside the forms toward a meta-empirical source.

A second problem is the ongoing question of whether or not the sacred is a universal class. There are religious expressions from numerous parts of the world that clearly *evident* the kind of structure of religious consciousness considered above. It is especially *appropriate* of some aspects in the religion of nonliterate societies, the ancient Near East, and some popular devotional aspects of Hinduism. There is, though, a serious question regarding the practicality of this structure in understanding a large part of Chinese religion, the social relationships (*dharma*) in Hinduism, the effort to accomplish superconscious cognizance in Hinduism (Yoga), Jainism, Buddhism (Zen), some forms of Daoism, and some modern and goal fundamentally dissimilar from human existence. If one takes the notion of sacred as something above (beyond, different from) the religious structure subjugated by divine or *transcendent* activity (described above), then this proposes that the notion of sacredness should not be partial to that structure. Thus, some scholars have found it perplexing to utilize the notion of sacred as a universal religious quality, for it has been acknowledged by many religious people and by scholars of religion as denoting to only one (though important) sort of religious consciousness.

The 20th-century discussion of the nature and manifestation of the sacred embraces other methods than those of scholars in the comparative study of religions. For example, *Sri Aurobindo*, a Hindu mystic-philosopher, speaks of the supreme reality as the "Consciousness-Force;" and Nishida Kitaro, a Japanese philosopher, expresses his uneasiness of universal reality as that of "absolute Nothingness." "Martin Heidegger," a German philosopher, expresses "the holy" as that dimension of existence through which there is the illumination of the things that are, though it is no absolute Being prior to existence; rather it is a *creative* act at the point of engaging the Nothing (*Nichts*). In contrast, the Protestant theologian Karl Barth discards philosophical reflection or mystical insight for capturing the sacred, and asserts that personal acceptance of God's self-revelation in a particular historical form, **Jesus Christ**, is the place to begin any consciousness of what philosophers call "ultimate."

Sociologists who study religion have, since Durkheim, usually recognized the sacred with social values that claim a supernatural basis. However, the sacred has been recognized mainly as found in the social occasions (festivals) that interrupt the common social order (by Caillois), or as the strengthening of social activities that protected an assumed social structure (by Howard Becker). During the 1960s, though, the usual definition of religion as those sacred activities which claimed a transcendent source was quizzed by some empirical scholars. For example, "Thomas Luckmann," a German-American sociologist, defined the sacred in modern society as that "strata of significance to which everyday life is ultimately referred"; and this definition embraces such themes as "the autonomous individual" and "the mobility ethos."

SACRED TODAY

The problems of defining and examining *religion* stated above are already expressive of the shifts in modern consciousness concerning the sacred. Both the physical and social sciences have given modern man a novel image of himself and techniques for refining his present life. The acceptance of cogent and critical viewpoints for judging the claims of religious ruling classes in Europe since the 18th century, plus the advancement of historical *criticism* and a sense of historical relativism, has donated to the affirmation of man as essentially a *secular* person. The once absolute authorities in the West (the **Bible**, priest, rabbi) are no longer the main sources for one's self-identity. To a rising extent the *cultures* in the East are also undergoing a loss of their traditional authorities. Some efforts have been made to re-sacralize modern cosmology, history, and personal experience by:

1. spreading the opportunity of religious apprehensions to "secular" areas such as politics, economics, personality development, and art; and

2. adapting theological positions, *ethical* norms, and liturgical forms to integrate novel modes of expression and to experiment with novel styles of living.

A significant 20th-century progress in religious life has been the relaxed flow of information between religious communities on different continents. This has offered a chance for experimenting with religious forms from outside the traditionally satisfactory forms in a culture. During the 1950s and 1960s, for example, Yoga and Zen meditation were serious religious options for some Westerners and a form of experimentation for large numbers.

The concern to experiment with personal experience and with styles of living during the 1960s in the West has itself been well-thought-out an important religious expression by some commentators. These years observed substantial examination in interesting experience with hallucinogenic drugs, many attempts to set up new communities for group living (communes)—though few survived more than a year—and a shift in the *values* of middle-class youth from a worry for personal economic security to social and experiential concerns. These current activities may be observed as efforts to retaking the experience of the sacred.

During the past hundred years a number of philosophers and social scientists have proclaimed the evaporation of the sacred and foretold the *demise* of religion. An investigation of the history of religions illustrates that religious forms transformed and that there has never been agreement on the nature and expression of religion. Whether or not man is now in a new state for emerging structures of ultimate values fundamentally different from those offered in the traditionally affirmed consciousness of the sacred is a vital question. The proposal that a fundamentally different kind of reality is conceivable is, of course, gobbledygook for those to whom the sacred already has been manifested once and for all in a particular form.

ANCIENT INTELLIGENT DESIGN

2000 years ago, Roman citizens and philologers have argued that the 'design' in plane nature pin point to a "Designer." Cicero (106–43 BC), Figure 7.21, a Roman writer, orator and statesman, in 44 BC, utilized the concept nature in his book *"De Natura Deorum"* - *"(On the Nature of the Gods)*. He composed his book to encounter the evolutionary ideas of other philosophers of his time.

GREEK GODS – FEAR OF DEATH

The main twin schools of way of life were Epicureanism and Stoicism. The "Epicureans" believed happiness stemmed through bodily pleasures and freedom from pain and anxiety. The two

Fig.7.21: Cicero (106–43 BC), UK Art

primary causes of anxiety were "fear of the gods" and "fear of death," so Epicurus pursued to annul both of these by teaching an "evolutionary atomic theory."

Epicurus deprived of that there was any aim in nature, as everything was composed of "particles" (atoma: atoms), Everything is deteriorating downwards. Also, he said that these occasionally instinctively 'swerved' to merge and form "bodies"—non-living, living, human, and divine. However, the gods were developed of finer atoms than humankind. Those gods did not create the world or have any control over it. These gods also were not disturbed with human affairs, and there was therefore no need for man to fear them. At death, the soul fragmented and became non-existent, so there was no necessity to be in terror of death or the outlook of judgment after death.

Cicero employed the "Stoic character" in his book to refute these ideas with arguments from design, intended to illustrate that the universe is ruled by an "intelligent designer." He argued that a "conscious" purpose was essential to express art (e.g., to make a picture or a statue). Accordingly, as nature was more perfect than art, nature illustrated purpose also. Cicero reasoned that the movement of a ship was steered by accomplished intelligence, and a sundial or water clock told the time by "design" rather than by "chance." Cicero said that even the barbarians of Britain or Scythia could not fail to understand that a model which illustrated the movements of the sun, stars and planets was the product of conscious intelligence.

Cicero sustained his challenge to the evolutionism of "Epicurus" by marveling that anybody could influence himself that "chance collisions of particles" could formulate anything as beautiful as the world. Also, he said that this was on a balance with trusting that if the "alphabet letters" were randomly thrown on the ground repeatedly enough they would spell out the *"Annals of Ennius." Annales is the name of a fragmentary Latin epic poem written by the Roman poet "Ennius" in the 2nd century BC.* Cicero questioned: if chance collisions of particles could make a world, why then cannot they build much less difficult objects, like a colonnade, a temple, a house, or a city?

RECENT DESIGN ARGUMENT

William Paley, (1743–1805) the most notable user of the design argument in the 18th century. In the book he authored, *"Natural Theology,"* he laid the case of someone finding a watch while walking in a barren countryside. From the functions which the numerous parts of the watch fulfil (e.g., spring, gearwheels, pointer), the solitary rational conclusion was that it had a maker who 'understood its structure and designed its utilization. Paley also deliberated evidence of design in the "eye" Figure 7.22—that as an instrument for vision it illustrated "intelligent design" in an identical way that telescopes, microscopes and spectacles do. And he elaborated on to discuss intricate design in many other human and animal organs, all pin pointing to the deduction that the existence of complex life point-toward an intelligent Creator.

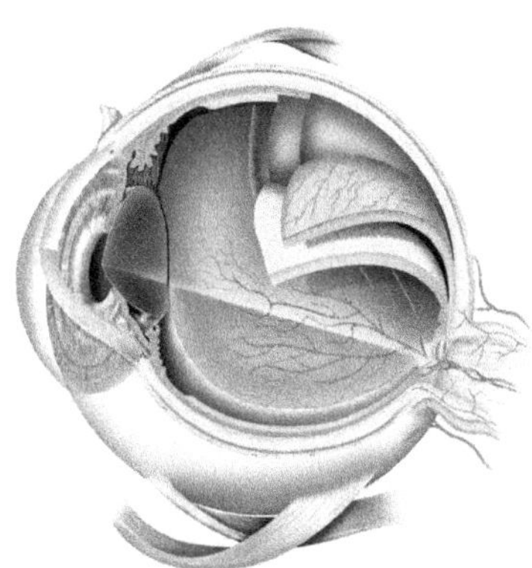

Fig.7.22: The Eye Anatomy Complex – Curtsy National Eye Institute

David Hume, the 18th century Scottish skeptical philosopher, attempted to counter the William Paley's watch argument by pointing out that watches are not living things which reproduce. Nevertheless, Paley wrote 30 years after Hume, and Paley's arguments are proof against most of Hume's objections. For example, a modern philosopher has countered Hume: 'Paley's argument about organisms stands on its own, regardless of whether watches and organisms happen to be similar. The point of talking about watches is to help the reader see that the argument about organisms is compelling.'

EVOLUTION AND WILLIAM PALEY

Charles Darwin was obligatory to read Paley during his theological studies at Cambridge (1828–31). He later said, '*I do not think that I hardly ever admired a book more than Paley's Natural Theology. I could almost formerly have said it by heart.*'

Nevertheless, Charles Darwin then employed up the rest of his life developing and promoting a theory to describe how 'design' in nature could happen deprived of God. Darwin proposed that small, convenient changes could happen by chance, and enable their possessors to survive and pass on these changes—natural selection. "Natural selection" would be achieved on even the tiniest progresses and, over immense ages, would allegedly accrue enough small changes to yield all the 'design' we observe in the living world.

Modern science vs Darwin

Evolutionists, counting the vociferously atheistic Oxford Professor "Richard Dawkins," still employ "Darwin's theory" to clash with the "design argument." Nonetheless, present day, evolutionist accept as true that "natural selection: acts on genetic copying mistakes (mutations), some of which are hypothetical to increase the genetic information content. But Dawkins' arguments have been brutally critiqued on scientific grounds. Dawkins' Neo-Darwinism has multiple flaws:

1. Natural selection necessitates self-reproducing entities. Constructing even the meekest self-reproducing organism by a chance combination of chemicals is even more unbelievable than constructing the "*Annals of Ennius*" by dropping letters on the ground. Living things necessitate long molecules with precise arrangements of lesser 'building blocks.' Not only will the 'building blocks' not combine in the right order, but they are improbable, by natural means, to construct up large molecules at all! Rather, large molecules incline to break down into smaller ones. Additionally, the 'building blocks' are unstable.

2. There is complex biological mechanism of which Darwin was merely ignorant. Biochemist "Dr Michael Behe" lists a number of examples: real motors, transport systems, the blood clotting cascade, and the complex visual machinery. Behe argues that they necessitate many parts or they would not function at all, so they could not have been constructed in small steps by "natural selection."

3. Biophysicist/information theorist "Dr Lee Spetner" points out that mutations have never been observed to add information, but only reduce it—this embrace even the rare helpful mutations. Also, Spetner points out that "natural selection" is deficient to accumulate slight advantages, as it would be too feeble to overcome the effects of chance, which would incline to eliminate these mutants.

THE BIBLE AND THE DESIGN

The Apostle Paul employed the design fight in *Romans 1:20,* where he affirms that God's everlasting power and divine nature can be comprehended from the things that have been made (i.e., evidences of design in nature). Apostle Paul says that because of this, the ungodly are 'deprived of excuse.' Nonetheless, Paul continues that people willingly reject this clear evidence.

This evidence of "design in nature" is sufficient to convict men, but it is not sufficient to save them. The Bible makes it distinctively clear that the preaching of the Gospel is also desirable to demonstrate how we are to come into a right relationship with the Creator.

Cicero lived in the century before Christ and perhaps had never heard of the God of Genesis; he employed design in support of the Greek pantheon of gods and goddesses of the Stoics. Nowadays, 'New Agers' may characterize design to "Mother Nature" or "Gaia," the Greek goddess of the earth.

EVANGELISM

When Christians employ design and other arguments from science, they are correctly engaging in pre-evangelism, i.e., they are looking for exposing the falsehood of the evolutionary presuppositions that blind the eyes of people today to the truth of the Word of God. This is illustrated by the Apostle Paul's experience in Athens. Paul 'preached Jesus and the resurrection' *(Acts 17:18),* which confronted both the "Epicurean and the Stoic philosophers" of his day—i.e., both Cicero's opponents and his fellow believers, Figure 7.23. Paul confronted their faulty ideas by pointing them to the one true God who had created everything. Nonetheless, Paul didn't stop with creation.

Fig.7.23: *Paul in action in Athens – Acts 17*

Paul counseled them to repent, and he said they could recognize there would be a Day of Judgment because God had appointed the Judge and given assurance of this by raising Him from the dead *(Acts 17:18–31).*

The solitary way to be saved is to believe in the Gospel of the Lord Jesus Christ *(Acts 4:12),* the Creator and Kinsman-Redeemer (Isaiah 59:20), who died and rose again to pay the penalty for mankind's sin. We should follow the way Paul offered the Gospel in *1 Cor. 15—N.B. verses 1–4, 21–22, 26, 45,* which make intelligent sense only with a literal Genesis—a literal Creation, Fall, death penalty for sin, etc.

John the Evangelist inscribed his Gospel *'so that you might believe that Jesus is the Christ, the Son of God, and that believing you might have life in His name' (John 20:31).* Nevertheless, he began his Gospel by declaring that Jesus is the Creator *(John 1:1–3),* the Second Person of the Trinity, who took on human nature *(John 1:14).* Thus, evangelism must present Christ as Creator or it is deficient—if Christ is not God, then He cannot be our Savior *(Isaiah 43:11).*

DNA-SCIENTIFIC DESIGN ARGUMENT

All the design in living things is programmed in a type of recipe-book with much information. Information designates the complexity of a sequence—it does not subject to the substance of the sequence. It could be a sequence of ink molecules on paper (book)—nevertheless, the information is not confined in the molecules of ink but in the patterns. Information can also be stored as sound wave patterns (e.g., speech), However, once more the information is not the sound waves themselves; electrical impulses (telephone); magnetic patterns (computer hard drive).

The anti-theistic physicist "Paul Davies" admits: '*There is no law of physics able to create information from nothing*' (Quantum leap of faith). Information scientist "Werner Gitt" has established that the "laws of nature" relating to information illustrate that, in all recognized cases, information necessitates an intelligent message sender, a conclusion rejected by Davies on purely philosophical (religious) grounds. Thus, a modern version of the design argument encompasses detecting high information content. In actual fact, this is exactly what the SETI project is all about—the Search for Extraterrestrial Intelligence involves trying to detect a high-information radio signal, which they would consider as proof of an intelligent message sender, even if we had no idea of the nature of the sender.

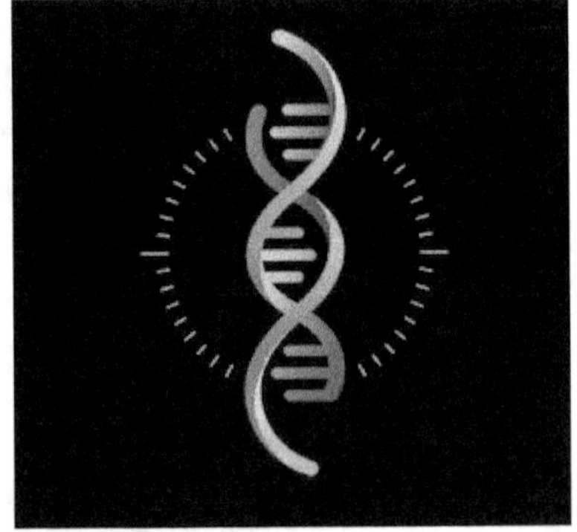

Fig.7.24: *Intelligent Design in DNA Structure*

In living things, information is all stored in patterns of DNA, Figure 7.24, which encode the instructions to make proteins, the building blocks for all the machinery of life. There are four types of DNA 'letters' called nucleotides, and 20 types of protein 'letters' called amino acids. A group (codon) of 3 DNA 'letters' codes for one protein 'letter'. The information is not contained in the chemistry of the 'letters' themselves, but in their sequence. DNA is by far the most compact information storage/retrieval system ever known.

Now consider if we had to write the information of living things in book form. "Dawkins" admits, '*There is enough information capacity in a single human cell to store the Encyclopedia Britannica, all 30 volumes of it, three or four times over. Even the simplest living organism has 482 protein-coding genes of 580,000 'letters'.*'

Let us assume we had the technology to go the other way, and store books' information in DNA—this would be the ideal computer technology. The amount of information that could be stored in a pinhead's volume of DNA is equivalent to a pile of paperback books 500 times as tall as the distance from Earth to the moon, each with a different, yet specific content. Phrasing the concept in another way, a pinhead of DNA would have a billion times more information capacity than a 4-gigabyte hard drive.

Just as letters of the alphabet will not write the "*Annals of Ennius*" by themselves, the DNA letters will not form meaningful sequences on their own. And just as the Annals would be meaningless to a person who did not understand the language, the DNA 'letter' arrangements would be worthless deprived of the 'language' of the DNA code.

Deprived of the message of design and the Creator, 'gospel preaching' lacks foundation. Deprived of Christ, the design argument cannot save. We must present a full Gospel, starting with creation by the Triune God, and combine it with the message of Christ's death for sin and His Resurrection.

THE INTELLIGENT DESIGN

The argument of 'intelligent design' (ID) has an extensive history going back to the ancient Greeks and Romans, as stated in our previous chapter. It was convincingly expressed by "William Paley (1743–1805)," who put forward the argument of a contingent divine "Watchmaker" in his book "*Natural Theology* (1802)." Contemporary biblical creationists have also employed the design argument in their disapproval to evolution, Figure 7.25. However, the works of modern researchers such as "Michael Denton" (*Evolution: A Theory in Crisis,* 1985) and "Phillip Johnson" (*Darwin on Trial,* 1991) have commanded the establishment of an association of scientists and other scholars, which has turn out to be recognized as the 'Intelligent Design Movement' (IDM or ID movement).

Fig.7.25: *"Watchmaker" in his book "Natural Theology (1802)" William Paley*

Many of "Intelligent Design" supporters have frequently enquired about the creationist's position on the ID subject. Accordingly, this chapter is primarily in response to that. Additionally, while this chapter is not envisioned to be an antagonistic assessment by any means to my opponents, in particularly to those oppose the 6-day creation, Figure 7.26. My response is also tailored also to my faculty-staff and colleagues who support naturalistic processes, and oppose creation in any form. Albeit, I have an amiable relationship with my faculty-staff, colleague and post graduate students, who oppose creation and the faith in my infinite Intelligent Designer.

Fig.7.26: *6-Day Creation*

The modern concept of intelligent design has been merely expressed as the belief that certain biological lines of evidence (e.g., the 'irreducible complexity' of features such as the bacterial flagellum) are evidence for a designer and against blind naturalistic processes.

INTELLIGENT DESIGN IDEOLOGY

The Intelligent Design ideology is mainly focused on the desire to challenge the blind acceptance of the materialistic, godless, naturalistic philosophy of "Darwinian evolution." The proponents of Intelligent Design are motivated to confront many of the philosophical underpinnings of today's evolutionary thinking. The opponent of Intelligent Design are unwilling to align themselves with biblical creationism.

The fundamentals of the ID has attracted a number of faithful believers and evangelical Christians, including believers in literal 6-day creation in Genesis, who see it as a helpful strategy to crack the foundation of evolution, which undergirds most of the world's cultures and schools.

Indication of ID's upward involvement was the effort to add the Santorum amendment to the 2002 US education bill; an amendment that stimulated schools to inform students about the 'continuing controversy' over 'biological evolution'. ID supporters have also been vigorous in enduring efforts to include ID in the educational standards in all the states of USA.

Exploring Intelligent Design language in Genesis 2 and Acts 17—yatsar and *poieō*

Divine Intention – Devine Action

Two examples of design language in the Bible, Figure 7.27.

- One is from the Hebrew Old Testament yatsar יָצַר *(Genesis 2:7),* the other
- from the Greek New Testament poieō ποιέω *(Acts 17:24-29).*

This illustrates that Scripture does express of God as a craftsman, conceivably as the potter of Adam in *Genesis 2,* and the author of life in *Acts 17.* The account of *Genesis 2* then emphasizes a discontinuity between Adam and the animals that is a problematic for theistic evolution. However, Scripture is vigilant not to separate divine intention from divine action, as Intelligent Design arguments from time to time seem to do.

A brief examination of Paul's speech in Athens also supports us to advance a biblically based apologetic. Paul's yearning to encounter idolatry mirrored Isaiah's concern, and his method balanced negative analogical intellectual with positive attributes of God's attributes.

Fig.7.27: The Bible – God's Words

This section of my writing will look at two examples of design language in the Old and New Testaments that express of God's creative activity in relationships to human craftsmanship. Given the method of the "Intelligent Design" movement, which pursues to discovery evidence of "divine intention" as opposed to "divine action" or contemplation of biblical timescales, it is essential to study how Scripture grips design language. Accordingly, this study may provide some vision for comprehending the legitimacy of "Intelligent Design" arguments. It will be demonstrated that some sustenance may be congregated from Scripture, however, care must be considered not to detach too strongly "design" from the act of "creation" because of a holistic approach to creation originate in Scripture.

- First for consideration is the Hebrew word "*yatsar*" (Hebrew יָצַר), employed for the formation of Adam. This will be traced from the Genesis account of creation in the Hebrew Masoretic text, and then carefully observe how it is translated into the Greek Septuagint (LXX) . On instances this is translated "*poieō*" (Greek ποιέω), but it is distinguished that "*poieō*" is also a valid translation of "*bara.*"

- The second consideration will be given to how Paul employs "*poieō*" in the Greek New Testament in *Acts 17* and *Romans 1*, and this may provide valuable visions for methods to Christian apologetics.

YATSAR IN GENESIS

In the 2nd chapter of Genesis God is defined as forming (Hebrew *yatsar* יָצַר) man from the dust of the earth *(Genesis 2:7).*

"The Lord God formed (*yatsar* יָצַר) the man from the dust of the ground and breathed into his nostrils the breath of life, and the man became a living being."

Figure 7.28 - The Hebrew word *yatsar* speaks metaphorically and analogically of God as the potter of Adam, however, also declares the literal creation of the first man from the dust of the earth.

The same word, *yatsar*, is parenthetically employed for the creation of the animals and birds in *Genesis 2:19-20*, with Adam assumed the task of naming his "fellow creatures." The metaphor employed in *Genesis 2* for the creation of Adam, then, might appear to be one of God acting as a potter, where the imagery is that of God taking raw matter from the earth and forming man directly from the ground and giving him the breath of life. Scholar "Unger" and scholar "White" point out that *yatsar* is a technical Hebrew term employed in the pottery trade, Figure 7.28, while it does have other related utilization. It does nevertheless, offer a fascinating possibility for comprehending God's creation of Adam. Therefore, even if the potter metaphor is difficult to establish conclusively in this passage, the word *yatsar* still expresses of God

Fig.7.28: The Hebrew word yatsar speaks metaphorically and analogically of God as the potter of Adam

in relationship of the master craftsman, forming Adam from the dust of the ground. This section will now consider the strength of the potter metaphor in *Genesis 2*.

In the wider Old Testament passages the verb *yatsar* is from time to time precisely employed to denote to the work of a potter in relation to God's activity in shaping and forming the people of Israel. It appears as a noun to refer to the pottery of Israel that had been *formed* by God in *Isaiah 29:16* (also *Jeremiah 18:4-6*) and God is frequently defined as the one *forming* Israel throughout Isaiah (64:8 ,49:5 ,11-9 :45 ,44:25 ,44:2 ,27:11). The feared sea monster was *formed* to play in the ocean (*Ps. 104:26*).

Correspondingly, in Isaiah (*44:9-12*) the word *yatsar* is employed to define the activity of those who form idols; it is employed for the work of an ironsmith who fashions, by cutting and hammering, a piece of hot metal into a shape believed suitable for worship. *Yatsar* likewise explicitly refers to God's decisive intention (*Isaiah 22:11; 46:11*), and in *2 Kings 19:25* it is employed to refer to God's action in 'planning' the fate of fortified cities. In *Genesis 6:5* the word is employed to define the evil 'imagination' of human beings who plan wrongdoing, and in *Psalm 94:20* where evil rulers are said to "*frameth* mischief by a law" (KJV). It may then on times express primarily of "divine intention," as opposed to "divine action," nonetheless not always. Though, it may appear difficult establishing a potter metaphor in the *Genesis 2* passage merely on a study of the language because the word *yatsar* can be employed in a number of different ways; instead, consideration of the context is essential.

Similarly, there are other words employed in the Genesis account for creation. The word "*bara*" means to "make" or "create" and is employed precisely with God as the subject. It typically refers to the act of creation out of nothing (*ex nihilo*) where God is capable to bring things directly into existence. Other words employed for creation have both God and human beings as their subjects; for instance, "*asah*" is normally employed as a verb in the Old Testament as a word meaning to make, or do. It is occasionally employed as a synonym when in conjunction with "*bara*." It is through, a more general word than "*bara*" and often has human agents as its subject. In *Genesis 1:7* "*asah* is" employed to refer to the making of the firmament "*raqiya*."

It is essential now to examine whether the formation of Adam from the dust of the ground should be read "literally." When read in this way it speaks of a "direct act" by God with a discontinuity between Adam and the other animals, which goes in contradiction of the position of theistical evolutionists who pursue to maintain "biological continuity;" i.e., scholar "Alexander's" belief that Adam was a federal head as "*Homo divinus*" called out from other early "*Homo sapiens*." Nevertheless, if formed from the ground, then Adam could not have had parents, as theistic evolution requires. Scholar "Walton" also believes that mention of the formation of Adam from the dust of the ground in *Genesis 2* should be read as archetypal and not as prototypal of humanity. That is, Adam is observed as just a representative of other people alive at the time, and the dust refers to Adam's mortality, and not to material substance. He thinks the passage is therefore, employing functional language about human frailty and not speaking materially about a literal creation. This position holds that Adam may have had a human mother and father and there is no material discontinuity as suggested by a literal reading of the text. However, the language of God as a potter, forming Adam from "*pre-created*" matter through the application of "divine intention" and "divine action," would break down if that were so.

Alexander and scholar "Walton's" position raises many theological difficulties as deliberated in the creationist literature; for instance, the fact that death would have existed before Adam's sin and what that means for the Gospel. There is no enough space to go over all that material here, however, just to note that the traditional position has read this passage literally, with Adam formed "*de novo*" (newly, at once) as even theistic evolutionists such as scholar "Lamoureux" have admitted. Scholar "Irenaeus," for instance, spoke of Adam as the 'protoplast' of humanity, 'the first-formed,' out of virgin soil; and it followed that Jesus also needed to be born without a human father in a virgin's womb in order to act as the last Adam.

Furthermore, a phenomenological reading, favored by "Lamoureux," would render a non-literal reading untenable. The simple statement that "*for dust you are and to dust you shall return*" *(Genesis 3:19)* is a literal statement based upon the original creation of Adam from the dust of the ground. Affirmatively, people die functionally, but they also die literally; in the same way Adam was created functionally, nonetheless, also formed literally.

While the language may be phenomenological in some sense as "Lamoureux" suggests, the Genesis creation account is really written from God's perspective; the traditional Hebrew reading has, for instance, painstaking that the Torah was dictated to Moses directly where God spoke *"face to face, clearly and not in riddles" (Numbers 12:8).*

REAL INTENTION – REAL ACTION

Subsequently, there are good reasons to read the text of *Genesis 2* literally where the creation of Adam by God involved "real intention" and "real activity." There is the intimation that employ of the word *yatsar* goes outside metaphor and is more akin to analogy where the difference is one of scale between the "human potter" and the "divine creator" of mankind; that it is univocal language as opposed to equivocation.

God literally formed Adam from the dust of the ground. However, a metaphor remains because clearly God did not create Adam in the same way a potter shapes a pot from clay placed on a wheel. God did not literally sit at a potter's wheel to form Adam with his hands, but he took dust and employed it to form the shape and person of Adam, in whatever way a spiritual being creates materially.

The word *yatsar* also has additional meaning relating to the pressures and stress of human existence; for instance, being in distress and the stress and suffering experienced in life. Scholar "Walton" employs this fact to build his own case for a functional creation account; he remarks, for instance, that *"yatsar"* need not refer to a 'sculpting process' involving matter, however, it has broader meanings concerning human experience. Nevertheless, I think his priority is incorrect here as he is possibly reading the prophets back into Moses. The metaphor for God's work as a potter in shaping individuals, and the nation of Israel, arises from the original creation of Adam that then speaks symbolically and perhaps prophetically of God's dealing with people and the nation. There are representational messages throughout Scripture that reveal theological truths about God's dealing with Israel and humanity, and this is frequently focused upon the Messiah, as "Augustine" and "Irenaeus" for instance believed. Nonetheless, it is symbolism that is grounded first in real people and events. Consider, for instance, the call of Abraham to sacrifice Isaac, and then the release from this call with the divine provision of a ram, this representative of the sacrificial work of Jesus many centuries later.

Scholar "Hamilton" has also probed the strength of the potter metaphor in *Genesis 2*. He has suggested that the metaphor of a potter is not a direct one, and is dependent upon the context of the passage since the various ways *"yatsar"* may be employed. The reason offered is that the word employed for dust is *"aphar"* עָפָר in *Genesis 2:7* and not the more typical Hebrew word for clay, which he points out is *"chomer"* רחֹמֶ. A potter metaphor, for instance, appears in *Isaiah 64:8*, where God is the potter who formed *"yatsar"* Israel out of clay *"chomer"*. Scholar "Walton" also proposes that clay would be a more probable ingredient for the formation of Adam. This is because dust, as a dry, loose, and lifeless material, is difficult to mold.

In response, it is pertinent to note that the word *"aphar"* is occasionally employed in the Old Testament to refer to dry dust, occasionally to clods of earth, and even in places to wet plaster (*Lev. 14:42*). Nevertheless, this on its own doesn't explain why the passage of *Genesis 2:7* refers to dust and not to clay if the potter metaphor is appropriate. But in support of use of the word *"aphar"* in relation to the potter metaphor, Augustine, in the *"City of God"* has provided a additional likely understanding. He gives an explanation, which he claims arises from others, founded upon the connection between verse 6 and 7. In verse 6 a mist is said to have arisen that watered the land, and then God took the wet *"aphar"* (or mud) to form Adam. If that is a correct interpretation, then we may see the formation of Adam involved water and dust as opposed to dry dust alone. Dry dust is of course very difficult to mold into anything as scholar "Walton" notes. On the other hand, clay is a mixture of water and dry earth, and the *Genesis 2* account possibly highlights these raw materials as opposed to the *"chomer"* mixture. This provides one likely explanation why dust is referred to in the passage and not directly to clay, and offers support to the potter metaphor. The notion of creation out of the dust of the ground also speaks of Adam's/mankind's frailty in life, and therefore to spiritual dependency upon God.

Therefore, there would seem to be a metaphor of God as a craftsman in the word *"yatsar"* in the *Genesis 2* account, and, from the contextual evidence, it may express metaphorically and analogically of God as a potter who shaped the human form. In this shaping there is the formation of a higher level of order out of the elementary chemical matter originated in the ground; the material first created out of nothing. There is also the inference that in molding and framing the form of man there is the application of planning and purpose as well as the art of the craft itself. Consequently, there is something of the concept of design in the word *"yatsar,"* but it is not just in a scientific intellect, but also in an intellect of the work of an artist. However, care does need to be taken not to separate the concept of design from the physical act of creation because of the scriptural approach to creation, which is holistic.

TRANSLATION OF "YATSAR"

It is essential now to reflect how the Hebrew word *"yatsar"* interprets from the Old Testament into the New. As noted, the word *"bara"* is employed in *Genesis 1*, and all over Scripture, to refer to God's activity as the creator, often referring to creation through the spoken divine word. *"Yatsar"* on the other hand is considered less sacred and is employed for both God's activity and that of man. In the Septuagint (LXX) *"bara"* is translated employing the Greek verb *"poieō"* (Greek ποιέω, in English = to do, to make) in *Genesis 1:1*. The word *"yatsar"* in *Genesis 2* is translated employing the Greek word *"plassō"* (Greek πλάσσω), meaning to form, mold, or shape, as an artist working in clay in *Genesis 2:7*. This is also how it is employed by Paul in *1 Tim. 2:13* in reference to the creation of Adam (i.e. Greek πρῶτος ἐπλάσθη). Another Greek word for art or skill is *techne* (Greek τέχνη), but this usually refers to a human art, craft, or trade, in the same way we use the English word 'technology'.

However, the (GREEK - SEPTUAGINT TRANSLATION OF THE OLD TESTAMENT (Greek - Septuagint translation of the Old Testament LXX)) is not consistent in employ of words in translation, and in Isaiah (44:2, 44:3, 44:28) *"yatsar"* is translated employing the Greek *poieō*. In the LXX *epoiēsen* (Greek ἐποίησεν) is employed in *Isaiah 40:19* in relation to man-made idols; in *Isaiah 44:2* in relation to the formation of Israel, and *epoiēso* (Greek ἐποίησω) in *Isaiah 44:3*; and *epoiēsei* (Greek ποιήσει) is employed in *Isaiah 44:28* in relation to the rebuilding of Jerusalem and the Temple. But importantly in *Isaiah 42:5* *"bara"* is translated as *epoiēsas* (Greek ποιήσας) with reference to the creation of the heavens and earth. Therefore, we find that the Greek word *poieō* is employed to translate both *bara* and *yatsar* in the (Greek - Septuagint translation of the Old Testament LXX), and this usage carries over into the New Testament.

Therefore, how is *poieō* employed in the Greek language? As a verb it means to make or do, whether for routine tasks or for distinct tasks such as writing or poetry; as *poiēma* it is employed as a noun to refer to the product or thing that is made; as *poiēsis* it refers to the action, as in a accomplishment or deed, and as *poiētēs* refers to the one who makes, performs, or does, but more specifically to a poet. Paul in fact employs the language of "divine workmanship: in his preaching and teaching; the two most notable places are recorded in *Acts 17:16-34* and in *Romans 1:19-23*. In *Romans 1:20* we find the word *poiēmasin* (Greek ποιήμασιν) referring to "what has been made". Scholar "Vine" notes that Paul employed *poieō* to refer to an action that was 'complete in itself' as a manner of expressing 'thoughts and feelings'; thus, the creation was observed as a completed act that articulated thought and intention. The Greek word *poieō* then is employed to translate both *"bara"* and *"yatsar"* and has diverse meanings in Greek; from the routine to the poetic. Therefore, on its own there is difficulty in saying much more about the translation of *"yatsar"* into Greek. However, in order to try to comprehend how Paul employed the word *poieō*, a satisfactory translation of both *"yatsar"* and *"bara,"* it is pertinent to consider the context in his preaching to the Greeks in Athens.

PAUL IN ATHENS

Fig.7.29: Paul Preaching in Athens by Raphael (1515).

Figure 7.29, St Paul Preaching in Athens by Raphael (1515). Paul's speech, recorded in *Acts 17:16–34*, provides lessons to Christians in how to present the Gospel to modern society with its Greek influence.

Luke records in *Acts 17:16-34* that Paul was intensely disturbed by the idolatry evident in Athens. He was then challenged by the "Epicurean" and "Stoic" philosophers to present the case for the gospel message at the "Areopagus" on "Mars Hill," Figure 7.29. Luke's account is evidently a refinement of a longer speech, and as Paul was stopped by his hearers after mentioning the "Resurrection" it may be probed whether the speech was finished. However, it is conceivable to get a feel for Paul's style in this passage and the fact that the "Resurrection" is declared suggests he did in fact preach the full gospel in Athens. Scholar "Dibelius," however, advocates that the speech was unusually Hellenistic for Paul, and questions whether it is authentic, maybe being an elaboration on the part of Luke. However, Paul was quite happy to articulate his preaching to the audience by becoming 'all things to all men' for the sake of the Gospel (*1 Cor. 9:19-23*), and scholar "Bruce" points out that the Hellenized apology need not have compromised his Christian faith. Scholar "Marshall" also proposes that scholar "Dibelius's" view is in fact too extreme and contemptuous in rejecting the accuracy of the account. While "Paul" molds his speech to the Greek mindset, he presents a message of material discontinuity in relation to the creation and "Resurrection" that would have challenged the Greek ways of thinking. Consequently, creation, including Adam, did not arise through self-generation from lower life-forms, as the works of scholar "Homer" suggested; rather, the whole of creation owes its existence to a special act of the divine will.

The foundation for Paul's speech seems to resemble with passages in the Old Testament book of Isaiah concerning the foolishness of idolatry, for instance *Isaiah 44:6-20*. Scholar "Fudge" points out that there is a stronger link between *Acts 17:24-25* and *Isaiah 42:5-6* with a quote employed from the (Greek - Septuagint translation of the Old Testament LXX).

"This is what God the Lord says—the Creator (Greek ὁ θεὸς ὁ ποιήσας) of the heavens, who stretches them out, who spreads out the earth with all that springs from it, who gives breath to its people, and life to those who walk on it:" (Isaiah 42:5).

"The God who made (Greek ὁ θεὸς ὁ ποιήσας) the world and everything in it is the Lord of heaven and earth … because he himself gives all men life and breath and everything else" (Acts 17:24-25).

Paul's quotation of the Isaiah passage relates to his own stated commission that he was called as a *"light to the gentiles"* (Acts 13:47); a verse taken from *Isaiah 42:6*. Thus, there is a link between Paul's earlier calling and the Athenian speech, as scholar "Fudge" notes. Thus, there appears to be cognizance in Paul's mind that in preaching to the Athenians he is fulfilling his calling to enlighten the Gentiles and lead them out of idolatry. The chapters of *Isaiah 42-49* also express of the person and servanthood of the coming Messiah, and Isaiah goes on to express of the futility of worshipping idols made by human hands (*44:12-13*).

Acts 17 may also have some resemblance with the Book of Proverbs (13:1-5), which again expresses in contradiction of idolatry and accentuates the enormity and beauty of God through analogical reasoning and apophatic or negative theology, as scholar "Pelikan" suggests—*"For by the greatness of the beauty, and of the creature, the creator of them may be seen, so as to be known thereby"* (Proverbs 13:5). God is presented here as so much greater than the works of human art and design. Nonetheless, in this Wisdom passage the Creator is described as the work master; (Greek *technítin* τεχνίτην) (v. 1) and creator (Greek *ktismáton* κτισμάτων) (v. 5), but *poieō* does not seem to be employed in these few verses.

The speech in *Acts 17* appears to be organized as an appeal to the pantheistic "Stoics," and a challenge to the atheistic "Epicureans," by utilizing an apologetic the Stoics would have had some affinity for, and the "Epicureans" would comprehend. Paul cited, for instance, from two Greek poets; "Epimenides in *Cretica* (about 600 BC)," *"In him we live and move and have our being"*, and "Aratus (315-240 BC)" in *Phaenomena*, *"We are his offspring"* (Acts 17:28). The "Epicureans" may have shared Paul's loathing for superstitious idolatry, while the "Stoics" believed in a pantheistic divine agent as the *logos spermatikos*. The fact that Athens was full of idols disturbed Paul. However, he praised the men of Athens for their measure of religion, but noted that it was in ignorance, and that they did not go far enough in acknowledging the one true and living God. Pagan Greek thought was dualistic and tended to be escapist, being focused upon the next life with less apprehension for the present, which is why they protested so strongly to the idea of a "physical resurrection" of the dead as Paul had preached. Why, they thought, would anyone want to come back into this world of suffering?

While the speech denotes to Greek literature, as noted there are also themes that are in harmony with the Old Testament, predominantly Paul's desire to challenge idolatry that mirrored Isaiah's concern. Scholar "Dibelius" notes that Paul's employment of *poieō* in *Acts 17* refers to the act of creating and follows the (Greek - Septuagint translation of the Old Testament LXX) that often translates the Hebrew word *"bara"* as *poieō* (*Genesis 1:1, 27, 31*). This is a perfectly valid comment, especially in light of Paul's use of the creation quotation from *Isaiah 42:5* in the Athenian speech (in *Acts 17:24*), which does precisely that, however, as noted it is also evident that the speech has some correlation with other parts of Isaiah that raise concerns against idolatry and translate *"yatsar"* as *"poieō."* There is evidently a degree of fluidity in how *"poieō"* is employed in translation in the (Greek - Septuagint translation of the Old Testament LXX) for both *"bara and yatsar."*

In the Athens speech Paul appears to accentuate the language of poetry and workmanship by comparing analogically the works of mankind with that of God, but also highlights the relational aspect with mankind identified as 'God's offspring'. God may then be known as father as a positive cataphatic approach to apologetics that balances the apophatic analogy that can elevate God so high that he becomes unknowable. Referring to God as the author or poet of life is also consistent with aspects of the creation account where God is presented as speaking life into existence. To highlight the analogy between human poetry and the divine author of life, it is worth now looking at how *"poieō"* is employed in the text of *Acts 17*.

Acts 17: [24] The God who made [*poiēsas* and *cheiropoiētois*] *the world and everything in it is the Lord of heaven and earth and does not live in temples built by hands* [*xeiropoiētois* χειροποιήτοις].

Acts 17: [25] And he is not served by human hands, as if he needed anything, because he himself gives all men life and breath and everything else.

Acts 17: [26] From one man [of one blood ἐξ ἑνὸς αἵματός][33] he made [*epoiēsen ἐποίησέν*] every nation of men, that they should inhabit the whole earth; and he determined the times set for them and the exact places where they should live.

Acts 17: [27] God did this so that men would seek him and perhaps reach out for him and find him, though he is not far from each one of us.

Acts 17: [28] 'For in him we live and move and have our being.' As some of your own poets [*poiētōn* ποιητῶν] have said, 'We are his offspring.'

Acts 17: [29] Therefore since we are God's offspring, we should not think that the divine being is like gold or silver or stone—an image made by man's design [*technēs* τέχνης] and skill" (*Acts 17:24-29*).

When reading this speech with consideration of the Greek it is possible to get a feel for the idea that God's workmanship in creation is a form of poetry—God as the author of life, and this is something outside human skill or trade. In Paul's mind there is something more sacred about the creation than the work of human crafts. God is also seen as the performer of the formed creation, which points to direct divine activity in creation, as opposed to the claims of those who hold that God must act indirectly through natural processes. In the context outlined in the passage, there is a comparison between the work of human poets and the divine poetry of life, but God's handiwork is considered far greater.

From this there is an allusion to another aspect; that is, who can rightly claim that the work of Greek authors and poets might have arisen by accident? The "Epicureans," like their modern "Darwinian" counterparts, held that life arose through chaotic forces involving random movement of atoms. Then, Paul's analogy between human poetry and the divine authorship of life points to the inevitability of intention and action of a divine agent, especially given the much greater complexity of organic life, and the presence of rational human beings who can write poetry. And this mirrors Isaiah's point that it is pure folly to worship images formed by human hands because the one who forms must be greater than that which is formed (*Isaiah 44:9-20*). But then again, Paul presents God in a relational manner; someone we may come to know as father. There are lessons in Paul's speech for Christians looking for respond to New Age pagans and atheists.

But are there lessons for Christian methods to "Intelligent Design?" In Genesis the word "*yatsar*" appears to be a metaphor for God as a craftsman, and arguably the idea of God as the potter of Adam. Later "*yatsar*" was translated in the (Greek - Septuagint translation of the Old Testament LXX) and New Testament employing a number of diverse Greek words, one being "*poieō*," which, when read in context, can be employed as a metaphor for God as a poet. Nonetheless, it is also true that "*poieō*" can have more mundane meanings. It is also a valid translation for "*bara*" and formed the basis for use of *poieō* in *Acts 17:24*. "*Yatsar*" and "*poieō*" both infer divine intention, but still retain the notion of the physical formation of life. While, there is a place for comparing aspects of creation to God's design, these can only be incomplete metaphors for God's greater and holistic work in creating life. Paul does appear to accentuate divine intention by using the word "*poieō*," nevertheless, in this there is also the implication of divine action in creation. We do necessity to be careful not to separate too strongly the two aspects of creation; that is the mental thought processes and the physical work of shaping and forming as the work of a potter, or the author of life.

This writing has emphasized the manner in which the Hebrew word "*yatsar*" has been employed in the Old Testament, translated on occasions into the Greek of the (Greek - Septuagint translation of the Old Testament LXX) as *poieō*, and then employed in the New Testament, although *bara* is primarily translated as *poieō* in the Greek version of the Old Testament.

When we turn to the New Testament, it is evident that Paul's speech to the Athenians compared God's workmanship in creation with human craftsmanship as part of his apologetic strategy. In other words, just as a finished poem is an expression of the human mind, so also is the created order a finished expression of the mind of God. Who, for instance, would say that a poem could arise through purely random processes apart from mental input? Paul's apologetic does seem to have an artistic dimension, but he appears to have inferred the need to account for mental causation and information in terms of a mind, instead of merely giving an explanation for physical causation.

Therefore, how does this relate to our apologetic work? On occasions Scripture speaks metaphorically and analogically of God as the potter of mankind, as the author of life, or as a divine poet. There is an artistic dimension that needs to be addressed in our work, which goes beyond rational arguments about the physical mechanisms of creation. Furthermore, it is evident that an accessible Gospel presentation to modern Greeks (i.e., New Age pagans and atheists) needs to start in creation, and while God may be presented as the author or poet of life, it is necessary to highlight the positive relational aspect with God revealed as father.

We can see as well in the language employed in *Genesis 2* that there is a discontinuity implicit in the Hebraic account of the creation of Adam from the rest of creation. This discontinuity is evident in the teaching of Paul, who saw human life stem from one man (for instance *Acts 17:26*), as did early Christian theologians such as Irenaeus, who saw Adam as the protoplast or prototype of human kind and not an archetype. The Apostles and many of the Church Fathers found it necessary to challenge Greek idolatrous ways of thinking present in the world that came against the Church and identifiable in the modern idea of evolution.

Intelligent Design

Propagating Intelligent Design (ID) is a intimidating task. ID was hurled as a movement in large part by Phillip Johnson's landmark bestseller "*Darwin on Trial.*} Subsequently. its publication in 1991, ID literature has been multiplying exponentially. The literature critical of ID was unhurried to score some successful, nonetheless, it is currently, widespread as well.

Intelligent Design has always ensued against evolution in a two-step progression.

- First, design supporters attack the sufficiency and accuracy of the ruling paradigm, Darwinism. Nonetheless, the fact that Darwinism is deficient and imprecise does not, by itself, create the conclusion of design. Therefore,

- as a second point to the argument for ID, design supporters argue that a scientifically more acceptable clarification is design by an intelligent designer. Even within a single argument, the two-point approach is frequently distinguishable. For example, in Michael Behe's formulation of irreducible complexity, the negative argument claims that the Darwinian process is powerless of forming an irreducibly complex biological structure. The positive argument states that an irreducibly intricate biological structure is characteristic of design.

The two-point nature of the argument is frequently lost on the evolutionists, but it is significant since it means that Intelligent Design's argument is not an argument from ignorance, as evolutionists frequently suggest. In this "*Politically Incorrect Guide,*" there is a clear dividing line between the anti-Darwinist and pro-ID sections.

ANTI-EVOLUTION

Many UD proponents are enthusiastic to demolishing the trustworthiness of Darwinism, investigating the fossil record, embryological recapitulation and genetic phylogenies, among other things. Most of the themes will be acquainted to anyone with even a passing understanding with ID and creationism. Thus far, lest you've been reading the literature frequently and diligently, this writing may have numbers of new scoops on any specified issue even for those who are relatively acquainted with the debates. Particularly in this writing you would think "Darwin is as perceptive as a modem god," exploding the Darwinist claim that evolution is essential to science.

Proponents to Intrigant Design noted that speciation is not a problem to Intelligent Design or young earth creation, but became preoccupied into hairsplitting over whether any speciation had actually been *observed*. This is certainly an interesting question worthy of inquiry, but not helpful insofar as it reinforces the (false) public impression that creationists and Intelligent Design advocates believe in fixity of species. 6-Day creationists made the effort to qualify their discussion properly, but once they have stated that speciation is not a problem, why go through a long, and on the whole, unconvincing, argument to show that we haven't observed speciation actually *happen*?

PROPONENTS OF INTELLIGENT DESIGN

Proponents of Intelligent Design stride from the negative case in contradiction of evolution to the positive case for "Intelligent Design," starting with the toughest of all ID arguments to explain at a popular level: "Dembski's" descriptive filter. He does a good job of explaining this mathematically intricate concept to laymen, before touching on to discuss the design arguments from biological information --focusing on "Stephen Meyer," from irreducible complexity, --focusing on "Michael Behe", and from cosmic fine-tuning, --focusing on Guillermo Gonzalez. This is much to cover. Proponent of "Intelligent Design intelligently" avoid trapped into niceties, hitherto, succeeds to provide enough detail to make the discussion feel considerable. Usually, a successful debate may incorporate the bacterial flagellum to interact peacefully with the critics.

Fig.7.30: Deformed Animal – Intelligent Design!

Why Deformed Design Argument Against Intelligent Design is Unsound?

The examples of life-forms that seem to be both ill designed and poorly modified to their environments can be clarified by the reality that balance in the natural world must exist in order for life to exist. Darwin's claimed evolution by natural selection is illustrated to ultimately cause the extermination of all life, Figure 7.30. Consequently, life must have built-in limits to indemnify that natural balance is upheld and that one animal does not become too successful numerically. The example of cancer is utilized to venture on the ensuing result of supposed natural selection; namely, death of all life. 'Ill-design' features are a consequence of design limitations needed by the necessity for a balanced ecology, or originate from God's Curse upon the world, and the introduction of death due to the Fall, resulting in worsening of the original created order.

ILL DESIGNED ORGANISMS

In examining the natural world, one becomes conscious of many examples of animals and plants that look as if to be both ill designed and poorly adapted to their environments. Besides, many animals are able to endure only in a very slim set of conditions, and necessitate a rigid ecological niche. Trivial changes in environmental conditions can frequently be mortal to various animals, and may even result in extinction of an animal kind. Many organisms are extremely demanding in their dietary necessities and, if these organisms were designed, the design appears ill. If a trivial change in their food requirement occurs, they are not able to survive. Thus, the query arises, 'Does evolution provides a better elucidation than creationism for what seems to be these "ill design" features in nature?' Reflect the microbial world with respect to nutrient supplies. Some bacteria can endure sufficiently on only a few types of nutrients, and others, such as the Spiro plasmas, are so finicky that they necessitate some 80 ingredients to survive. Some bacteria necessitate a diet comprising all 20 amino acids, yet other bacteria require only a few or no amino acids in their diet.

Humans, in comparison, require only 10 amino acids. No pattern of primitive, simpler, less evolved to more evolved can be distinguished. One bacteria type, which yields large, reddish-colored colonies, received 'The Guinness Book of World Records' 'world's hardest bacterium' honor. Called "Deinococcus radiodurans," it was revealed in 1956 in a can of decayed meat at the Corvallis, Oregon, Agricultural Experiment Station. The bacterium had endured the radiation used to sterilize the food, and has subsequently been verified to 'tolerate one thousand times the radiation level that a person can' (and can even survive in the powerful radiation of a nuclear reactor)! "Radiodurans" like certain other bacteria has 'the extraordinary capability to readjust its radiation-shattered pieces of genetic material and, employing enzymes to convey in new nucleotides and stitch together the pieces, repair the damage.'

Cancer!

An enquiry that must be posed is, 'If this organism has evolved this critically significant ability—which offers it a foremost endurance advantage—why is this mechanism not more common?' This mechanism could effectively abolish cancer and other genetic diseases among humans and animals. Why would it have been lost in the alleged macroevolutionary process?

The Koala

At the macroscopic level, a good example of a tremendously demanding animal is the koala, which survives on a diet of only eucalyptus leaves. When eucalyptus leaves are few and far between, many koalas will die, even if other categories of food are in abundant supply. Some koala species are even more particular in that they consume only certain species of gum leaves, Figure 7.31.

Fig.7.31: The Koala

Dodo Bird

One of the best-known examples of an animal that is frequently claimed to be 'ill designed' for survival is the "dodo" bird. Its lifestyle and anatomy made it virtually certain that it would become extinct, if it comes across any aggressive, large, predatory animal. The "dodo" (and all other non-flying birds) lays its eggs on the flat ground, in its place of in a safer location. Laying eggs on the open ground exposes them to hundreds of ground-dwelling animals, and consequently, the eggs are often consumed. The ground egg-laying trait is a significant reason why some birds are today threatened with extinction. If they produced a large number of eggs,

7.32: Could the dodo bird be brought back from extinction?

survival would be less of a problem, but many ground-laying birds lay only a few eggs, or even one egg, Figure 7.32.

Giant Panda

Yet another well-known example is the giant panda, Figure 7.33. These animals are so clumsy at reproducing that only about a thousand pandas are left in the world in spite of a 30-year, multi-million-dollar campaign to hearten their breeding success by leading animal experts. The reasons why they are endangered with extinction embrace the fact that, although they can survive on a bambooless diet 'for a while at least', they normally survive only on a single species of bamboo.

Fig.7.33: *Giant Panada*

Inept Panda Reproduction

Their reproduction methods are also so inept that, even under idyllic conditions, they seldom reproduce very successfully, and under most conditions they don't reproduce at all.

One would imagine that ***millions of years of evolution*** would have improved their reproductive system to the point that they could efficiently reproduce at least in their natural environment. Any minor improvement, no matter how small, would be selected for, and only a few changes would have made them much more appropriate. The same could be said of the koala and the dodo. Although the giant panda has survived until modern times, their numbers were never large, even during the most favorable time of their existence, and loss of their habit may yet result in their extinction.

NATURAL SELECTION FAILED

The Crocodile and the Deer

Crocodiles usually capture their prey by going to the water's edge, and then clutching and drowning their victim. Certain sorts of deer-like mammals frequently drink by the water's edge, disregarding the local crocodiles that typically can easily pull a deer into the water, and then kill it by drowning before consuming it. After the millions of years appealed by evolutionists, it would appear that animals coming to water holes to drink where crocodiles feed would be able to sense the crocodile's presence better. Those animals that are even slightly more aware of the crocodile's presence would be more probable to live and pass

Fig.7.34: *Crocodile Capturing a Dear*

this trait on to their descendants, Figure 7.34. In the end, the whole population would similarly be more active in avoiding crocodiles. Neo-Darwinism also would predict that, as a deer evolved to be more sensitive to crocodile noise, sight and smell, the crocodile would evolve to be much more discrete than it is now. Yet this has not occurred. The deer are unusually unconscious to the crocodiles, and the crocodiles need only to swim to where the deer are and attack. As long as there are deer, the crocodiles will have plenty of easy meals.

Copulatory Organ

A current example of what seems to be ill adaptation, that would be strongly selected against, is the male bean weevil's copulatory organ. It is a spine-covered structure that slashes the female's copulatory organ. For understandable reasons, females characteristically fight potential mates by kicking with their hind legs. Consequently, of the injury, females that never mate have a much longer lifespan—about a month—while those that mate once live an average of only ten days, and the twice-mated females live a mere nine days. Selection would cause the female to advance a more robust copulatory organ.

Been Weevil

Any small mutation or genetic variant that lessens the stiffness or size of the spines would hypothetically be selected for as the female would be less probable to reject this mate. It would appear that millions of years of evolution would have eradicated this major obstruction to reproduction. In the end, a mutation or other genetic variation would have happened that produced the male spines to be less rigid, or would have caused their forfeiture all together, Figure 7.35. This adapted weevil would have augmented its chances of mating significantly, and consequently this male would be more possible to have more offspring than a bean weevil with the wild-type, rigid, spine-covered copulatory organ.

Fig.7.37: *Bee Weevil – Unadopted Process Contrary to Darwinian Theory*

Guinea Pigs

Another example is the need for dietary vitamin C, a critically significant composite obligatory for many body functions, not the least of which is this antioxidant's ability to aid neutralize free radicals. Guinea pigs, anthropoid apes and humans are the only known species that cannot synthesize vitamin C. Because it is frequently problematic to obtain sufficient in the diet, the ability to synthesize vitamin C would confer a main survival advantage. Many so-called primitive organisms have this capability, but numbers of higher animals deficient of it. Evolutionists claim that it was vanished during the evolution into higher life-forms. They point to evidence for a pseudogene (an sedentary or spoiled gene) involved in vitamin C production found in one sample (so far none has been found in any other primates). Up till now, if this is the case, the initiation of the gene would be highly favored. An animal that had the aptitude to manufacture vitamin C would permit them to continue in a far wider set of circumstances. Absence of vitamin C is predictable today as the main cause of a wide diversity of diseases. No longer would a vegetarian diet high in vitamin C be essential, but an animal could do very well on a much poorer-quality diet.

Additional example is the human species, Homo sapiens, evidently the most highly evolved animal on Earth. Bearing in mind body weight, humans are about 10 times more susceptible to most toxins than are many experimental animals. The variance is owing partially to more active biotransformation systems that detoxify poisons in many so-called lower animals, and is a vital reason why humans necessity to rigidly control their environment in order to survive.

Many animals also enjoy behavior traits that are frequently lethal—a well-known example is that dogs, and many other animals, generally consume animal excrement. The reason why this behavior can be deadly is that around 40% of the dry weight of most mammal excretory matter is bacteria, many kinds of which are pathogenic.

As dumping raw sewer water in drinking water can be catastrophic, so, too, are the ways of many animals. Some Darwinists claim that coprophagy (eating excrement) can be beneficial as a means of removing the markings of a contestant for territorial reasons. Yet coprophagy hardly removes the scent, and most dogs do not limit their coprophagy to any one territory. Some investigations show that coprophagy consequences from chronic stress and if not is normally uncommon.

One investigation discovered it in about 9% of dogs, many of which were in a stressful situation. In the wild (where dogs face much stress), coprophagy is obviously very common, but irrespective of how common it is, coprophagy is still very damaging health-wise, and is not a functional response to stress. Confidently, natural selection would have removed this trait after millions of years (or would never have selected for it). In spite of this, dogs have continued living very well in the wild and in captivity. However, their coprophagy behavior results in significantly higher rates of morbidity and mortality, as any dog lover knows.

CREATION OR EVOLUTION

These and other examples of ill fitness, either of biochemistry or behavior, have been employed by evolutionists as indication that life was not created. They reason: 'Why would a creator create animals that were so perceptibly marginally, or ill, adapted and could survive only in a very thin ecological niche or environment?' Darwin claimed that these examples were evidences of ill design that mitigated against an intelligent-design worldview. On the other hand, since evolution is alleged to be an undesigned, undirected, and unplanned process, Darwinists reason that if evolution were true, it would not be unexpected to discover many examples of ill design in nature. The problem with this reasoning is:

'To find fault with biological design because it misses an idealized optimum, as "Stephen Jay Gould" frequently does, is therefore gratuitous. Not knowing the purposes of the designer, "Gould" is in no position to say whether the designer has come up with a defective compromise among those objectives.'

These examples of less-than-optimal design argue against the efficacy of mutation/natural selection paradigm, but creationists also have come short of a good clarification for these observations, excluding pointing out that intelligent design is not essentially optimal design, and that flaws in the creation are predictable as a result of the biblical Fall.

Evolution concludes that the reason ill design exists universally is because what evolves is a result of chance, time, and the constraints of natural law. If an adaptation works sufficiently well to survive, the animal will not become extinct. However, actually, evolution has main problems clarifying what is normally observed: millions of years of evolution should not have formed the many ill adapted animals. If an animal cannot effectively compete or survive, the trait will not be passed on to its offspring. Natural selection should therefore consistently select for the variations that can contest and function better. In the words of "Timms and Read":

'The features that compel niche expansion lie at the heart of a key problem in evolutionary ecology: why are there so many different kinds of species? Why is there not an ultimate organism adapted to exploit all ecological niches? ... Why are there no parasite species exploiting all the members of large taxa such as mammals or birds?'

Evolutionists attempt to answer this question by, for example, perceiving factors limiting a species' range (water barriers, for example). This may explain a small number of incidents, but another factor may be more important. As "Timms and Read" note: 'We have unusually little comprehending of the relative significance of these alternatives in limiting host range in natural parasite populations.

Impossible Neo-Darwinism Evolution – Cancer illustrates Why

Let us consider cancer as a specimen of the basic mechanism of evolution through natural selection that exemplifies why organisms require to be less than optimally designed in order for life to survive. The progress of cancer necessitates a series of mutations which enable differential survival of that cell compared to other cells. If sufficient mutations happen to progress the survivability of a cell even slightly, then that cell will have an advantage compared to other cells. It is clear that the development of cancer is an example of classical, idealized, neo-Darwinian evolution, requiring both mutations and natural selection to happen. Additionally, the circumstances is frequently stated in this way in the literature. For example, research has found that prostate cancer progresses from a localized condition to a widely circulated malignancy and that each step along this progression pathway involves multiple genetic changes that convey a survival advantage to the tumor cell over its regular counterparts and may confer resistance to therapy.

In the case of cancer cells, each mutation that permits the cancer cells to replicate, even to some extent faster than the neighboring normal cells, upsurges the relative number of cells that encompass those mutants. Among the mutations that support the development of cancer cells are mutations in proto-oncogenes, genes that have a character in the cell that frequently is compared to an automobile accelerator. A mutation, in essence 'jams the accelerator' in the 'on' position, enabling unrestrained cell division (what actually occurs is that the mutation converts a proto-oncogene into an oncogene). The cell encompasses many systems designed to restore DNA damage, as well as tumor-suppressor genes (that halt the cell cycle so repair can occur) and various repair systems (such as proofreading and excision repair). If these repair systems are spoiled so that they no longer are functional, a mutated oncogene will not be repaired, and as a result a DNA-damaged cell is permitted to reproduce.

Cancer is fundamentally a survival-of-the-fittest fight connecting the mutated cell competing with the body's normal cells for food, nutrients and space. Each mutation that permits or heartens unrestrained cell division (the jammed accelerator example) similarly, is favored in the natural selection of cells. Likewise, mutations that contribute to the loss of the cell's capability to control reproduction, as well as mutations to tumor-suppressor genes, proto-oncogenes, DNA and other cell-repair genes, telomerase (an enzyme that adds base pairs to DNA that permit it to survive beyond the average number of cell divisions), and apoptosis coding genes (a complex mechanism that causes the cell to self-destruct) will be favored.

Few hypotheses in history have been so articulately and dynamically reinforced through empirical observation as has Neo-Darwinism in the case of cancer. Cancer research laboratories are accurately coming up with new evidence monthly and the similarity of cancer to evolution has been noted by many researchers. "Weinberg" noted that the detection that human tumor development caused from a set of gene mutations was:

'... enormously substantial because it reverberated a theme that had been echoing in the halls of science for more than a century. Tumor advancement illustrated outstanding parallels to the evolution of species. In the mid-nineteenth century, Charles Darwin had defined evolution in terms of nature's aptitude to select the fittest from among heterogeneous populations of organisms. After the detection of gene mutations in the 1920s and 1930s, Darwin's theory of natural selection was advanced and prolonged. Now scientists understood that randomly occurring mutations generated genetically heterogeneous populations of organisms, and that natural selection selected among these, preferring the survival and reproduction of those organisms that occurred to carry the most favorable gatherings of genes.'

"Weinberg" then contended that an analogous process exists within human tissues, specifically between individual cells:

'A cell that occurred to withstand a mutation changing one of its growth-regulating genes might have a growth advantage over its genetically normal neighbors. It would spawn a host of offspring which would accrue in disproportionate numbers in the tissue. Later, another mutation happening in one of these offspring would produce a cell having even greater growth potential, permitting this cell to create a more violently growing flock. These cells would be even more active in elbowing out their neighbors, outcompeting them for the limited space and nutrients within a tissue.'

"Weinberg concluded" that the cell evolution occurring within a living body that resulted in cancer is different from Darwinian evolution only '… in one important respect: The continual genetic enhancement of the evolving population would ultimately compromise its own long-term capability by destroying the atmosphere that cultivated it. Sooner or later, evolving cancer cell inhabitants would kill the host organism that was vital to their own survival.'

In fact, if Darwinism were true, the same result would also occur at the multi-celled-organism level. It must be stressed that cancer cells are not better cells as a whole, even though they can replicate more efficiently. Cancer cells, like mutations, cause degeneration. For example, they do not illustrate a 'gain of information, but normally illustrate a loss' or more disorder of functions. This is another example of evolutionists distinguishing advancement where only variation exists, often that causing degeneration.

CANCER ULTIMATE EFFECT

The eventual consequence of cancer is destroyed cells that have a reproduction advantage, and as a result, multiply faster. As these cells accrue more and more mutations, which allow them to divide faster and faster, ultimately runaway cell reproduction consequences. Then crowding and nutrient deprivation distressing normal cells ultimately harvests death for the whole organism. In this event, the end result of mutations and natural selection is at all times the death of the organism. Similarly, if Darwinism were true, an evaluation of the natural world discloses that the same prospect also would happen with all life: one of the animals in competition with another would ultimately win out. Ultimately, a 'super-animal' would evolve that could run faster than most others, endure in a wide diversity of temperature conditions, and be capable to ingest and digest a wide diversity of foods.

This animal would out-compete most other animals, and would ultimately dominate the earth, triggering most other animals to become extinct. This event would occur recurrently until, ultimately, only one super-animal remained. The disturbance of the ecosystem produced by the extinction of the other animals and life-forms, and the damage of biodiversity that resulted, similarly would ultimately cause the super-animal itself to become extinct, just as cancer evolution would cause the death of the organism. For this reason, an animal cannot have too great an advantage over other animals. The rivalry, in other words, must frequently be somewhat close to equal, accomplishing what is recognized as an ecologically balanced system.

ENDLESS IMPROVEMENT LEADS EXTINCTION ALL LIFE

The universal equilibrium in the natural world is accomplished by a extensive diversity of means that frequently cannot be considered by the theory of natural selection. Natural selection aids more to keep the animal numbers constant than to reason the development of mechanisms that serve to knowingly upsurge its population numbers. For example, if an animal species is endangered with only few predators, it inclines to have a 'natural' short lifespan, few offspring, or both. If it has many predators, it inclines to have a longer lifespan, many offspring, or both.

An animal that is troubled with a large number of predators will also typically possess many multifaceted protective/survival mechanisms. For example, those animals that cannot flee from predators frequently possess some resourceful means of defending themselves, such as quills for a porcupine, or ferocity in some rodents. Animals that have a high death in their young also incline to have more offspring. But in the case of the 'higher' animals (such as mammals)—most of which have few offspring—small numbers limit natural selection, plummeting the likelihood of development of survival-facilitating organs and structures. The fact of omni-balance forces in nature (unless humans drastically upset them) has been recurrently exemplified and highlighted by many researchers.

Another example is little water bears, crustaceans less than a millimeter long that are part of "Phylum Tardigrada." The over four hundred species that have been recognized inhabit a variety of niches extending from high mountains to the ocean abyss, and from the Arctic to the Antarctic. They can endure in temperatures ranging from higher than that of boiling water to those that are as low as 0.0008 Kelvin, or close to absolute zero. These shellfish survive environmental immoderations by going into a profound dormancy state in which they are unaware to hunger for hundreds of years, then emerging like Sleeping Beauty. They can also endure radiation a thousand times above the deadly dose for humans. They are in numerous ways enormously resilient, yet are inept in other ways, such as the capability to protect themselves against predators. Hsü concluded that:

'If the ability to endure a crisis is the substratum criterion of fitness, then little water bears are the fittest of us all, and that is the direction, the purpose, and the perfection to which natural selection should have inclined. Luckily, it has not.'

Essentially, the best example of a super-animal may be human beings. We now have the capability to cause many, if not most, animals to become extinct. Most current extinctions were triggered by humans or natural disasters (such as the Ice Age), not by other animals as a consequence of natural selection competition. This power so far has not been fully exercised, partially since humans distinguish that their life is contingent on the existence of a balanced ecosystem. Humans also, conferring to some research, usually have an inborn visceral love of animals, particularly baby animals–though no hesitation much of this is acquired through culture.

Thus, knowledge, culture and perhaps this assumed internal instinct aid as a brake to enable humans to control their drive, so as not to reach the state whereby all, or most, animals become **extinct**

NATURAL SELECTION – LIMITED LIFE SURVIVAL

Granting, most animals have their ecological niche, the widely held believe, however, face some competition. This rivalry, though, must be measured so that the familiar 'balance in nature' is upheld. If it is vanished, it must soon be re-established or extinction results. To survive, consequently, natural selection cannot purpose to significantly upset the balance among numerous forms of life, since so doing will sooner or later cause the extinction of the competition (and ultimately of all life). Accordingly, just as many human inventions have built-in weak points that spur-of-the-moment under pressure and avoid other points from deteriorating; similarly, built-in feebleness must exist in all life in order to safeguard that the balance of nature continues to exist. This built-in feebleness can be understood as essential, in order to uphold a balance in nature; i.e., natural selection at finest trims out the substandard and weaker individuals, reducing the amount of devolution.

In industry, many machines encompass a built-in designed weak point that will fail first, averting more impairment from happening to other parts of the unit. The best example is a fuse or circuit breaker, which are designed to fail before the internal wires overheat to the point of generating a fire or destroying the electrical components. Fuses and circuit breakers, undoubtedly, have prohibited millions of fires, and reduced or averted the damage of multiplied millions of electrical and electronic equipment units.

Circuit breakers, well-thought-out one of the most significant inventions ever, prove intelligent design. Likewise, the 'circuit breakers' found in nature that avoid one lifeform from causing the extinction of other lifeforms also validate intelligent design. This supplementary exemplifies the observation that intelligent design necessity not be optimal design, in terms of maximizing the survival of a specific species. This significant, built-in limitation in life clarifies the main inconsistency between the reality of nature and natural selection, and the balance discovered to exist in virtually all areas, as revealed by the investigations of ecology. What seems to be less than optional design in nature is essential to protect that one form of life does not dominate and consequence in the death of other forms and ultimately their extinction.

The set of observations studied here has significant consequences for both the creation and evolution worldviews. They explain the observation that many animals—even the most intelligent animals—normally manifest behaviors or dissimilarities that are inept from a existence viewpoint. This view clarifies the reasons designs, once adjudicated as inadequacies in the natural world (obviously a contradiction, as is calling a fuse or circuit breaker an imperfection), essentially have a critical function. This so-called imperfection is a essential design mandatory in order for life to endure in richness and diversity in the long term. Notwithstanding this built-in balance, occasion ally the balance is offset (frequently as a result of human intervention, and infrequently due to main natural catastrophes), compelling a novel equilibrium to be reached.

Beneficial mutations (motivating minimal information loss) are conceivable (although enormously rare), and natural selection has been documented to harvest a limited level of enhanced adaptation to the local environment. The problem that Darwinists must discourse is the origin of the variation, not the fact that certain variations cannot enable individual survival. Accordingly, variation within the created kinds (commonly misinterpreted as microevolution) is not a problem for creationists. Large-scale biologic transformation, or evolution, has never been made known to happen. If it could, such a change would be a momentous danger to the ecological balance of the biosphere.

The view argued here is that, in addition to genetic mechanisms, ecological mechanisms also occur to avoid evolution. This is since, as is the case with cancer, macroevolution would ultimately outcome in extinction of all life. These mechanisms, both genetic and ecological, embrace features of nature that have been disregarded by evolutionists as 'ill design'. Biblical creationists uphold that these 'ill design' properties are a result of either design limitations required by the essential for a balanced ecology, or emanate from God's Curse upon the world, and the introduction of death due to the Fall, consequential in the worsening of the original created order.

MALE REPRODUCTIVE SYSTEM

Perfect or flawed!

One of the up-to-date proofs of human evolution is the allegedly poor design claim, namely that an intelligent "Creator" would not design certain human body part in a particular manner. A sample is the human male reproductive system, Figure 7.36, which an Evolutionist listed as number four in his list of the top 10 design flaws in the human body.

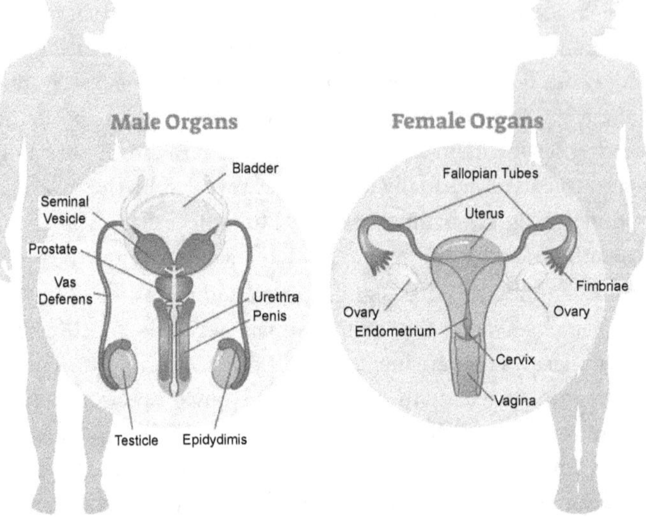

Fig.7.36: *Human Productive System*

The human male reproductive system unfortunate design claim emphases on the opinion that "if testicles were designed," then why didn't God "shield them better. Couldn't the Designer have placed them exclusively inside the body, or enclosed them in bone" like the brain which is walled by a hard skull?

Closing that a body edifice is poorly designed, as Oxford University Ph.D. Professor "Hafer" claims, as an alternative of asking *why* the existing design exists, is a science stopper. The 'why' question inspires research into the causes for the design. As this method was implemented to the human appendix, the tonsils, the backward retina, and the numerous putative other specimens of supposed poor design, good reasons for the prevailing designs were discovered in all cases. The same is true of the male reproductive system.

"Hafer" elucidated that when she was searching for novel methods to disprove Intelligent Design, she knew she "had a winner when … in the middle of an Anatomy and Physiology lecture" she determined that the male reproduction system "is a great first argument against Intelligent Design". She held that she also had a good "political-style argument" against ID. Her foremost argument is that since male testicles are outside of the body, they are disposed to damage. She augments that in numerous animals, counting cold-blooded reptiles, they are situated inside the body where they are fully safe guarded.

Reasons for Design

Male testicles exist outside of the body in humans and most mammals for numbers of significant reasons, counting operative directive of scrotal temperature for ideal spermatogenesis generation. Additional reason is to retain sperm comparatively inactive until they cross the threshold of warm boundaries of the female reproductive system. Even merely a few degrees above the optimal temperature is disadvantageous to both sperm generation, precisely in the later stages of spermatogenesis, and sperm maturation.

A low ambient temperature is indispensable for typical spermatogenesis in humans and *most* mammals since the enzymes essential for the process are denaturized if their temperature is not finely controlled.

One study in mice discovered temperatures of 37°C or higher caused "a momentous decrease in the percentage of motile sperm", creating an upsurge in the number of spermatozoa with plasma membrane damage. The mammal exclusions include monotremes, mammals that lay eggs as an alternative of giving birth to live young, and have intra-abdominal testes. Similarly, some placental mammals, such as insectivores (shrews, hedgehogs, and moles), as well as elephants and hippopotamuses, all have intra-abdominal testicles. One influence is externalized testes are found only in certain mammals whose lifestyle includes jumping, leaping, or galloping. In big size animals with this lifestyle this behavior would be predictable to place great pressure on the testicles and even expel their contents by creating concussive hydrostatic rises in peritoneal pressure.

Wikimedia Commons

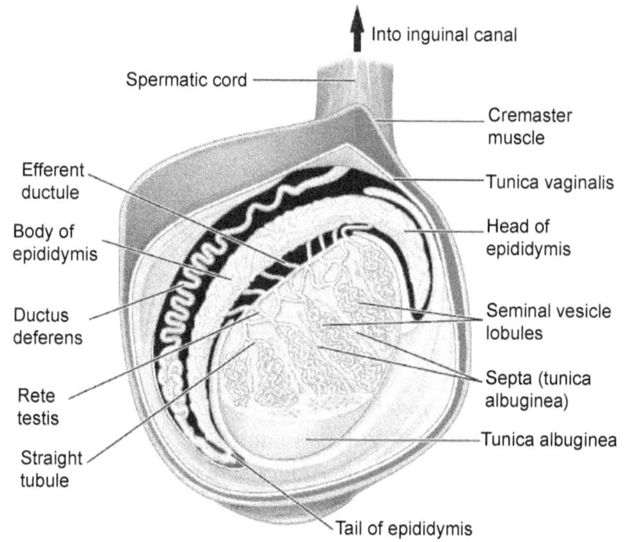

Fig.7.37: *Diagram of a Male Testicle Showing its Internal Structure.*

Equated to core body temperature, the typical temperature drop accomplished in the current design construction is idyllic, Figure 7.37. The sperm generation system is upheld at a temperature very close to 4°C cooler than the typical body temperature of 37°C. An upsurge in temperature by as little as 2°C unfavorably distresses sperm development. The result of this 2°C upsurge in humans embraces a lesser sperm count, and a significant surge in the number of uncharacteristic sperm.

If the testes were placed inclusively inside the body, the enzymes sperm necessitate to be healthy would be denaturized in a matter of hours. New sperm would have to continuously be generated to permit humans to be fertile year-round, as is ordinary for humans. This matter would not be an apprehension for most animals that are fertile only during very short windows each year.

Numbers of intricate mechanisms exist to safeguard that the 4°C variance is upheld at 37°C down to 33°C. Once testicle temperature drops below 33°C, a complex feedback system causes the cremaster muscle that surrounds the testicles to contract, which moves the testicles closer to the warm 37°C body in order to recompense for the heat loss. When their temperature increases above 33°C, the cremaster muscle relaxes, permitting the testicles to move away from the body. This ensures that the ideal male reproductive system temperature is upheld within a very slim tolerance of the 33°C ideal.

Their 33°C temperature is likewise upheld by increasing or decreasing the external area of the tissue surrounding the testicles, the scrotum, permitting faster or slower dissipation of their heat. It does this by expanding like a crumpled balloon does when air is pumped into it. Additionally, scrotal skin is very thin, permitting the testes to effortlessly lose heat into the surrounding environment. The air circulating around the scrotum sack added helps to enable the cooling of the scrotal skin, in sequence serving to cool the sperm development system.

Besides, to uphold the appropriate temperature, the arteries carrying blood into the scrotum run along with the veins that carry blood away. This sophisticated heat-exchanger mechanism drops the temperature of the blood supply travelling to the testicles. The warm arterial blood approaching from the abdomen drops heat to the cooler venous blood coming away from the testes. The consequence is that the blood is cooled slightly before even entering the scrotum. For these details, the existing system is an outstanding design, well recognized to engineers as *counter-current exchange*, to uphold optimal spermatogenesis temperature control. This design is prevalent in biological systems.

YEAR-ROUND FERTILITY

One main cause for the rigid temperature regulation is due to humans are fertile throughout the entire year, and effectively all animals with internal testicles, as well as all cold-blooded vertebrates and birds, are not. Most animals require to be fertile only during a very short period of time during their mating season. This is frequently when outdoor temperature permits maintenance of the proper temperature for spermatogenesis, such as is the case for reptiles. In agreement with this observation, "Freeman" found that "taxa with internal testes yield large volumes of low-quality sperm while taxa with scrotal testes yield smaller volumes of higher-quality sperm".

Additional reason for the unbending temperature regulation is that sperm have a very short lifespan and must be kept at lower than body temperature to keep them dormant longer. Sperm are kept in the epididymis where they mature, and the warmth of the female reproductive system attends to help stimulate them. If the testicles were located inside of the male body, the sperm would be triggered much sooner, and thus, given their short lifespan, measured in hours, bulky numbers of sperm would die before they could even enter the vagina.

Evidence for the temperature effect embraces the fact that semen quality is inferior in the warm summer months compared with the cold winter months. The semen volume does not vary significantly, but the total sperm count falls in the summer compared to winter, specifically in the northern hemisphere during the hottest summer months of July and August.

DESIGN PROTECTIONS

Several designs reduce the prospect of testicular damage, counting the left testicle typically dangling lower than the right one and, consequently, pressure causes one to slip past the other without pain or injury. Each testicle is contained in a strong fibrous outer casing called the "tunica albuginea," and an operative lubrication system permits the slippage to happen without pain or problems. Injury is rare, and the main source of injury is in sports, which is why it is recommended that sport contestants continuously utilize defensive gear, such as a jockstrap or hard cup, while playing.

CRYPTORCHIDISM

Another cause for the existing positioning of the human male reproductive system outside of the body is that the

"… postpartum testicle is designed to operate at this lower temperature. Failure of the testicles to descend into the scrotum, called cryptorchidism, causes a greater risk of malignancy and other major health problems. The process of testicle descent is also both complex and poorly understood."

Failure of the testicles to descend following birth of a baby leads to progressive anomaly in both the biochemistry and physiology of the testis, often causing infertility.

One example is that the abnormal biochemistry caused by descent failure hinders many essential reproductive system developments. Cases including the transformation of neonatal gonocytes into type A spermatogenesis, a step vital to produce viable sperm. This is one reason why failure of the testicles to descend is a foremost reason for male infertility.

SCROTUM EVOLUTION?

In what manner the various complex male temperature regulatory system parts could have evolved can only be ventured, and not be founded on observation and science. Accordingly, evolutionists must provide many just-so stories in an effort to explain their presence and function. In a word, a literature review discovered that "all of the current hypotheses concerning the origin and evolution of the scrotum" and external testicles are extremely problematic. Explanations include, assuming external testicle evolution from lower life forms with internal reproductive organs is problematic because it is "why the scrotum has been lost in so many groups, that should be explained." The evolutionists even speculate that the scrotum may have evolved before mammals did:

"… in concert with the evolution of endothermy in the mammalian lineage, and that the scrotum has been lost in many groups because descent in many respects is a costly process that will be lost in mammal lineages as soon as an alternative solution to the problem of the temperature sensitivity of spermatogenesis is available."

Clear evidence is obvious as the year-round reproductive cycles, in addition to the necessity that human sperm must be kept close to a constant temperature of 4°C below that of the core body temperature, efficiently explains the existing design of testicles. Men who have unremedied non-descended testicles are typically infertile and predisposed to to many other health problems, including cancer. In a nutshell, the existing intricate design is obligatory for many reasons, including fertility and health reasons. It is, consequently, clear that "Hafer's poor design claim," along with those of other evolutionists, is grossly reckless.

APPENDIX

The following dictionaries have been employed: Strong, J., 'New Strong's Concise Dictionary of the Words in the Hebrew Bible (HB), and New Strong's Concise Dictionary of the Words in the Greek Testament (GT), in *The New Strong's Exhaustive Concordance of the Bible*, Thomas Nelson, Nashville, TN, 1995; Unger M.F. and White, W., 'Nelson's Expository Dictionary of the Old Testament; in Vine, W.E., Unger, M.F. and White, W. (Eds.), *Vine's Complete Expository Dictionary of Old and New Testament Words*, Thomas Nelson, Nashville, TN, 1985; Vine, W.E, 'An Expository Dictionary of New Testament Words; in Vine, W.E., Unger, M.F. and White, W. (Eds.), *Vine's Complete Expository Dictionary of Old and New Testament Words*, Thomas Nelson, Nashville, TN, 1985.

REFERENCES

1. For instance; Behe, M., *The Edge of Evolution: The Search for the Limits of Darwinism*, Free Press, New York, p. 166, 2007.

2. Unger and White, ref. 2, p. 86. Return to text

3. There is another metaphor in the passage of Genesis 2, that of God as the gardener bringing forth plants and establishing the Garden of Eden. Return to text

4. Unger and White, ref. 2, pp. 51-52; Strong, ref. 2a, pp. 32, 111.

5. Alexander, D.A., *Creation or Evolution: Do We Have to Choose?* Monarch, Oxford, p. 237, 2008.

6. Walton, J.H., *The Lost World of Genesis One*, InterVarsity Press, Downers Grove, IL, pp. 69-71, 2009.

7. Walton, J.H., An Historical Adam: Archetypal Creation View; in: Barrett, M. and Caneday, A. (Eds.), *Four Views on the Historical Adam*, Zondervan, pp. 90-93, 2013. Return to text

8. Statham, D., Dubious and dangerous exposition, A review of *The Lost World of Genesis One: Ancient Cosmology and the Origins Debate* by John H. Walton, *J. Creation* 24(3):24-26, 2010; creation.com/genlostworld.

9. Lamoureux, D., Response from the Evolutionary View, to Walton's chapter 'An Historical Adam: Archetypal Creation View'; in: Barrett, M. and Caneday, A. (Eds.), *Four Views on the Historical Adam*, Zondervan, 2010.

10. Irenaeus, *Against Heresies*, Book III 21:10.

11. Hamilton, V.P., *The New International Commentary on the Old Testament: The Book of Genesis Chapters 1–17*, William B. Eerdmans, Grand Rapids, MI, p.156, 1990; *aphar* Strong HB6083, *chomer* Strong HB2563.

12. Augustine, *The City of God*, Book XIII, ch. 24.

13. Dibelius, M., *Studies in the Acts of the Apostles*, Greeven, H., (Ed.), Ling, M. (transl.), 1st English edn, SCM Press Ltd, London, pp. 26-83, 1956.

14. Bruce, F.F., *The Book of Acts*, Revised Edition, Eerdman, Grand Rapids, MI, pp. 341-343, 1988.

15. Marshall, I.H., *The Acts of the Apostles: An Introduction and Commentary*, InterVarsity Press, Leicester, pp. 281-283, 1980.

16. Charles, J.D., *The Unformed Conscience of Evangelicalism*, InterVarsity Press, Downers Grove, IL, pp. 151-155, 2002.

17. Fudge, E., Paul's Apostolic Self-Consciousness at Athens, *J. Evangelical Theological Society* 14:193-198, 1971. See also Isaiah 44:6-7 "Israel's King and Redeemer, the Lord Almighty: I am the first and I am the last; apart from me there is no God. Who then is like me? Let him proclaim it. Let him declare and lay out before me what has happened since I established my ancient people … "

18. Pelikan, J., *Brazos Theological Commentary on the* Douay-Rheims Bible, *Bible-Acts*, Brazos Press, Grand Rapids, MI, Pelikan, J, *Acts-Brazos theological commentary on the Bible*, Brazos Press: Grand Rapids, 2005, pp. 192–196 Wisdom 13:5 is from the, 2005Wisdom 13:5 is from the. The Wisdom passage is taken from the Douay-Rheims translation.

19. Not all manuscripts have αἵματός 'blood', as Vine, ref. 2, p. 71, for instance, points out.

20. Rowe, C., The 10 Design Flaws in the Human Body, *Nautilus*, 24 March 2016, p. 2.

21. Hafer, A., *The Not-So-Intelligent Designer: Why evolution explains the human body and Intelligent Design does not*, Cascade Books, Eugene, OR, p. 5, 2016.

22. Hafer, A., No data required: why intelligent design is not science, *The American Biology Teacher* **77**(7):507–513, 2015; Hafer, ref. 2 p. 2.

23. Wechalekar, H., Setchell, B.P., Peirce, E.J., Ricci, M., Leigh, C., and Breed, W.G., Whole-body heat exposure induces membrane changes in spermatozoa from the cauda epididymis of laboratory mice, *Asian J. Andrology* **12**(4):591–598, 2010.

24. Setchell, B.P., The effects of heat on the testes of mammals, *Animal Reproduction* **3**(2):81–91, 2006.

25. Sodera, V., *One Small Speck to Man—The evolution myth*, 2ndedn, Vija Sodera Productions, pp. 109–110, 2009.

26. Chance, M.R.A., Reason for externalization of the testis of mammals, *J. Zoology* 239(4):691–695, 1996.

27. Hutson, J.M. and Clarke, M., Current Management of the Undescended Testicle, *Seminars in Pediatric Surgery* 16:64–70, 2007; p. 65.

28. Ahmad, G. Moinard, N., Esquerre-Lamare, C., Mieusset, R., and Bujan, L., Mild induced testicular and epididymal hyperthermia alters sperm chromatin integrity in men, *Fertility and Sterility* 97(3):546–553, 2012.

29. Van Niekerk, E., Vas deferens—refuting 'bad design' arguments, *J. Creation* 26(3):60–67, 2012; creation.com/vas-deferens

30. Freeman, S., The evolution of the scrotum: a new hypothesis, *J. Theoretical Biology* **145**:429–445, 1990; p. 429.

31. Chen, Z., Toth, T., Godfrey-Bailey, L., Mercedat, N., Schiff, I., and Hauser, R., Seasonal variation and age-related changes in human semen parameters, *J. Andrology* 24(2):226–231, 2003.

32. Chung, E. and Brock, G., Cryptorchidism and its impact on male fertility: a state of art review of current literature, *Canadian Urological Association J.* 5(3):210–214, 2011.

33. Gallup, G.G., Finn, M.M., and Sammis, B., On the origin of descended scrotal testicles: the activation hypothesis, *Evolutionary Psychology* 7(4):517–526, 2009; Freeman, S., The evolution of the scrotum: a new hypothesis, *J. Theoretical Biology* 145:429–445, 1990.

34. Heyns, C.F. and Hutson, J.M., Historical review of theories on testicular descent, *J. Urology* **153**:754–767, 1995; Ivell, R., Lifestyle impact and the biology of the human scrotum, *Reproductive Biology and Endocrinology* **5**(15):1–8, 2007.

35. Werdelin, L. and Nilsonne, A., The evolution of the scrotum and testicular descent in mammals, *J. Theoretical Biology* 196(1):61–72, 1999; pp. 61–62.

36. Black, J., Microbiology Principles and Explorations, Prentice Hall, Saddle River, 1999.

38. Lewis, R., Human Genetics, McGraw Hill, New York, p. 157, 2001.

40. Burton, M. and Burton, R., Wildlife Encyclopedia, Funk & Wagnalls, New York, 1970. Actually, sometimes they can eat other leaves, even of non-Australian trees such as Monterey Pine. Koalas seem to have become dependent on addictive chemicals in the leaves. These chemicals pass through the mother's milk.

41. Morris, R. and Morris, D., Men and Pandas, McGraw Hill, New York, p. 151, 1966.

43. Hsü, K., The Great Dying; Cosmic Catastrophe, Dinosaurs and the Theory of Evolution, Harcourt, Brace, Jovanovich, New York, p. 49, 1986.

44. Schaller, G.B., The Last Panda, University of Chicago Press, Chicago, 1993.

46. Crudgington, H. and Siva-Jothy, M., Genital damage, kicking and early death, Nature 407:855, 2000.

47. Wentzler, R., The Vitamin Book, Gramercy Publishing Company, New York, 1978.

49. Pauling, L., Vitamin C the Common Cold and the Flu, W.H. Freeman and Company, San Francisco, 1976.

51. Evolution 14(9):333–334, 1999; p. 334. 19. Timms and Read, Ref. 18, p. 333. 20. Sullivan, G.F., Amenta, P.S., Villanueva, J.D., Alvarez, C.J., Yang, J.M. and Hait, W.N., The expression of drug resistance gene products during the progression of human prostate cancer, Clinical Cancer Research 4(6): 1393–1403, 1998; p. 1343. 21. Weinberg, R., One Renegade Cell; How Cancer Begins, Basic Books, New York, p. 55, 1998. 22. Weinberg, Ref. 21, p. 56. 23. Demick, D., Cancer and the Curse, Back to Genesis 145:a–c, January 2001. 24. Hsü, Ref. 5, p. 245. 25. Marchant, R.A., Man and Beast, Macmillan, New York, 1968.

52. Gitt, W., *In the Beginning was Information*, CLV, Bielefeld, Germany, 1997.

53. Dawkins, R., *The Blind Watchmaker*, W.W. Norton, NY, USA, p. 115, 1986.

54. *genitalium*, *Science* 270(5235):397–403, 1995; perspective by Goffeau, A., Life with 482 Genes, same issue, pp. 445–446.

55. Gitt, W., Dazzling design in miniature, *Creation* 20(1):6, 1997

56. A fictitious dialogue involving an Epicurean, a Stoic, and a speaker from the Academy (the philosophical school founded by Plato). Return to text.

57. Based on the teachings of Epicurus (341–270 BC).

58. Based on the teachings of Zeno of Citium (335–263 BC). The Stoics were pantheists. For them, happiness lay in emulating the calm and order of the universe by enduring hardship and adversity with fortitude and a tranquil mind. The name derives from the porch (Greek: stoa) where Zeno taught.

59. Derived from Democritus (460–361 BC). These philosophies were of Greek origin.

60. Cicero, *De Natura Deorum*, Book 2, sections 87–88.

61. Compare Grigg, R., Could monkeys type the 23rd Psalm? *Creation* 13(1):30–34, 1990; updated in *Apologia* **3**(2):59–64, 1994.

62. Paley, W., *Natural Theology*, first published 1802, republished by Bill Cooper as *Paley's Watchmaker*, New Wine Press, Chichester, England, pp. 29–31, 1995.

63. Sober, E., *Philosophy of Biology*, Westview Press, Boulder, CO, USA, p. 34, 1993, cited in Behe, Ref. 20.

64. C. Darwin to John Lubbock, Nov. 15, 1859, *Life and Letters of Charles Darwin*, D. Appleton & Co., **2**:15, 1911.

65. Wieland, C., Darwin's real message: have you missed it? *Creation* 14(4):16–19, 1992.

66. Gitt, W., Weasel words, *Creation* 20(4):20–21, September 1998.

67. Bohlin, R.G., Up the river without a paddle—Review of *River Out of Eden: A Darwinian View of Life*, *Journal of Creation* 10(3):322–327, 1996.

68. Sarfati, J.D., Review of *Climbing Mount Improbable*, *Journal of Creation* 12(1):29–34, 1998.

69. Truman, R., The problem of information for the Theory of Evolution: Has Dawkins really solved it? 14 July 1999; trueorigin.org/dawkinfo

70. Sarfati, J., Origin of life: the polymerization problem, *Journal of Creation* 12(3):281–284, 1998.

71. Sarfati, J., Origin of life: Instability of building blocks, *Journal of Creation* 13(3):124–127, 1998.

72. Behe, M.J., *Darwin's Black Box*, The Free Press, NY, USA, p. 217, 1996. He calls this property: 'irreducible complexity'.

73. Spetner, L.M., *Not by Chance*, The Judaica Press, Brooklyn, NY, USA, 1997; see A review of *Not by chance!*

74. Matthew 28:18–20; Mark 16:15; Luke 24:47; Romans 10:13–15.

75. For further reading: Morris, H.M., Design is not enough! *Back to Genesis* No. 127: a–c, July 1999; icr.org/article/859/321.

8

MOLECULAR INTELLIGENT DESIGN

Life did not arise by physics and chemistry without intelligence, Figure 8.12. The intelligence needed to create life, even the simplest life, is far greater than that of humans; we are still scratching around trying to understand fully how the simplest life forms work. There is much yet to be learned of even the simplest bacterium. Indeed, as we learn more the 'problem' of the origin of life gets more difficult; a solution does not get nearer, it gets further away. But the real problem is this: the origin of life screams at us that there is a superintelligent Creator of life and that is just not acceptable to the secular mind of today.

God is the master engineer. He did not just create life; He designed life to robustly respond to the environment. He overdesigned living things so that they could withstand, so far, thousands of years of mutational accumulation. His designs were holistic, in that He accounted for what each kind would need in the future. He built redundancy in the genomes to ward off the effects of decay, but He also designed specific aspects of the genome to change via mutation.

ORIGIN OF LIFE

The origin of life is also recognized as *abiogenesis* or sometimes *chemical evolution*.

Life is founded on long information-rich molecules such as DNA and RNA that contain instructions for making proteins, upon which life depends.

DNA: Stands for deoxyribonucleic acid, (abbreviated DNA). It provides the code for the cell 's activities.

RNA: Stand for Ribonucleic acid (abbreviated RNA). It converts that code into proteins to carry out cellular functions.

THE MARK OF LIFE

However, the reading of the DNA/RNA to make proteins, and the replication of DNA or RNA to make new cells (reproduction, the mark of 'life') both depend on a large suite of proteins that are coded on the DNA/RNA. Both the DNA/RNA and the proteins essential to be present at the same time for life to begin.

Thus, the origin of life is a frustrating problem for those who insist that life arose through purely *natural processes* (physics and chemistry alone, deprived of God).

Some evolutionists claim that the origin of life is not a part of evolution. However, probably every evolutionary biology textbook has a section on the origin of life in the chapters on evolution. The University of California, Berkeley, has the origin of life included in their 'Evolution 101' course, in a section titled "*From Soup to Cells—the Origin of Life.*" High-profile defenders of '*all-things-evolutionary*', such as P.Z. Myers and Nick Matzke, agree that the origin of life is part of evolution, as does Richard Dawkins.

THE SPECIAL THEORY OF EVOLUTION (STE)

A well-known evolutionist of the past, G.A. Kerkut, did make a distinction between the *General Theory of Evolution* (GTE), which included the origin of life, and the *Special Theory of Evolution* (STE) that only dealt with the diversification of life (the supposed topic of Darwin's 1859 book).

It is only recently that some defenders of evolution have tried to divorce the origin of life from consideration. It's probably because the hope of finding an answer is rapidly fading, as one scientific discovery after another of sophisticated machinery in even the simplest living cells makes the problem of a naturalistic origin ever more difficult.

So, what do we need to get life? We can break the problem of the origin of life into a number of topics in an attempt to explain to non-scientists what is involved (although it still might be mind-stretching).

THE FREE-LIVING CELL

What is it that we have to acquire to produce a **living** cell?

A living cell is capable of obtaining all the resources it needs from its surroundings and reproducing itself. The first cell had to be free-living; that is, it could not depend on other cells for its survival because other cells did not exist.

Parasites cannot be a model for 'first life' because they need existing cells to survive. This also rules out viruses and the like as the precursors to life as they must have living cells that they can parasitize to reproduce themselves. Prions, misshaped proteins that cause disease, have nothing to do with the origin of life because they can only 'replicate' by causing proteins manufactured by a cell to become misshaped.

The first things needed are the right ingredients. It's almost like baking a cake; you can't make a apple pie if you have no apples or flour.

THE RIGHT INGREDIENTS

Right here there is a major problem for chemical soup approaches to the origin of life: all the components have to be present in the same location for a living cell to have any possibility of being assembled. But necessary components of life have:

carbonyl (>C=O) chemical groups

that react destructively with amino acids and other amino:

$$(-NH_2) \text{ compounds.}$$

Such carbonyl-containing molecules include sugars, which also form the backbone of DNA and RNA. Living cells have ways of keeping them apart and protecting them to prevent such cross-reactions, or can repair the damage when it occurs, but a chemical soup has no such facility.

Cells are incredibly complex arrangements of simpler chemicals. I am not going to cover every chemical that a first cell would need; it would take a book and some to cover it. I am just going to highlight some of the basic components that have to be present for any origin of life scenario.

1. Amino Acids

Living things are loaded with *proteins*; linear strings of *amino acids. Enzymes* are special proteins that help chemical reactions to happen (catalysts). For example, the enzyme amylase is secreted in our saliva and causes starch molecules from rice, bread, potatoes, etc., to break up into smaller molecules, which can then be broken down to their constituent glucose molecules. We can't absorb starch, but we are able to absorb glucose and use it to power our bodies.

Some reactions necessary for life go so slowly without enzymes that they would effectively never produce enough product to be useful, even given billions of years.

Other proteins form muscles, bone, skin, hair and all manner of the structural parts of cells and bodies. Humans can produce well over 100,000 proteins (possibly millions; nobody really knows exactly how many), whereas a typical bacterium can produce one or two thousand different ones.

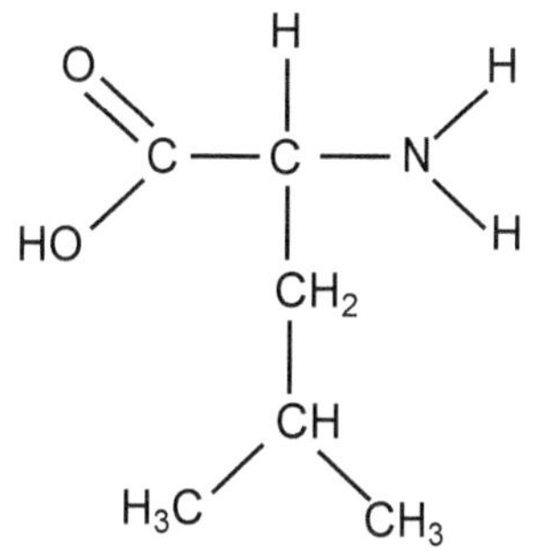

Proteins are made up of 20 different amino acids, Figure 8.1 (some microbes have an extra one or two). Amino acids are not simple chemicals and they are not easy to make in the right way without enzymes (which are themselves composed of amino acids);

Fig.8.1: Leucine, the most common amino acid, which is a specific arrangement of atoms of carbon (C), hydrogen (H), oxygen (O), and nitrogen (N).

The 1953 Miller–Urey experiment, which almost every biology textbook still presents, managed to make some amino acids without enzymes. It is often portrayed as explaining '*the origin of life*', but that is either very ignorant or very ignorant or very deceitful.

Although *tiny* amounts of *some* of the right amino acids were made, the conditions set up for the experiment could never have occurred on Earth; for example, any oxygen in the 'atmosphere' in the flask would have prevented anything from forming. Furthermore, some of the *wrong types* of amino acids were produced, as well as other chemicals that would 'cross-react', preventing anything useful forming.

The amino acids necessary for functional proteins could never have been made by anything like this experiment in nature. When Stanley Miller repeated the experiment in 1983 with a slightly more realistic mixture of gases, he only got trace amounts of glycine, the simplest of the 20 amino acids needed.

There appears to be rehabilitated enthusiasm in some quarters for injecting new life into the Miller-Urey experiment: "Ouellette, J.," Scientists recreated classic origin-of-life experiment and made a new discovery, "arstechnica.com", 29 October 2021. This commentary hypes a finding of scientists that silica leached from the borosilicate glass flasks utilized by Urey contributed a catalytic function to the apparatus. They also detected a wider range of organic compounds than Miller-Urey. Nevertheless, these discoveries do not change the conclusion that the Miller-Urey experiment is unconnected to the origin of the first cell, for the reasons stated herein.

The origin of the correct mix of amino acids remains an unresolved problem, Figure 8.2.

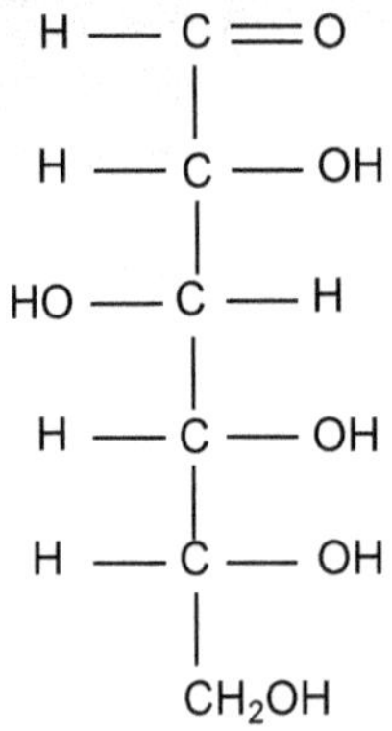

$$
\begin{array}{ccc}
H & - \; C = O \\
& | \\
H & - \; C - OH \\
& | \\
HO & - \; C - H \\
& | \\
H & - \; C - OH \\
& | \\
H & - \; C - OH \\
& | \\
& CH_2OH
\end{array}
$$

Fig.8.2: Glucose, linear form.

2. Sugars

Some sugars can be made just from chemistry without enzymes (which are only made by cells, as explained above). Sugars are theoretical to have fashioned from naturally occurring formaldehyde in the presence of alkali by the "*formose* (or *Butlerov*)" reaction. Nevertheless, the very same alkaline conditions that are required for this reaction also destroy sugars such as *ribose* and *glucose* that are crucial for life.

The *formose* reaction that is proposed for the formation of sugars also necessitates the *absence of nitrogenous compounds*, such as amino acids, because these react with the formaldehyde, and the sugars, to produce non-biological chemicals.

Ribose, the sugar that forms the backbone of *RNA*, and in modified form *DNA*, a vital part of all living cells, is specifically problematic. It is an unstable sugar (it has a short half-life, or breaks down quickly) in the real world at near-neutral pH (neither acid nor alkaline).

3. The Components of DNA and RNA

How can we get the nucleotides that are the chemical 'letters' of *DNA* and *RNA* without the help of enzymes from a living cell?

The chemical reactions need formaldehyde ($H_2C=O$) to react with hydrogen cyanide ($HC\equiv N$). However, formaldehyde and cyanide (especially) are deadly poisons. They would *destroy* critically important proteins that *might* have formed!

Cytosine, Figure 8.3, one of the five indispensable nucleotide bases of DNA and RNA, is very difficult to make in any realistic pre-biotic scenario and is also very unstable.

DNA and RNA also have backbones of alternating sugars and phosphate groups. The problems with sugars are explained previously. Phosphates would be precipitated by the abundant calcium ions in sea water or cling strongly onto the surfaces of clay particles. Either scenario would prevent phosphate from being used to make DNA.

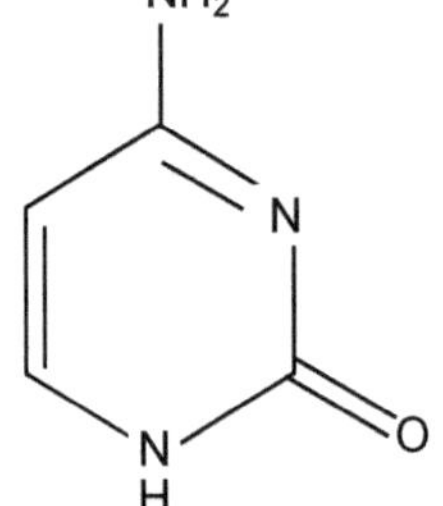

Fig.8.3: Cytosine, one of the simpler of the five nucleotides that make up DNA and RNA. In this form of chemical diagram, each unlabeled bend in the ring has a carbon atom at the bend.

4. Lipids

Lipids ('fats') are vital for the formation of a *cell membrane* that contains the cell contents, as well as for other cell functions. The cell membrane, embrace of several different complex lipids, is an vital part of a free-living cell that can reproduce itself.

Lipids (fats) have much higher energy density than sugars or amino acids, so their formation in any chemical soup is a problem for origin of life scenarios (high energy compounds are thermodynamically much less likely to form than lower energy compounds).

The fatty acids that are the primary component of all cell membranes have been very difficult to produce, even assuming the absence of oxygen (a 'reducing' atmosphere). Even if such molecules were produced, ions such as magnesium and calcium, which are themselves vital for life and have two charges per atom (++, i.e., divalent), would combine with the fatty acids, and precipitate them, making them unavailable. This process similarly hinders soap (essentially a fatty acid salt) from being useful for washing in hard water—the same precipitation reaction forms the 'scum'.

Some popularizers of abiogenesis like to draw diagrams showing a simple hollow sphere of lipid (a 'vesicle') that can form under certain conditions in a test-tube. However, such a 'membrane' ***could never lead to a living cell*** because the cell needs to get things through the cell membrane, in both directions. Such transport into and out of the cell entails very complex protein-lipid complexes known as transport channels, which operate like electro-mechanical pumps.

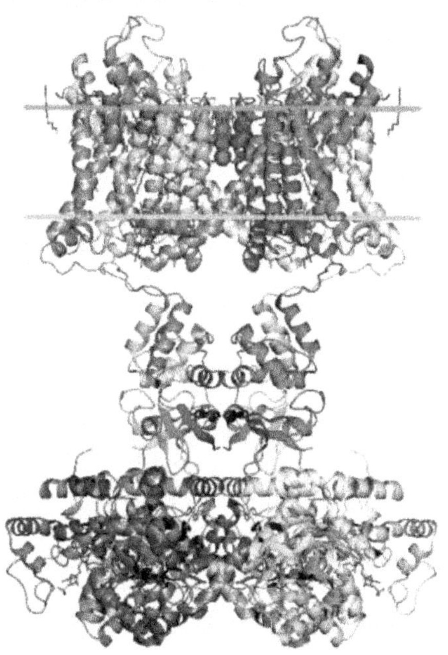

Fig.8.4: *A Potassium Transport channel. The red and blue lines show the position of the lipid membrane and the ribbons represent the transporter, which includes a number of proteins (different colors). To give some idea of the complexity, each loop in each of the spirals is about 4 amino acids.*

They are specific to the various chemicals that must pass into and out of the cell (a pump that is designed to move water will not necessarily be suitable for pumping oil). Many of these pumps use energy compounds such as ATP to actively drive the movement against the natural gradient. Even when movement is with the gradient, from high to low concentration, it is still facilitated by carrier proteins, Figure 8.4.

The cell membrane also enables a cell to maintain a stable pH, necessary for enzyme activity, and favorable concentrations of various minerals (such as not too much sodium). This requires transport channels ('*pumps*') that specifically move hydrogen ions (protons) under the control of the cell. These pumps are highly selective.

Transport across membranes is so important that "20–30% of all genes in most genomes encode membrane proteins." The smallest recognized genome of a free-living organism, that of the parasite *Mycoplasma genitalium*, codes for 26 transporters amongst its 482 protein-coding genes.

A pure lipid membrane would not allow even the passive movement of the positively-charged ions of mineral nutrients such as *calcium, potassium, magnesium, iron, manganese*, etc., or the negatively-charged ions such as *phosphate, sulfate*, etc., into the cell, and they are all essential for life. A pure-lipid membrane would repel such charged ions, which dissolve in water, not lipid. Indeed, a simple fat membrane would prevent the movement of water itself (try mixing a lipid like olive oil with water)!

Membrane transporters would appear to be essential for a viable living cell.

In the 1920s the concept that life began with soapy bubbles (fat globules) was popular (Oparin's 'coacervate' hypothesis) but this pre-dated any knowledge of what life entailed in terms of DNA and protein synthesis, or what membranes have to do. The ideas were naïve in the extreme, but they still have an airing today in YouTube videos showing bubbles of lipid, even dividing, as if this were applicable to explaining the origin of life.

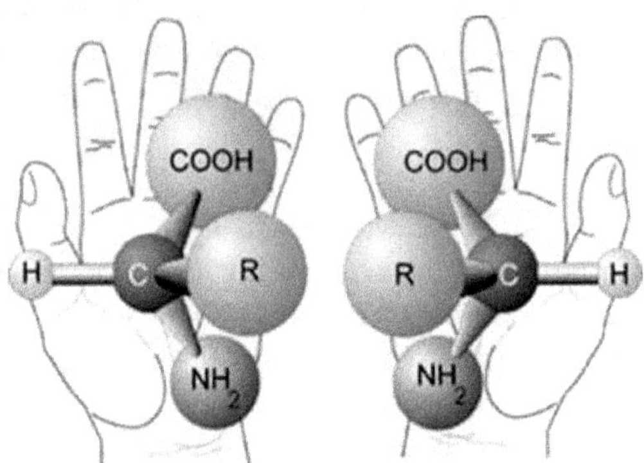

Fig.8.5: The chirality of typical amino acids. 'R' represents the carbon–hydrogen side-chain of the amino acid, which varies in length. R=CH$_3$ makes alanine, for example.

5. Handedness (Chirality)

Amino acids, sugars, and many other biochemicals, being 3-dimensional, can usually be in two forms that are mirror images of one another; like your right and left hand are mirror images of each other. This is called handedness or *chirality*, Figure 8.5.

Now living things are founded on biochemicals that are pure in terms of their chirality (homochiral): left-handed amino acids and right-handed sugars, for example. Here's the rub: chemistry without enzymes (like the Miller–Urey experiment), when it does anything, produces mixtures of amino acids that are both right-and left-handed. It is likewise with the chemical synthesis of sugars (with the *formose* reaction, for example).

SCIENTISTS BATTLE – THE ORIGIN OF LIFE

Origin-of-life scientists have battled with this problem and all sorts of potential solutions have been recommended but the problem remains unsolved. Even reaching 99% purity, which would require some totally artificial, unlikely mechanism for 'nature' to create, doesn't solve it. Life needs 100% pure left-handed amino acids. The reason for this is that placing a right-handed amino acid in a protein in place of a left-handed one results in the protein having a different 3-dimensional shape. None can be tolerated to obtain the type of proteins vital for life.

MINIMUM REQUIREMENTS FOR CELL TO LIVE

A minimal free-living cell that can manufacture its components employing chemicals and energy obtained from its surrounding environment and reproduce itself must have the followings:

1. **A Cell Membrane**. This separates the cell from the environment. It must be capable of maintaining a different chemical environment inside the cell compared to outside (as stated above). Without this, life's chemical processes are not possible.

2. **A Way of Storing** the information or specifications that instructs a cell how to make another cell and how to operate moment by moment. The only known means of doing this is *DNA* and any proposals for it to be something else (such as *RNA*) have not been proven to be viable—and then there has still to be a way of changing from the other system to DNA, which is the basis of all known life.

3. **A Way of Reading** the information in item (2) above to make the cell's components and also control the amount produced and the timing of production. The major components are *proteins*, which are strings (polymers) of hundreds to thousands of some 20 different *amino acids*. The only known (or even conceivable) way of making the cell's proteins from the DNA specifications involves over 100 proteins and other complex *co-factors*. Involved are:

    a   **nano-machines** such as *RNA polymerase* (smallest known type has ~4,500 amino acids),

    b. *gyrases*, which twist/untwist the DNA spiral to enable it to be 'read' (again these are very large proteins),

    c. *ribosomes*, sub-cellular 'factories' where proteins are manufactured, and

    d. at least **20 transfer-RNA molecules**; these select the right amino acid to be placed in the order specified on the DNA (all cells that we know of have at least 61 because most amino acids are specified by more than one DNA three-letter code). The transfer-RNAs have sophisticated mechanisms for making sure the right amino acid is selected according to the DNA code.

    e. **There are also mechanisms** to make sure that the proteins made are folded three-dimensionally in the correct way that involve *chaperones* to protect the proteins from mis-folding, plus *chaperonin* folding 'machines' in which the proteins are helped to fold correctly). All cells have these.

    That's just the basics!

    A greatly simplified animation of protein synthesis, which includes the action of RNA polymerase, ribosomes, transfer-RNAs, chaperonins, and chaperones. All living cells have this system of protein synthesis.

4. **A Means of Manufacturing** the cell's biochemical needs from the simpler chemicals in the environment. This includes a way of making "*ATP*, the universal energy currency of life." All living cells today have *ATP synthase*, a phenomenally complex and efficient electric rotary motor to make ATP (or in reverse to create electric currents that drive other reactions and movement both inside and outside the cell), Figure 8.6.

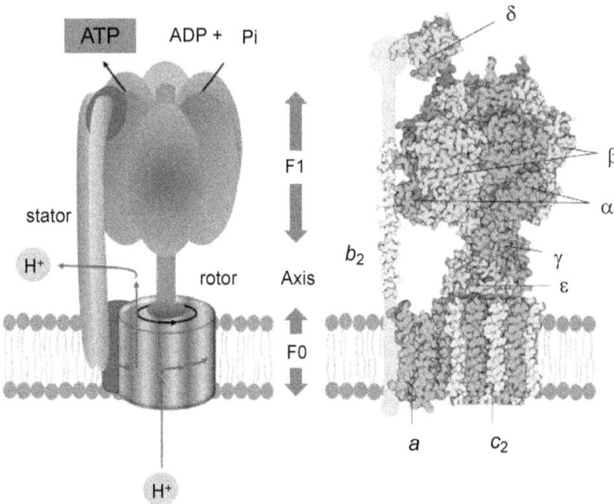

Fig.8.6: *Structure and Mechanism of ATP synthase.*

5. **A Means of Copying** the information and passing it on to offspring (reproduction). A recent simulation of one cell division of the simplest known free-living bacterium (which 'only' has 525 *genes*) **required 128 desktop computers working together for 10 hours.**

This gives some indication of what needs to happen for the first living cell to *live*.

An interesting project began some years ago to ascertain what could be the minimal cell that could operate in a free-living manner; that is, not dependent on another living organism. Though, it did have available a nutrient-rich medium that provided a wealth of complex organic compounds such that the cell did not have to synthesize many of its needed biochemicals. This *minimal cell* is now known to need over **400 protein and RNA components**, and of course that means that its DNA needs to be loaded up with the specifications for making these. That is, the DNA needs to have over 400 'genes.'

POLYMER FORMATION

Life is not just composed of amino acids or sugars but it is loaded with *polymers*, which are strings, or chains, of simpler compounds joined together. A polysaccharide is a polymer of sugars. A protein is a polymer of amino acids and DNA and RNA are polymers of nucleotides. Polysaccharides are the simplest, where the links in the chain are typically the same sugar compound, such as glucose (making starch in plants or glycogen in animals). Proteins are much more complex, being chains of amino acids where each link in the chain can be one of 20 different amino acids. And there are four different links in DNA and RNA.

Now water is a vital ingredient of living cells; characteristic bacteria are about 75% water. Being the 'universal solvent', water is a essential carrier for the various components of cells; it is the milieu in which it all happens

There is a huge problem for origin-of-life scenarios: when amino acids are joined together, for example, a water molecule is released. This means that in the presence of water, the reaction is pushed in the wrong direction, backwards; that is, proteins will fall apart, not build, unless the water is actively removed. A cell overcomes this by protecting the reaction site from water (inside ribosomes) and providing energy to drive this and the polymer formation. Thus, the formation of proteins of more than a few amino acids is a huge problem for all origin-of-life scenarios (and adding more time does not solve the problem; they just fall apart more).

Polymer formation also requires that the ingredients (monomers) that are joined together are *bi-functional*. That simply means that the amino acids for making proteins (or sugars for making polysaccharides) have at least two active sites that will allow another amino acid (or sugar) to be joined to each end. A protein-forming amino acid will have at least one amino group ($-NH_2$) and one carboxyl group

Fig.8.7: Creating life in a test–tube would not demonstrate that life could have made itself without intelligent input.

($-COOH$), with the amino group of one amino acid joining to the carboxyl group of another, thus growing the chain. A compound with only one active site (*mono-functional*) would terminate the formation of the chain. The problem for origin-of-life scenarios is that any proposed chemical reactions that produce some amino acids also produce mono-functional ones that terminate protein formation.

Nucleic acids such as DNA and RNA are based on a sugar-polymer backbone. Again, the presence of some sugars that are mono-functional would terminate the formation of these and the presence of water also drives this reaction in the wrong direction as well (to fall apart), Figure 8.7.

ORIGIN OF LIFE PROGRAMMING – NOT CHEMISTRY

The provided information, this far would be enough to eliminate notions of the naturalistic origin of life, but we have not covered the most important problem, which is the origin of the programming. Life is not founded merely on polymers but polymers with specific arrangements of the subunits; specific arrangements of amino acids to make functional proteins/enzymes and specific arrangements of nucleic acid bases to make functional DNA and RNA.

As astrobiologist 'Paul Davies,' now director of the *Beyond Center for Fundamental Concepts in Science* at Arizona State University, said:

"To explain how life began we need to understand how its unique management of information came about.

"The way life manages information involves a logical structure that differs fundamentally from mere complex chemistry. Therefore, chemistry alone will not explain life's origin, any more than a study of silicon, copper and plastic will explain how a computer can execute a program."

Davies' precision on this point ought not to be a surprise to his fellow evolutionists, given his similarly plain-speaking public utterances for well over a decade previously. E.g., *"It is the software of the living cell that is the real mystery, not the hardware." And: "How did stupid atoms spontaneously write their own software? … Nobody knows …".*

ORIGIN OF LIFE – ORIGIN OF INFORMATION

Any attempt to explain the origin of life without explaining the origin of the information processing system and the information recorded on the DNA of a living cell is circumventing the issue. We just have to look at the simplest free-living cell possible to understand how the origin of the information is an insoluble problem for scenarios that depend on physics and chemistry (*that is, no intelligent design allowed*).

Sir "Karl Popper," Figure 8.8, one of the most prominent philosophers of science of the 20th century, realized that:

What makes the origin of life and of the genetic code a disturbing riddle is this:

The genetic code is without any biological function unless it is translated; that is, unless it leads to the synthesis of the proteins whose structure is laid down by the code. But … the machinery by which the cell (at least the non-primitive cell, which is the only one we know) translates the code consists of at least fifty macromolecular components which are themselves coded in the DNA [ed: we now know that over 100 macromolecular components are needed]. Thus, the code cannot be translated except by using certain products of its

Fig.8.8: *Sir Carl Popper (1902 – 1994) – Prominent Philosophers of Science of the 20th Century*

translation. This constitutes a baffling circle; a really vicious circle, it seems, for any attempt to form a model or theory of the genesis of the genetic code.

"Thus, we may be faced with the possibility that the origin of life (like the origin of physics) becomes an impenetrable barrier to science, and a residue to all attempts to reduce biology to chemistry and physics."

ORIGIN OF THE DNA CODE

The *coded* DNA information storage system as described by Sir Carl Popper cannot arise from chemistry, but demands an intelligent cause. If we think of other coding systems, such as the Morse code or a written alphabetical language, where symbols were invented to represent the sounds of speech, such coded systems only arise from intelligence. It is an arbitrary convention that 'a' is usually pronounced as in 'cat' in English; nothing about the shape of the letter indicates how it should be pronounced. Likewise, there is just no conceivable possibility of explaining the DNA coding system from the laws of physics and chemistry because there is no physical or chemical relationship between the code and what is coded.

Furthermore, if the origin of any DNA code were not a big enough problem, the DNA code turns out to be, of the many millions possible, "at or very close to a global optimum for error minimization: the best of all possible codes." This error minimization in the code is possible because there are potentially 64 '*codons*' for 20 amino acids, so that nearly all amino acids have more than one codon (a few common amino acids, such as leucine, have six). These multiple codons are sometimes called 'redundant', often taken to mean 'extra to needed' or 'superfluous.' However, the extra codons are optimized such that the most likely single-letter mistakes (mutations) in the coding are more likely not to change the amino acid, or at least to change it to a chemically similar one (thus being less disruptive to the structure of the protein manufactured).

The extra codons are also involved in sophisticated control of the amount of protein synthesized, through 'translation level control.' This control system operates in bacteria and higher organisms.

There is no possibility that a coding system can be generated in successive stages to be optimized. If any workable coding system did come into existence by some incredible fluke, no significant change in the basic code could thenceforth occur because the code and the decoding system (reading machinery) would have to change at the same time (there are some very minor variations in the basic code in some bacteria, for example, where one of the three normal 'stop' codons codes for an extra amino acid to the normal 20). Therefore, the *optimized* code cannot be explained except as another incredible fluke of 'nature,' right at the supposed beginning of life.

CODING AND INFORMATION

Not only does the origin of the coded information storing system need to be explained, the information or specifications for proteins, etc., stored on the DNA has also to be explained. Revisiting the *simplest* cell, derived by knocking out genes from a viable free-living microbe to see which ones were 'essential', this minimal cell needs over 400 protein and RNA components. Specifications for all these have to be encoded on the DNA, otherwise this hypothetical cell cannot manufacture them or reproduce itself to make another cell. It would take a large book to print this information coded in the four 'letters (ATGC)' of the DNA.

As per the Paul Davies analogy, the problem is similar to a computer program.

How do we explain the existence of a program?

There is first the programming language (Python, Fortran, C++, Basic, Java, etc.) but then there is the actual set of instructions written in that language. The DNA problem is likewise two-fold; the origin of the programming language and the origin of the program.

Proposals for something simpler that 'evolved' into this simplest cell need to demonstrate the route from their hypothetical simpler start to the first living cell. Enthusiasts for abiogenesis frequently appeal to 'billions of years' as a hand-waving approach to solving the problems, but this provides no *mechanism*. Reactions that are going in the wrong direction are not going to reverse and go in the correct direction by adding more time.

NECESSITY GENE REGULATION

Genes in viable living cells are regulated; the cell controls the activity of each gene to match the cell's needs.

Production of proteins, such as enzymes to digest food, requires energy as well as the raw ingredients (amino acids). If a gene is unregulated, the protein it codes for will be produced up to the limit of the energy and/or amino acid supply; whichever is exhausted first. Multiply this by the 400+ genes in the minimal viable free-living cell and you have ultra-chaos and non-viability. A lack of gene regulation would obviously be unworkable.

Furthermore, many products of enzyme-catalyzed reactions are toxic if manufactured in too great a quantity ('the dose makes the poison'). For example, Down Syndrome problems are caused by overexpression of genes on the tripled chromosome 21.

It appears that all processes in cells are regulated and in very precise ways. Not surprisingly, failure of individual gene regulation causes many diseases in humans.

Gene regulation involves sensing a need and producing the right amounts of proteins to meet that need.

A foremost form of gene regulation involves proteins that bind to a gene or a DNA sequence that activates a gene or suite of genes. Binding of the protein can stop the production of the protein(s) or promote it. Binding or not-binding can be determined by the presence of a substrate or a product of the reactions that are catalyzed by the protein/enzyme(s) produced. Thus, the amount of enzyme produced matches the need.

Another form of gene regulation encompasses methylation. This involves the addition of methyl ($-CH_3$) groups to some of the nucleic acid bases that make up a gene. This then influences the activity of the gene (whether it is transcribed to mRNA and produces a protein or not). mRNA is Messenger RNA. It tells your DNA how to make specific proteins. The mRNA contains instructions that your body can read to create a special type of protein. When the J. Craig Venter Institute produced their 'synthetic' cell, Syntia, in 2010, they found that it was not feasible unless they copied the DNA methylation pattern of the bacterium that they were copying (*Mycoplasma mycoides*).

There are many types of gene regulation, even in the simplest living single cells. How could gene regulation arise, as it would need to be present when a new gene came into existence?

Without regulation a new gene would create chaos. The origin of any new gene is difficult enough, but the need to add its regulatory network at the same time adds further to the impossibility of abiogenesis. Abiogenesis, is defined by Evolutionists as the idea that life allegedly arose from nonlife more than 3.5 billion years ago on Earth. Abiogenesis proposes that the first life-forms generated were very simple and allegedly, through a gradual process became increasingly complex.

LIFE NEEDS ERROR-CORRECTING SYSTEMS

Molecular biology has discovered that cells are phenomenally complex and sophisticated, even the simplest ones. The information, as stated, is stored on the DNA. However, DNA is a very unstable molecule. One report says:

There is a general belief that DNA is 'rock solid'—extremely stable," says "Brandt Eichman," associate professor of biological sciences at Vanderbilt, who directed the project. "Actually, DNA is highly reactive. On a good day about one million bases in the DNA in a human cell are damaged.

Therefore, all cells must have systems for correcting faults that develop in the structure of the DNA or in the coded information. Without these error-correcting systems, the number of errors in the DNA sequence accumulate and result in the demise of the cell ('*error catastrophe*'). This feature of all living cells adds *yet another 'impossible'* to origin of life scenarios.

Any information that happened to arise on a theoretical DNA molecule in a primordial soup would have to be reproduced accurately or the information would be lost due to copying errors and chemical damage. Without an already functioning repair mechanism, the information would be degraded quickly. However, the instructions to build this repair machinery are encoded on the very molecule it repairs, another vicious circle for origin of life scenarios.

When scientists revealed bacteria that live in extreme conditions, such as around hydrothermal vents in the sea, they were heralded as '*primitive life*' because some origin-of-life researchers had proposed that life might have started in such places. However, these '*extremophiles*', as they have been called ('*liking extremes*'), have quite sophisticated error-correcting systems for their DNA. For example, *Deinococcus radiodurans* is a bacterium that can withstand extreme doses of ionizing radiation that would kill you or me, or other bacteria. It does sustain DNA damage where the DNA is fractured into many pieces. However, about 60 genes are activated to repair the breaks and reconstruct the genome in the hours following the damage.

Hydrothermal vents are hot, inhospitable places and the DNA of microbes that live there is continually being damaged, such that the microbes must have sophisticated error-protecting and correcting systems to survive. They are not at all simple and do not provide any sort of viable model for explaining the origin of life.

Moreover, all bacteria, not just the 'extremophiles', must have sophisticated error-correcting systems that involve many genes, and when the error correction is inactivated by mutations the bacteria become non-viable. This provides yet another problem for the origin of life.

ORIGIN OF LIFE SITUATIONS

Did life originate in a warm pond (as speculated by Darwin), near a deep-sea vent, on clay particles, or somehow/somewhere else?

The number of scenarios proposed, with no winner, suggests that they all have major deficiencies.

A major problem with warm pond and deep-sea vent ideas is the presence of water, which prevents many of the reactions needed; to get polymers, for example. Furthermore, the heat in deep sea vents would speed up the *breakdown* of any lucky chemical formation.

Because of these problems with the presence of water, physical chemist and origin-of-life researcher, Graham Cairns-Smith proposed that clay surfaces were involved in facilitating some of the needed reactions.

However, experiments in warm volcanic ponds have shown that clay particles bind amino acids, DNA and phosphate, essential components of life, so strongly that the clay prevents any necessary reactions from occurring.

The origin of a whole cell including the DNA, proteins and RNA needed for it to reproduce will never happen by an accident in a chemical soup, as demonstrated above. Therefore, advocates of abiogenesis have tried to imagine situations whereby life began with simpler necessities and then progressed to life as we know it today.

PROTEINS FIRST

Most effort has gone into a '*proteins first*' approach, whereby proteins supposedly formed first and the DNA sequences to make the needed proteins and the RNAs necessary to make proteins from the sequences of DNA came later. However, other than the problem of getting the correct set of optically pure amino acids and the problem of polymerization to make the protein chains of amino acids, few proteins can act as templates to make copies of themselves. Also, a fundamental problem is that there is **no mechanism for creating the DNA sequence for a protein from the protein itself**, as pointed out by information theorist Hubert Yockey.

RNA FIRST

In the 1980s, some RNA molecules were discovered that have the ability to catalyze some chemical reactions; these were dubbed 'ribozymes' (from *ribo*nucleic acid en*zymes*). This finding stimulated a lot of excitement and so a lot of effort has gone into RNA-first scenarios, or the 'RNA world'. At least there are enzymes that can generate DNA code from RNA code; that is, if you could get the RNA, you might be able to imagine a scenario for getting the DNA.

However, the enzyme complexes that can make a DNA copy of an RNA sequence are phenomenally complex and themselves would never arise by natural processes. And there are many other seemingly insurmountable problems with the RNA-first scenarios, 19 of which have been enumerated by "Cairns-Smith." Moreover, RNA is much less stable than DNA, which itself is very unstable, as documented previously.

The multiplicity of scenarios proposed reinforces the conclusion that scientists really have little idea how life could have '*made itself*'. There is no viable hypothesis as to how life could start off simpler and, step-wise, progress to become an actual living cell. Neo-Darwinism (mutations and natural selection) is often invoked to try to '*climb mount impossible*' but this cannot help, even hypothetically, until there is a viable self-reproducing entity, aka a *cell*, the minimum requirements for which I set out earlier ('*What are the minimum requirements for a cell to live?*').

LIFE FROM OUTER SPACE

Fig.8.9: *Probability calculations for the origin of life*

Francis Crick, co-discoverer of the DNA double helix structure, is a well-known proponent of '**life from space**'. He proposed that aliens sent life to earth, known as '*directed* panspermia'. Another form of this idea, simply 'panspermia', is that life arose somewhere else in the universe and came to earth as microbes on meteorites or comets; Earth was 'seeded' with life in this manner. Either version of panspermia effectively puts the matter beyond the reach of science. About the only element of panspermia that is testable is the ability of microbes to survive riding on/in a meteorite to earth. And this has been tested and found wanting; **microbes don't survive**.

A lot of the motivation for the search for extra-terrestrial intelligence (SETI) and extra-solar planets comes from a desire to find evidence that life might have formed '*out there.*' But even allowing the whole universe as a laboratory does not solve the problem; life would never form, as the following section reinforces.

PROBABILITY CALCULATIONS FOR THE ORIGIN OF LIFE

Many attempts have been made to calculate the probability of the formation of life from chemicals, but all of them involve making simplifying assumptions that make the origin of life even possible (i.e., probability > 0), Figure 8.9.

Mathematician Sir "Fred Hoyle" stated in various ways the extreme improbability of life forming, or even getting a single functional biopolymer such as a protein, Figure 8.10. Hoyle said, "Now imagine 10^{50} blind persons [ed: standing shoulder to shoulder, they would more than fill our entire planetary system] each with a scrambled Rubik cube and try to conceive of the chance of them all simultaneously arriving at the solved form, Figure 8.11. You then have the chance of arriving by random shuffling of just one of the many biopolymers on which life depends. The notion that not only the biopolymers but the operating program of a living cell could be arrived at by chance in a primordial soup here on earth is evidently nonsense of a high order. Life must plainly be a cosmic phenomenon."

Fig.8.10: *Mathematician Sir Fred Hoyle (1915–2001)*

Indeed, we can calculate the probability of getting just one small protein of 150 amino acids in length, assuming that only the correct amino acids are present, and assuming that they will join together in the right manner (polymerize). The number of possible arrangements of 150 amino acids, given 20 different ones, is $(20)^{150}$. Or the probability of getting it right with one try is about 1 in 10^{195}. Lest someone protest that not every amino acid has to be in the exact order, this is only a small protein, and only one of several hundred proteins needed, many of which are much larger, and the DNA sequence has to arise as well, seriously compounding the problem. Undeniably, there are proteins that will not function at all with even a small alteration to their sequence.

At that time Hoyle argued that life must therefore have come from outer space. Later he realized that even given the universe as a laboratory, life would not form anywhere by the unguided (non-intelligent) processes of physics and chemistry:

"The likelihood of the formation of life from inanimate matter is one to a number with 40,000 naughts after it … It is big enough to bury Darwin and the whole theory of evolution. There was no primeval soup, neither on this planet

Fig.8.11: *"Fill our entire planetary system – each with a scrambled Rubik cube and try to conceive of the chance of them all simultaneously arriving at the solved form!*

nor any other, and if the beginnings of life were not random, they must therefore have been the product of purposeful intelligence."

Does a figure of 1 in $10^{40,000}$ make the origin of life somewhere in the universe impossible without purposeful intelligence?

Can we say that?

The total number of *events* (or 'elementary logical operations') that could have occurred in the universe since the supposed big bang (13.7 billion years) has been calculated at no more than 10^{120} by MIT researcher "Seth Lloyd." This sets an upper limit on the number of experiments that are theoretically possible. This limit means that an event with a probability of 1 in $10^{40,000}$ would *never happen*. Not even our one small protein of 150 amino acids would form.

However, biophysicist "Harold Morowitz" came up with a much lower probability of 1 in $10^{10,000,000,000}$. This was the chance of a minimalist bacterium being assembled from a broth of all the basic building blocks (e.g., theoretically obtained by heating a brew of living bacteria to kill them and break them down to their basic constituents).

As an atheist, "Morowitz" argued that therefore life was not a result of chance and posited that there must be some property of available energy that drives the formation of entities that can use it (aka 'life'). This sounds much like the idea of Gaia, which attributes pantheistic mystical properties to the universe.

More recently the atheist philosopher "Thomas Nagel" proposed something similar to account for the origin of life and mind.

Anything but believe in a supernatural Creator, it would appear.

The different probabilities calculated arise from the difficulty of calculating such probabilities and the differing assumptions that are made. If we make calculations using assumptions that are most favorable to abiogenesis and the result is still ridiculously improbable, then it is a more powerful argument than using more realistic assumptions that result in an even more improbable result for the materialist (because the materialist can try to argue against some of the assumptions with the latter approach).

However, all calculations of the probability of the chemical origin of life make unrealistic assumptions in favor of it happening, otherwise the probability would be zero. For example, Morowitz's broth of all the ingredients of a living cell cannot exist because the chemical components will react with each other in ways that will render them unavailable for forming the complex polymers of a living cell, as explained above.

The origin of life is about as good as it gets in terms of scientific 'proof' for the existence of God. High profile information theorist Hubert Yockey (UC Berkeley) realized this problem:

The origin of life by chance in a primeval soup is impossible in probability in the same way that a perpetual motion machine is in probability. The extremely small probabilities calculated in this chapter are not discouraging to true believers … [however] A practical person must conclude that life didn't happen by chance."

It is interesting to learn that in Yockey's calculations, he generously granted that the raw materials were available in a primeval soup. But in the previous chapter of his book, Yockey showed that a primeval soup could never have existed, so belief in it is an act of 'faith.' He later concluded, "the primeval soup paradigm is *self-deception* based on the ideology of its champions."

MORE ADMISSIONS

Note that Yockey is not the only high-profile academic to speak plainly on this issue:

"Anyone who tells you that he or she knows how life started on earth some 3.4 billion years ago is a fool or a knave. Nobody knows."—Professor Stuart Kauffman, origin of life researcher, University of Calgary, Canada.

"...we must concede that there are presently no detailed Darwinian accounts of the evolution of any biochemical or cellular system, only a variety of wishful speculations." —"Franklin M. Harold," Emeritus Professor of Biochemistry and Molecular Biology Colorado State University.[46]

"We are almost as much in the dark today about the pathway from nonlife to life as Charles Darwin was when he wrote, 'It is mere rubbish thinking at present of the origin of life; one might as well think of the origin of matter.'"—

"Paul Davies," director of BEYOND: Center for Fundamental Concepts in Science at Arizona State University.

"The novelty and complexity of the cell is so far beyond anything inanimate in the world today that we are left baffled by how it was achieved."—

"Kirschner, M.W." (professor and chair, department of systems biology, Harvard Medical School, USA.), and "Gerhart, J.C." (professor in the Graduate School, University of California, USA).

"Conclusion: The scientific problem of the origin of life can be characterized as the problem of finding the chemical mechanism that led all the way from the inception of the first autocatalytic reproduction cycle to the last common ancestor. All present theories fall far short of this task. While we still do not understand this mechanism, we now have a grasp of the magnitude of the problem."

"The biggest gap in evolutionary theory remains the origin of life itself... the gap between such a collection of molecules [amino acids and RNA] and even the most primitive cell remains enormous."—

"Chris Wills," professor of biology at the University of California, USA.

Even the doctrinaire materialist "Richard Dawkins." admitted to "Ben Stein" (*Expelled,* the movie documentary) that no one knows how life began:

Richard Dawkins: *"We know the sort of event that must have happened for the origin of life—it was the origin of the first self-replicating molecule."*

Ben Stein: *"How did that happen?"*

Richard Dawkins: *"I've told you; we don't know."*

Ben Stein: *"So you have no idea how it started?"*

Richard Dawkins: *"No, nor has anybody."*

Fig.8.12: If a scientist managed to create a machine that could make life, it would only go to show how much intelligence was needed to create life!

"We will never know how life first appeared. However, the study of the appearance of life is a mature, well-established field of scientific inquiry. As in other areas of evolutionary biology, answers to questions on the origin and nature of the first life forms can only be regarded as inquiring and explanatory rather than definitive and conclusive." [*emphasis* added]

Life did not arise by physics and chemistry without intelligence, Figure 8.12. The intelligence needed to create life, even the simplest life, is far greater than that of humans; we are still scratching around trying to understand fully how the simplest life forms work. There is much yet to be learned of even the simplest bacterium. Indeed, as we learn more the 'problem' of the origin of life gets more difficult; a solution does not get nearer, it gets further away. But the real problem is this: the origin of life screams at us that there is a *super-intelligent* Creator of life and that is just not acceptable to the secular mind of today.

The origin of life is about as good as it gets in terms of scientific 'proof' for the existence of God.

SPECIES CHANGE – THE HAND OF GOD

People of faith in the Infinite Creator do not, and have never, taught that God created all species as we see them today. Undoubtedly, people in the past believed in 'the fixity of species' but this was more due to the inescapable influence of the pagan Greek philosopher Aristotle than to anything in the Bible. One of Darwin's principal influences, the lawyer-turned-geologist Charles Lyell, taught fixity of species. He also believed that species were placed in "center of creation." In other words, species were more or less created in their current locations.

Many of Darwin's points in the *Origin of Species* and later books were designed to refute Lyellian and Aristotelian arguments, although we can see their influences in his earlier writings.

The biblical Creation/Flood/Dispersion model actually has nothing to do with Darwin's objections. He was not arguing against biblical teaching. Actually, there is an excellent example in the Bible itself that illustrates how fast species can change: the episode where Jacob bred spotted and speckled animals from all-white flocks *(Genesis 30)*. Clearly, species are not fixed, Figure 8.13. The Bible talks about a worldwide dispersion of the animals on the Ark after the Flood. Thus, in no way does the Bible suggest that all species were created where they are found today. Darwin observed that many species on the Galápagos islands were clearly related to those on the South American mainland, but they were subtly different as well. This refuted Lyell's "centers of creation" model, yet the observations are even better explained by post-Flood migration paths.

Fig.8.13: *Birds in Different forms –
Still Birds*

CREATURES DESIGNED TO CHANGE

The predominant idea within creationist circles is that God created distinct "kinds" that were front-loaded with a large amount of genetic diversity that could then be used as the kinds turned into our modern species. Dr. "Peer Terborg" used the term "pluripotent baranomes".

Pluripotent describes something that can turn into many other things. A pluripotent stem cell, for example, can be used to grow any organ in the body. A baranome is a play on words. The suffix "–ome" is a made-up root word used by biologists to denote a total collection of objects. Therefore, a genome is a collection of all the 'genes' found in an individual or species, the proteome is all the proteins found in an individual or species, and the microbiome is the total collection of microbes in a human or animal body. *Bara* is the Hebrew word for "create". Therefore, in the same way that a *genome* represents all the genes in a single person, a *baranome* represents all the genomes in a single baramin (that is, a created kind). These were filled with reams of useful information, therefore, the original horses, whatever they may have looked like, could turn into all the varieties we see both today and in the fossil record.

When the Infinite, Omniscient God designed life, He foreknew how far livings things could be pushed. Actually, he wrote a potential-tolerance code into the design specs of each created kind. Think of the great variety we observe in the various kinds. Horses, for example, range from very small to very large. They can be shaggy or short-haired. Their colors range from white to red to brown to black, with many **potential** combinations in the same animals. Much (but not all) of this was coded into the first horses.

INITIAL CONDITIONS–FRONT–LOADED INFORMATION

Some of this initial information was in the form of heterozygous alleles (i.e., genetic variation potentials). We are not told how many individuals in each kind God created, except for humans. Given an initial population, God could have engineered any amount of genetic diversity into the genomes of the created kinds. Even humans could have had more initial diversity than is normally carried in two people because there was nothing stopping God from engineering a different genome into every one of Adam and Eve's reproductive cells.

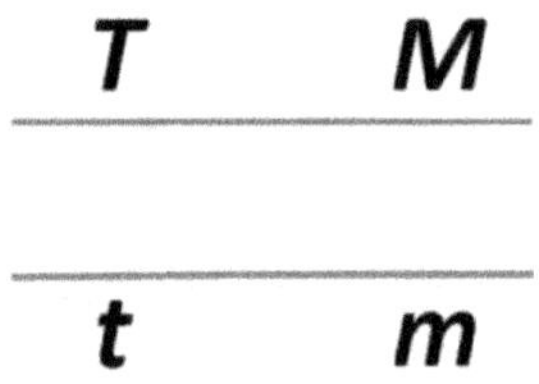

Alternatively, Adam and Eve could have been created with normal reproductive cells but with a lot more genetic diversity than modern humans currently have. In either case, the amount of diversity we observe among humans today would depend, in large part, on how many children they had.

*Fig.8.14: Two genes that are close together in the genome and that were initially found in the pairs **TM** or **tm**.*

Some of that initial information could also have been in dormant form. God did not have to hard-code every single future trait ever seen in every species but could have allowed for the *possibility* of certain changes to take place. For example, retrotransposons ('jumping genes') can turn genes on or off depending on where they are located in the genome. Whole new phenotypes could arise, depending on where these genetic elements are, or are not, located in the genome. No "new" information needs to be created if this was part of the original design criteria for life. Yet, "new" characteristics ('phenotypes') could certainly arise.

The four-dimensional structure of the genome also plays a role in how the genome can change over time.

Retroviruses like HIV enclosure themselves into the genome. Indeed, they *change the genome*. However, this is not a random process. They are much more expected to integrate in areas that are open, not areas that are deeply buried in coils of DNA. They are also more likely to join in areas that affect the structure of DNA itself. The structure of DNA also impacts the timing, rate, and types of mutation (discussed below) that will happen in the genome.

Thus, when God programmed those linear strings of DNA we call 'the genome', He knew how it would fold up into a 3-D shape and how that 3-D shape could be modified in the 4th dimension (i.e., time). Part of this high-level organization elaborate forward planning. By putting a gene in a specific place, the infinite omniscient God was setting up specific possibilities for future modifications.

Another method to create new phenotypes is through *recombination*. Chromosomes get scrambled each generation. This brings genetic variants together in new and unique ways. Let's say that two genes that affect height and muscularity can be found near each other in the genome of some species. Let us call the first one *T*, for tall. We'll call the second one *M*, for muscular. Let's say they each have a recessive form. We'll call them *t* (short) and *m* (thin) and that God originally set it up so that any one chromosome had either *T-M* or *t-m*, Figure 8.14.

Using dogs as an example, since individuals carry two copies of each chromosome, they could be:

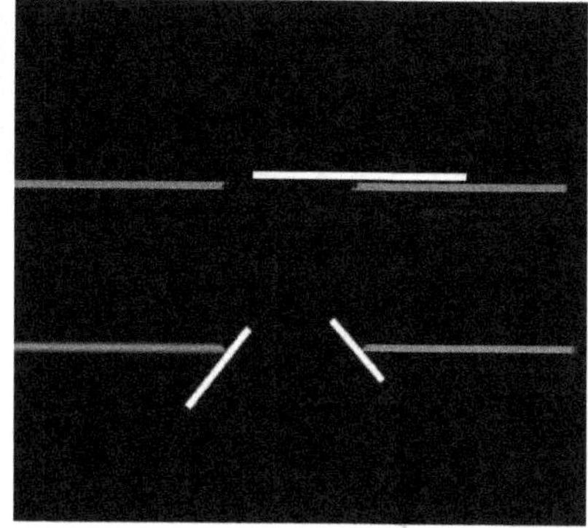

Fig.8.15: *Recombbcination leads to new gene pairs.*

- *TTMM* (tall and muscular, like a mastiff)
- *TtMm* (also tall and muscular, since *t* and *m* are recessive)
- *ttmm* (short and thin, like a Chihuahua)

Nonetheless, what is lacking? No individual can be *TTmm* or *ttMM*. It is not possible to get an animal that is tall and thin (like a greyhound) or short and muscular (like a bulldog). However, what if a recombination occurred between the two genes, Figure 8.15?

Suddenly, we have two new combinations, Figure 8.16. We started with only three possibilities. We now have five (*TTMM, TTmm, TtMm, ttMM,* or *ttmm*). Given that each species has thousands of genes, recombination could be a continual source of new phenotypes, especially early on.

Yet another way to create new phenotypes is through gene gain and gene loss, *Figure 8.16*. Genetics has thrown us a few curve balls over the years. Take the lowly *E. coli* bacterium. It has a genome of 4.6 million letters and contains 4,288 protein-coding genes. Only that's not really true, for different bacteria in that same species can carry different sets of genes. If something they need is being produced by another species living nearby, for example, deleting that gene will be beneficial. Why spend time and effort maintaining that section of DNA and making proteins when the thing you need

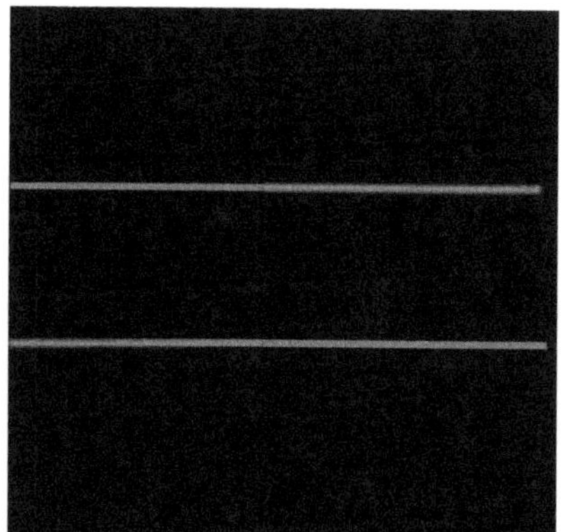

Fig.8.16: *The new gene pairs, never seen before.*

is free? Because of this, members of the same species carry different genes depending on the environment in which they are living. If you add up all the genes carried by all *E. coli* in the world, you end up with what we now call a *pangenome*, with up to 5,500 genes. The pangenome concept gives the infinite omniscient God a way to create a phenomenal amount of potential diversity. The difference in gene content from one organism to another in the same kind could have been pre-engineered or could have arisen through mutation.

ENGINEERED MUTATIONS

Clearly, the infinite omniscient God could engineer a *lot* of potential changes into creation. However, there is something else we have to consider: mutations.

God was not naïve about mutations. When He wrote out each genome, He knew how that code could be mutated. In effect, part of the design specifications of life included an understanding of how DNA would be modified over time.

Engineered Error-Checking and Repair Systems

Yet, God also knew that unchecked mutation would not be a good thing, because most mutations are harmful, simply because there are many more ways to break something than to make it. Therefore, He also created multiple amazing, complex, and well-engineered error-checking and repair systems. There is a different repair system for each type of mutation, from single-letter misincorporations to near-fatal double strand breaks.

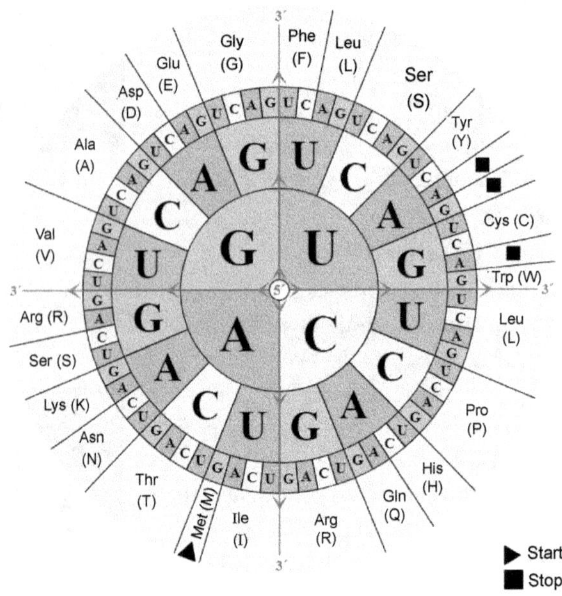

Fig.8.17: The amino acid code. Starting from the inside and working outwards, we can see that each amino acid is coded by three letters in the mRNA. However, what is not immediately obvious is that the table is also optimized to reduce the amount of amino acid changes in the case of mutation. Many single-letter mutations will code for the same amino acid, and those that don't will often code for an amino acid with similar chemical or physical characteristics.

Even so, some mutations still slip through this system, therefore, God added yet another level of complexity. You may have heard that three letters of DNA codes for one amino acid in a protein. Because there are four bases, there are 64 possible 3-letter codons (4^3) but only 20 amino acids, meaning some amino acids are coded by more than one codon, Figure 8.17. However, what they usually forget to tell you in school is that the 'redundancy' in the amino acid code is optimized to slow the rate of change. There are millions of ways to assign 20 amino acids to 64 codons, but God picked one in which the genetic code helps to minimize the effect of mutation.

When a mutation does occur, the amino acid is often unchanged, and when an amino acid is changed, it is often changed to a *similar* amino acid. Most other arrangements are not nearly as effective at reducing the potential effects of mutation.

If a mutation has little to no effect, it is effectively neutral. There is no problem with thinking this and so we can take a clue from the evolutionary community, many of whom believe that most mutations are harmless, because some certainly are. Yet, mutations are not truly random.

Some mutations are much more likely than others. Thus, given any initial starting point, any genetic system will change in a preferred direction. This is completely contrary to the evolutionary idea, where all possibilities must stay on the table. Instead, we can see specific types of changes building up in different things, like the high number of C to T and G to A mutations seen in the human H1N1 influenza virus since 1917, Figure 8.18. We documented a 1% change in the number of A's and G's and about a 0.5% change in the number of U's and C's between 1917 and 2009. This was due to the simple fact that not all mutations are equally likely, because some bases are more thermodynamically stable than others. When infinite omniscient God wrote out the DNA sequences of the original baranomes, He certainly foreknew the direction in which they would change. This was all part of His perfect initial plan.

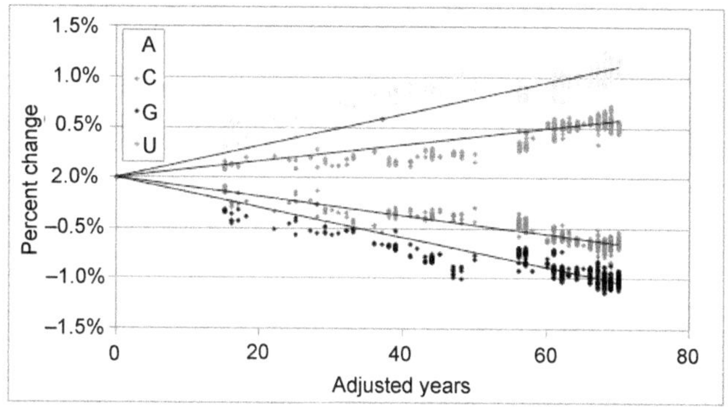

Fig.8.18: *Changes in the letters of the human H1N1 influenza virus over several decades. Some mutations are simply more common than others. These changes are mainly driven by a high number of C to T and G to A mutations. After Carter (2014).*

To that end, some genes are more protected than others, with robust and continual error checking. Other genes are much more likely to mutate than others. For example, the genes that control melanin production in mammals are highly mutable. There are also examples of genetic systems that seems to be designed to rapidly change in order to exploit new resources. An example of this would be the ability of certain bacteria to be able to digest nylon, a man-made fiber never seen in nature. Instead of being an example of evolution, these bacteria clearly have a *designed* ability to adapt. In fact, geneticists are moving away from the old, now outdated, idea that mutations are inherently random.

Non-randomness could be due to differences in the chemical code (i.e., the C to T change is much more common than T to C, and some genes are enriched in C), the position within the nucleus when the genome is in its normal 3D configuration, the presence of duplicate copies of a gene, the presence of repetitive DNA close by, the near presence of non-B-DNA (non-B (including cruciform loops, Z-DNA, and quadruplex looping), and many other factors. God knew this when he wrote the initial code. He *knew*.

Mutation is not a surprise to God. He is the one who wrote the laws of chemistry, and He knew how oxygen and water would continually react with DNA when He chose to use DNA as the information storage mechanism in the cell. He designed the DNA repair systems and engineered a known fault tolerance level into them. He knew that the DNA polymerases that copy DNA would occasionally make mistakes and He designed them with this error rate in mind. If He wanted more faithful replication, He would have designed the system more stringently. If He wanted sloppy DNA copying, this would have been easy to do. Again, He *foreknew*.

Another factor that God deliberately designed into living things is something called epigenetics. In our cells, there are all sorts of switches that can turn genes on and off. Genes that are not needed can be turned off. The cell can stick extra carbon atoms on the DNA in that region (a process called methylation) or it can modify the histone proteins the DNA wraps around. Unlike mutations, many of which don't have any measurable effect, epigenetic changes are *designed* to produce phenotypic change. Yet, this depends on a specific set of letters being in place. If those letters mutate, the epigenetic switch can be broken, either in the 'off' position or the 'on' position.

Of course, most of this applies after the Fall of Adam and Eve. There is some small scope for mutations prior to sin and death entering the world, but these would have to be carefully orchestrated by God. In the same way that Adam would never have been able to trip over a root and plummet headlong off a cliff before the Curse, mutations would never have been able to create deformities in living things. But this does not mean that there was no chance for any change in any letter in any baranome in all of creation.

NATURAL SELECTION

As organisms within each kind reproduce, it is natural to think that some will be better fit for some environments while others would be better suited to other environments. This is the essence of natural selection. Much has been written about this already, and it is not a threat to the creation model. In fact, in some ways natural selection would even apply to the time before the Fall. After the Fall, however, selection would go into overdrive. As the environment, especially after the Flood, changed, the phenotype of specific organisms would be very important. Death also reigned. Survival became an all-important factor as organisms now struggled with environmental conditions that were never seen before (like the Ice Age). It is good to remember that natural selection is not a creative force. It is, however, something that happens automatically as organisms reproduce over time. This is one of the many factors that combine to create change over time. Notice that it was also saved for last. Darwin once thought natural selection was the preeminent force driving change over time. Instead, it is a minor component among a diverse array of factors that God put into his creation.

GOD DESIGNED LIFE

God is the master engineer. He did not just create life; He designed life to robustly respond to the environment. He overdesigned living things so that they could withstand, so far, thousands of years of mutational accumulation. His designs were holistic, in that He accounted for what each kind would need in the future. He built redundancy in the genomes to ward off the effects of decay, but He also designed specific aspects of the genome to change via mutation.

But let's leave off with a verse that tells of the greatness of our Creator, *Romans 11:33–36*:

"Oh, the depth of the riches and wisdom and knowledge of God! How unsearchable are his judgments and how inscrutable his ways! "For who has known the mind of the Lord, or who has been his counselor?" "Or who has given a gift to him. For from him and through him and to him are all things. To him be glory forever. Amen."

Fig.8.19: A Male Cardinal
(Cardinalis cardinalis)

SPECIES DESIGNED TO CHANGE

Limitation of Change

The infinite omniscient God designed life with a large amount of information, up front. Some of this was in the form of genetic diversity and some was in the form of latent (dormant) information that could arise through the process of recombination. We also saw that he *foreknew* the effects of mutation and incorporated this into His perfect design. God was not caught by surprise here. He *foreknew, even before the universe began.*

We now broaden this discussion to include population-level changes that lead to the rise of new species. Nonetheless, first, we must define what a species *is*. This is not easy to do, and many different definitions exist. Essentially, if a group of organisms breeds '*true to type*' we call that a species. The problem is that many 'species' can interbreed, producing hybrids. Worse, in the fossil record we see skeletons of all sorts of deceased organisms, yet we know nothing about which ones could have interbred. In the end, the word species is just a term of convenience. For example, it is helpful to separate the North American cardinal (*Cardinalis cardinalis*), Figure 8.19, from the pyrrhuloxia (or desert cardinal, *Cardinalis sinuatus*), Figure 8.20, even if they share a lot of physical features, have identical songs, and can interbreed.

The easiest definition to understand is the '*biological species concept.*' Fundamentally, if two organisms can successfully breed (i.e., not make sterile hybrids), they belong to the same species. If they can't, they are separate species. Yet, the lines between 'species' are often quite blurry. This is perhaps not unexpected in the creation model. If you take a group of organisms from a single created kind and spread them out across the earth, it is not hard to imagine each of the resulting subpopulations changing in different ways while maintaining the ability to interbreed with other subpopulations.

Numerous 'species' can interbreed. In fact, there are multiple examples of species from different genera and even from different families being able to successfully interbreed. Therefore, they belong to the same biological species by definition. Yet, just because two things cannot interbreed today does not mean they do not belong to the same biblical kind. We have to take each case separately and study it in detail.

Given the pluripotent baranomes, the infinite omniscient God put into His creation, with a rich diversity of genes, we would naturally expect *rapid* diversification. As that diversity was slowly lost, mutated, and pigeonholed into discrete species, the rate of change should have slowed over time. However, hybridization can lead to a burst of change because mixing the genes of two species partially restores the original diversity. Evolutionists have been slow to realize how much hybridization among species has always existed. In fact, they are now saying that new species *often* arise due to hybridization. Of course, they are putting a much longer timescale on these events, but what is clear is that standing variation is a rich source of rapid change and the rise of new phenotypes.

Fig.8.20: *Pyrrhuloxia or Desert Cardinal (Cardinalis sinuatus)*

SPECIATION

Speciation can be driven by reproductive incompatibility, but it can also be driven by mate choice. In songbirds, the females will often choose to mate with a male that has a song very similar to her father's. In the nest, he imprinted a specific song into her mind. Likewise, the males will grow up singing the song they heard their father sing. Cross-species mating can sometimes occur if the nests of two songbird species are too close together.

MALE IMPRINT SONGS ON BABIES

In this case, one male can imprint his song onto the babies in the other nest. In a long-term study of Galápagos finches, the team of "Peter and Rosemary Grant" noticed that a strange bird had arrived on the small island of Daphne Major. He was himself a hybrid of a medium ground finch (*Geospiza fortis*) and a common cactus finch (*Geospiza scandens*). He was larger than any of the other closely related finches on the island but managed to mate with a G. *fortis* female and raise a clutch of eggs. One of those birds flew off to mate with another G. *fortis* individual, but every other bird mated with a sibling. There are not a lot of these birds, but they have been on that island for over three decades. They are genetically and morphologically different from any other species on the island, and they sing a different song. In short, we saw the rise of a new species, through hybridization.

Frequent hybridization among species caught the evolutionists flat-footed. They resisted the thought for many years because it did not fit well with the whole evolution concept. If species can mingle, and if species have been around for a very long time, giving them plenty of time to 'mingle', why do discrete species even exist? Perhaps they don't.

CHANGE IN POPULATION-LEVEL

As we observed above, there are things that happen on the population level that can affect the way organisms look and act. For example, if you take a population and divide it into subgroups, each little group might not carry all the genes in the original population, Figure 8.21. Yet, the future individuals born into those subpopulations can only carry the genes of their sires. This is called the *founder effect*. If the founders of the subpopulation carried more "tall" genes, or "yellow" genes, or "fast" genes than average, the resulting subpopulation would be *different*. We would expect such a thing to be in play in many of the biblical kinds as they were spreading out on the earth after the Flood.

Think about any post-Flood kind (dogs, bears, birds, or whatever). As they travel further away from where the Ark came to rest, they would have to cross rivers, mountain ranges, and deserts.

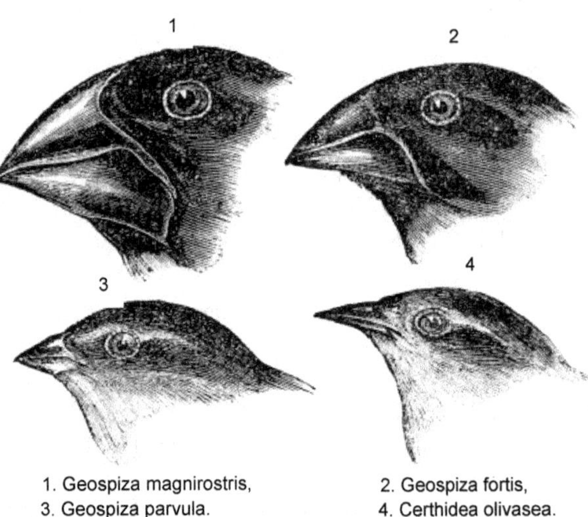

1. Geospiza magnirostris,
3. Geospiza parvula.

2. Geospiza fortis,
4. Certhidea olivasea.

Fig.8.21: Darwin's finches display a range of beak sizes and shapes, but many of these species are known to hybridize.

It would become more and more difficult to interact with other members of the original population. Given founder effects, the appearance of mutations unique to each subpopulation, genetic drift removing genetic diversity at random, and things like that, the different groups would quickly change. This easily explains how the pair of the Ursid created kind on the Ark could give rise to grizzlies, brown bears, black bears, sun bears, polar bears, and the extinct cave bear and short-faced bear. We don't need evolution to explain speciation. All we need is an initially high genetic diversity, a way to scramble that diversity, and the isolation of subpopulations.

The rise of unique phenotypes after isolation is called *allopatric speciation*, and mountainous regions (such as Ararat!) are well known for it. Actually, many types of bears can still interbreed, such as the grizzly and polar bears having a pizzly cub together, so all bears are really a single *biological* species, by definition.

HUMAN STAYED IN BABEL A CENTURY

As an aside, we observe a similar process for humans, but it was delayed by a century. That is, humans disobeyed God's command to fill the earth and intentionally stayed gather in one place (Babel). God confused their languages to force them to spread out in the days of Peleg. They were divided into small, isolated subpopulations. This is largely responsible for the small differences between 'races.' Nonetheless, no speciation has occurred. All humans (including Neanderthals and Denisovans) are or were interfertile, so are one species, by definition. We started off as one group and there has been considerable mixing throughout history, so people can often share things like blood groups and Y-chromosomes with people of supposedly very different races. Thus, 'inter-racial' organ transplants are more common than many people might assume. Why? Because you can be more similar to a person from another 'race' than to people from your own.

How Fast is the Change

It Appears rational that change would have happened more quickly in the past. Thus, when Darwin observed small changes in some species over a few years, he was witnessing the end game. He drew grand conclusions from a very limited sampling in time. Had he considered what God would have initially created, perhaps he would have come up with a different theory.

Once living things exist, they can change. However, as the species derived from each original kind aged, they lost some of that original, good, God-created diversity and started to accumulate mutations. Because of this, it is like modern species are poised on the edge of a knife. You cannot change them too much without killing them. Their ancestors were more robust than the inbred, high-mutation-load individuals that exist today. Plus, due to the way species adapt to different environmental niches, genetic pigeonholing happens.

Natural selection doesn't create genes, but instead culls them. Once they go down one path, they can no longer hop to another. They are stuck.

Change is (mostly) a one-way street. Do we believe in 'hyper-evolution', as some evolutionists claim, dutifully followed by old-earth compromisers like "Hugh Ross?" Not at all. representatives on the Ark. If we have to account for the rise of millions of species in just a few thousand years, it sure sounds like we are talking about evolution, right?

Actually, no. We do not have to account for much change at all. It depends on where you draw the line. First of all, we do not have to account for anything but **the** *nephesh chayyah* **that were on the Ark, which generally equate to the land-based vertebrates (i.e., mammals, reptiles, and birds).** Thus, it does not matter how many species of beetles are in the world. Yet, for the animals on the Ark, we would have a major problem if "kind = species." Nonetheless, what if we draw the line at the family level? Of the 151 mammalian families, for example, some have upward of 400 species, but most have less than 20.

The families with more species tend to include the smaller species, with high reproduction rates, short generation times, and large populations (in other words, the ones with more opportunities for speciation processes to operate). We have had 4,500 years of post-Flood history to account for within-kind speciation. Even a modest speciation rate can account for the numbers of species we see, particularly if the "kind" can divide into two species, which divides into four, etc. Just because a subpopulation split off from the main population does not mean it cannot split again and again, especially in those early years. I fully expect that the rate of speciation has slowed down over time.

Allowable Change

The next question we must face is one of magnitude. Given pluripotent baranomes and several thousand years of individual and population-level changes, how much 'change' can we acquire? Clearly, there is a limit. God may have engineered things to dynamically adapt to new environments, but we have only so much time for mutation, selection, drift, and recombination. We also have limits to the design itself. An organism cannot exceed its design constraints, or it perishes.

Some creationists have pushed for much more potential change than we are comfortable with. For example, Dr "Kurt Wise" once wrote:

Finally, some of the animals which are aquatic or marine today may not have been aquatic at the time of the Flood. The marine and sea otters, for example, are members of the mustelid (weasel) family and their aquatic character is likely to have been revealed after the Flood. The whales might turn out to be another example. Only when including the legged *archaeocetes* (and/or possibly the terrestrial order Acreodi) do the whales have a fossil record continuous with the Flood. Vestigial legs and hips in modern whales confirm legged ancestors of the whales existed only a short time ago. It is possible that the purely marine cetaceans of the present were derived from semi-aquatic or even terrestrial ancestors as well.

Whales came off the Ark? At face value, this seems preposterous. There is nothing seriously wrong with thinking that the legless lizards once had legs, and the same goes for snakes. There is no problem thinking that wingless beetles once had perfectly functioning wings, or that blind cave creatures had sighted ancestors. In fact, there are all sorts of examples of *loss* that we can cite.

But whales can't be explained just by leg loss. In any case, the so-called vestigial legs are not what is often claimed in modern whales, and in the extinct whale *Basilosaurus* they appeared to be reproductive claspers. The evolutionary ancestors of whales could not nurse their young underwater. Their breathing and eating systems were still connected. They would have had no need for those strange mouth plates that baleen whales use to filter out krill from the water. They would not have had the ability to echolocate. There are entire suits of morphological and behavioral changes that are necessary to explain whales.

Modern whales are also found in the fossil record. There is debate among creationists about where to locate the Flood/post-Flood boundary. If the boundary is high up, whales died in the Flood, so they did not come from any animal on the Ark. If the boundary is lower, there is still not much room for the evolution of whales from land animals because whales are found about half-way between today and the end of the Cretaceous, where some creationists put the Flood/post-Flood boundary. This would require an incredible amount of change in a very small amount of time.

We are not talking about cat-like lions and cat-like tigers arising from a cat-like ancestor.

The amount of change permitted in the creation model is an interesting question. In some ways, it is still an open question. Yet, the line must be drawn somewhere. Generally, if the proposed changes necessarily involve the appearance of brand-new features that would have been useless to the original organism (i.e., a land-based predator with baleen instead of teeth), it is almost certainly excluded from consideration. If, however, it can be shown that the information for the new features could have been coded into the original baranome (either directly or in latent form), or if it could have arisen through the destructive process of mutation, then the resulting changes might be allowed in the creation model.

God is the master engineer. He did not just create life; He designed life to robustly respond to the environment. He overdesigned living things so that they could withstand, so far, thousands of years of mutational accumulation. His designs were holistic, in that He accounted for what each kind would need in the future. He built redundancy in the genomes to ward off the effects of decay, but He also designed specific aspects of the genome to change via mutation.

On the last page of *Origin of Species*, Charles Darwin wrote, "There is grandeur in this view of life," referring to his claim that all life evolved from primitive ancestors. How wrong he was! Our God, however, is an amazing, fantastic, super-intelligent, all powerful, and omniscient Creator. Consider the words of Paul: "For by him all things were created, in heaven and on earth, visible and invisible, whether thrones or dominions or rulers or authorities—all things were created through him and for him. And he is before all things, and in him all things hold together." *(Colossians 1:16–17).* There is grandeur in *that* view of life.

MYSTERIES OF LIFE

Life is not a naturalistic phenomenon with limitless evolutionary potential as Darwin proposed. It is intelligently designed, governed by undisputable laws, and survives only because it has a built-in *facilitated variation* mechanism for continually adapting to internal and external challenges and changes. The indispensable components are: functional molecular architecture and machinery, modular switching cascades that control the machinery and a signal network that coordinates everything. All three are required for survival, so they must have been present from the beginning—a conclusion that demands intelligent design. Life's built-in ability to adapt and diversify looks like Darwinian evolution, but it is not. Darwin's theory of speciation via natural selection of natural variation is correct in principle, but it cannot be extrapolated to universal ancestry.

What we see instead is different kinds of organisms having been designed for different kinds of lifestyles, with enormous potential for diversification built-in at the beginning, but with time this potential for diversification has become exhausted by selection and degraded by mutations so that we are now rapidly heading towards extinction. Intelligent design and rapid decay point to recent Creation and Fall, as the Bible tells us.

Biologists have long sought the laws that govern life, but it is only now that we see the molecular detail that these laws have appeared. What we discover is not a naturalistic phenomenon, but intelligent design. In this section, I shall briefly outline the most important laws and how they work together for the survival of individual organisms, and their diversification into different species. Life's evolutionary potential is not unlimited, as Darwin proposed, but limited to the transformations and combinations of what was built-in at the beginning. These limits are also explored.

Fig.8.22: The Ability to Produce Continual Variation Requires a Core of a Mechanically Stable Structure.

Pasteur's Law of Biogenesis

The first person to discover a universal law of life was French chemist Louis Pasteur. Since ancient times, life forms that had obscure reproductive stages were thought to arise via spontaneous generation from non-living matter. Fermentation—the central process in wine and cheese making—occurs spontaneously, and when Pasteur began to research the matter, he was able to show that microbes—not spontaneous generation—are the cause. He demonstrated in 1861 that microbes do not arise spontaneously from their growth medium, but from physical transmission of spores, some of which are carried in air. From this, he formulated a *law of biogenesis—**that life comes from life**—*a universal principle that has stood the test of time. Origin-of-life researchers continue to look for means of *abiogenesis* (life from non-life) but without success, Figure 8,22.

DARWINIAN EVOLUTION

In his 1859 famous book *The Origin of Species*, Charles Darwin proposed a universal law—that all species arise from a common ancestor through a collective struggle that leads to survival of the fittest. This produces diverse species when natural selection of random natural variations favors different survival strategies in different environments.

This theory is correct in principle within limits—the species we see around us today have indeed arisen from ancestral species via natural selection of natural variation. However, it is incorrect in its extrapolation to *all* of life, and is thus not a universal law. We shall see shortly that Darwin's mechanism is just part of the larger *law of survival.*

Neo-Darwinists proposed that natural variation occurred primarily via *mutation* of genetic information. However, recent discoveries show that **mutations are accidental damage** events and are sending us all to extinction on a remarkably short time scale due to what appears to be the universally deleterious effects of these mutations.

Life's Irreducible Structure

In order to understand life, we need to know what it is made of. It consists mostly of architecture and machinery made from long-chain molecules having a 'backbone' of carbon atoms tightly linked together, with hydrogen, oxygen, nitrogen, phosphorus and sulfur attached along the sides, Figure 8.23. Many other elements are involved in other special ways as well. What makes life work is the peculiar manner in which these molecules are structured, and the intricate and super-efficient ways they work together as an integrated system.

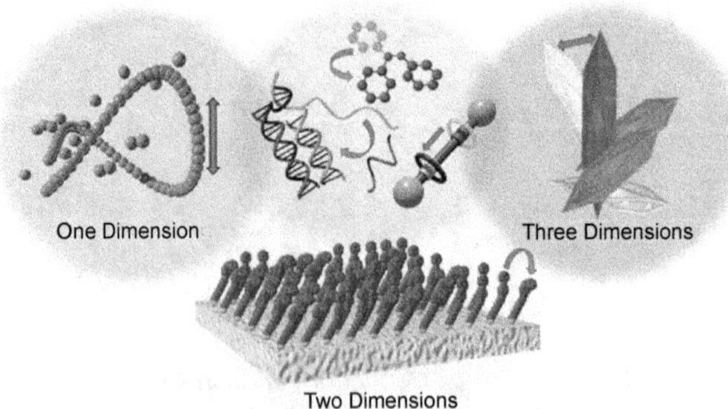

Fig.8.23: *Molecular Structure*

Famous European polymath Professor "Michael Polanyi" described the principle of life's *irreducible structure* in 1968, and it points unerringly to intelligent design. Polanyi argued that the special structure of life's machine-like components cannot be explained by (or reduced to) the properties of the atoms and molecules they are made of; something else is required. He did not speculate on what the extra ingredient might be because molecular biology was in its infancy. However, we can now say with great certainty that it is *coded information*. The very precisely 'engineered' structures in living things are crafted by the cell, one molecule at a time, by carefully following the instructions coded on the DNA molecule.

Polanyi's law of irreducible structure is a universal principle—life consists of **information-driven molecular machinery**.

Where does the information come from? According to Pasteur's **law of biogenesis**, if 'life comes from life', then life's information must come from its parent's information.

THE LAW OF SURVIVAL

Despite life's marvels, it all dies. Why? Theologically, because of the Fall—Adam and Eve's sin brought death into a perfect world. The biological question then is 'How?', and we shall see later that the answer is 'mutations.' How then do species survive? In the original creation, reproduction glorified God in filling the earth with His creatures. In this fallen world, reproduction now solves the problem of species survival. For a species to survive, it must be capable of reproducing itself and passing on its store of functional information to its offspring.

However, if an organism were a simple mechanical structure, it would become extinct at its first malfunction. An intelligent designer must therefore build-in to it a **self-repair mechanism.** Anticipating the need for repair would also logically lead to a complementary **routine maintenance system** to avoid the more obvious hazards. This combination of self-maintenance and self-repair would allow life to explore beyond its normal range of operations, because any damage incurred could be repaired. Life that survives beyond its normal operating range could then be said to have **adapted** to a new set of conditions when frequent repair turns into routine maintenance.

But repair and maintenance mechanisms are subject to the same damage hazards as the rest of life, so a longer-term solution is required if a species is to survive. Reproduction is the answer—it starts again using fresh materials and the original recipe, to build a new organism that has less accumulated damage.

Another life-challenge is that environments have changed dramatically during Earth's history, so an intelligent designer must build-in to the reproductive mechanism a system of **continual variation** that will produce a range of different capabilities in the offspring so that they might have a better chance of survival than the parents. *Continual* variation is necessary because if only a limited number of options were on offer, life would become 'stuck' on the most functional option, the alternatives would degenerate by mutation, and there would be no reserves to call upon during dramatic environmental change.

Therefore, we arrive at the *law of survival—life must vary and adapt or become extinct*. Life works only if it is *robustly flexible* (via self-maintenance and repair) to survive in this generation, and can *reproduce itself in continuously variable forms* to provide adaptability amongst its descendants. This condition is universal across the whole spectrum of life.

Some might object that 'living fossils' that are on the verge of extinction, such as the 'dinosaur pine tree' *Wollemia nobilis* are not showing any signs of genetic variation and thus contradict this law. Not so. This law applies to how the Creator made the *original baramins*, and not necessarily to the remnant populations we see today that have had their built-in original stores of variation exhausted by selection and depleted by mutation.

THE LAW OF FACILITATED VARIATION

Life, by and large, is beautiful and inspiring and wonderfully adapted to seemingly endless ways of making a living. There are some ghastly forms—parasites can do terrible things to the living bodies of their hosts. But if we put aside our squeamishness, we can still marvel at the fact that the parasite has developed astonishing ways to 'earn a living and provide for its family', often through multiple stages of very different life-forms in very different environments and/or host organisms.

Neo-Darwinists attributed all of this functional beauty to mutations and natural selection. But it has long been known that mutations produce defects and monsters, not beautifully functional adaptations to different ways of life. According to "Kirschner and Gerhart's" theory of *facilitated variation*, the enormously varied but functional beauty of life results from a combination of three fundamental components:

- robust core processes of cell structure, function and body plan,
- *modular regulatory mechanisms* that can be easily pulled apart (like Lego° blocks) and rearranged into new circuit-and-switch combinations that generate variation by activating different components in new times, places and amounts during embryonic development,
- *signaling systems* to coordinate everything.

Organisms have a built-in capability for variation which *facilitates* changes via an integrated modular structure that is able to maintain functionality in the face of internal and external challenges.

We know of no life form that does not have these three components ('Kirschner–Gerhart properties') so they constitute a universal principle, the *law of facilitated variation—species survival requires*:

- a robust core structure,
- a regulatory system that provides built-in capacity for variation through randomly rearranged module combinations, and
- a signaling system to coordinate and maintain the process.

Variation and Stasis

Species survival requires a balance between variation and stasis. To create life, there would be no point putting together a molecular machine that reproduced perfect copies of itself. It could not adapt to changing conditions, either in the current generation, or amongst its descendants. There is also no point putting together a machine that continually varies itself, because its variations would escalate into *error catastrophe*. Neither kind of machine will work on its own, because *both* kinds are needed *together*. Continuous variation has to be achieved in a manner that is compatible with—and thus limited by—the necessity for continuous function.

This principle is illustrated in games of chance with tossed coins or dice, or a spinning roulette wheel, Figure 8.22. A continuously variable outcome depends upon maintaining the rigid mechanical structure and continuous functional performance of the mechanism.

The law of Inverse Causality

The law of facilitated variation turns causality on its head and takes us out of the realm of physics and chemistry. To understanding this crucial statement, we need to look at the *law of **cause and effect***.

The Law of Cause and Effect

The law of cause and effect is one of the most fundamental in all of science. Every scientific experiment is based upon the assumption that the result of the experiment will be caused by something that happens during the experiment.

Now a naturalistic origin-of-life scenario must explain life in terms of a series of naturalistic causes. The first cause must produce the second step, which then causes the third step, and so on. The logic must be in the following order:

$$A \rightarrow B \rightarrow C \rightarrow \ldots Z \rightarrow \text{life}$$

where the arrow means 'causes' and **A** would be either chemicals (in a proteins-first scenario) or information (in an information-first scenario). All naturalistic theories are bound by this principle of causality because they do not admit any purpose in the process that could manipulate the steps towards a predetermined goal, nor is any intelligent agent able to help.

However, according to the *law of facilitated variation* we have to rule out such a scenario because none of the steps **A** to **Z** have the necessary properties to survive. The Kirschner–Gerhart properties define the necessary conditions for survival and they are not present in the **A** to **Z** series, only in the '*life*' at the *end* of the series.

If Darwin had conducted an origin-of-life experiment, he could justifiably have begun with chemicals in 'a warm little pond' because the molecular basis of life was not understood at that time. However, we now know too much to allow such a thing. Because we now know the *laws of survival* and *facilitated variation*, an intelligent designer would have to begin with that end in sight. To have anything less in mind would be to decide upon failure.

To have a particular end-product in mind is a case of *inverse causality*. In normal causality, the cause always comes *before* the effect. In biology however, we see the universal occurrence of *inverse causality*, recently acknowledged by Darwinian philosopher of science Michael Ruse. His chosen example was that stegosaur plates begin forming in the embryo but only have a function in the adult—supposedly for temperature control. Other examples include the adult's need for robustness in the face of environmental challenges that the zygote has not yet faced, and the adult's need for successful reproduction of variable offspring that is essential to its species' survival. All these lies in the zygote's *future* but must be present in the zygote.

For the zygote to have even arrived where it is, all of these events must have been in view in its progenitors' developmental program as well as in its own developmental program, otherwise life would have become extinct. This is why the ability to survive takes us out of the realm of physics and chemistry—and into the realm of **intelligent design**. In physics and chemistry, normal causality rules. In life, *inverse causality* rules. Although life *uses* the normal causality of physics and chemistry, it is not *bound* by its rules.

This characteristic is universal in all forms of life, so it constitutes a *law of inverse causality*—the Kirschner–Gerhart properties *inversely cause* the development of the adult from the zygote and they produce the adaptive variety necessary for survival. Since all this must be present at the beginning for species to survive, intelligent design is the only possible explanation.

The Law of Code Variation

How does life manage to transcend normal causality and move into inverse causality? The answer is *coded information*! It is through the coded information in our genomes that inverse causality rules biology. The adult end-product of development is coded into the zygote's genome. The genome guides the zygote to fulfil a destiny that was written down *before the event occurred*. It could be no other way, because life that lacked the Kirschner–Gerhart properties could not survive.

This leads us to another universal law. The *law of code variation,* which is that the Kirschner–Gerhart properties are encoded in the zygote, and it is through *the zygote's built-in ability to read, implement, **rearrange** and replicate this coded information* that the Kirschner–Gerhart properties inversely cause the development of the organism, the production of its variable offspring, and its ability to adapt and survive, Figure 8.24.

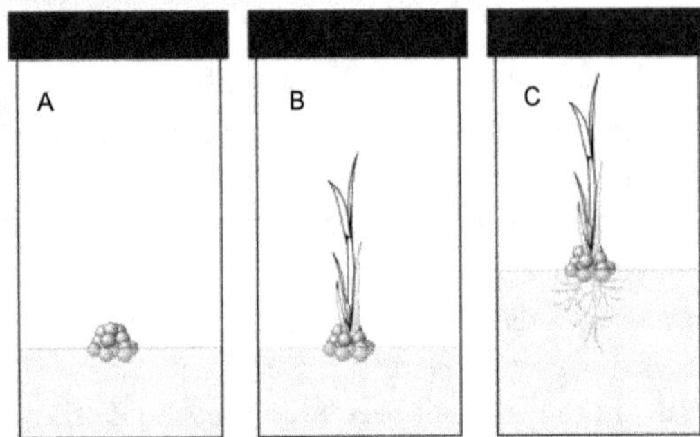

Fig.8.24: *Plant tissue culture. Plant stem cells grow in an amorphous mass in culture medium (**A**). When cytokinin is added, the cells begin to develop shoots, but no roots (**B**). When the rootless plant is transferred to a culture medium containing auxin, the root system begins to grow and a whole plant develops (**C**).*

Life does its own *natural genetic engineering*, illustrated most clearly in the fact that all *our* tools for genetic engineering have come from living organisms. It occurs in eukaryotes during the remarkable process of meiosis when the cell cuts the father's and mother's chromosomes up into segments and rearranges them so that the gene combinations get mixed up. A number of other enzyme-mediated changes can also occur—deletions, insertions, inversions, duplications, transpositions and retro-transpositions.

Microbes (bacteria and viruses) have the added ability to splice foreign DNA in and out of their genomes, thereby 'sampling' the genetic environment for potentially useful sequences. The (eukaryote) single-celled ciliate *Oxytricha trifallax*, displays an extraordinary talent for natural genetic engineering. It has two nuclei in its single cell, one large and one small. The large nucleus carries out the everyday activities of life, and the small nucleus remains quiet until it is time to reproduce. At reproduction, the small nucleus undergoes meiosis, but the chromosomes in the large nucleus are chopped up into hundreds of thousands of separate pieces. In the new daughter cells, all these fragments are then re-assembled into chromosomes in a new large nucleus.

COMPARTMENTS, MODULES AND SIGNALS

Development of the zygote into an adult is organized by the use of compartments, modules and signals of different kinds, and these can operate independently, yet in a cooperative manner so as to produce a functional whole organism. We can illustrate the basic principles in both plant and animal development with two simple experiments.

Many plants are easily grown in cell culture by taking some stem cells from a growing tip of a root or shoot, and placing them on a sterile growth medium, Figure 8.24. After a few days, they will multiply and produce a mass of undifferentiated cells. If a drop of *cytokinin* is then added, the cells will begin to organize into a shoot, with stem and leaves, but with no root. If the half-plant is then transferred to a new growth medium with *auxin* in it, a root system will develop and we will soon have a whole plant with both roots and shoots.

A comparable system in animals is seen in the planarian flatworm. Planarians are free-living freshwater animals that have the remarkable ability to regenerate themselves after being cut in half! The signaling system that controls head and tail regeneration was recently identified by experimentally interfering with signals to see what would happen, Figure 8.25. When the *β-catenin* signaling pathway was blocked, a head developed at the tail end of the head section, producing a two-headed flatworm, and a tail developed at the head end of the tail section, producing a two-tailed flatworm.

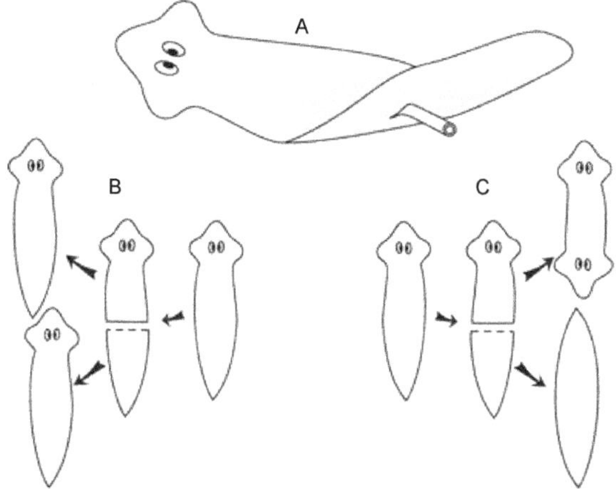

Fig.8.25: *A plananrian flatworm is a free-living freshwater creature with two eyespots at one end and a feeding tube at the other end (**A**). When cut in half, each of the two halves normally regenerates a complete organism (**B**). However, when the beta-catenin signaling system is blocked, the head end regenerates another head and the tail end regenerates another tail (**C**).*

Now *cytokinin*, *auxin* and *β-catenin* are just protein molecules. They do not carry any coded instructions like RNA or DNA, so they are unable to tell the cells what to do or how to do it. They are simply signaling molecules that carry a GO/STOP message.

The plant cells already have the built-in ability to produce a whole plant—it only needs to be switched ON. But notice that the top part of the plant can develop quite independently of the bottom part. Likewise in the planarian flatworm, the head end and the tail end can develop independently of one another. This is *compartmentation*. Organisms are arranged into compartments so that development can proceed in one compartment independently of what happens in an adjoining compartment. But notice also that adjoining compartments cooperate at the joining edges so that the development in each compartment is smoothly integrated with its neighbors. Compartments also exist at smaller scales, for example leaves, stems, flowers, etc. in plants, and limbs, head, alimentary tract, etc. in animals. The human embryo contains about 300 different compartments.

Within each compartment there are numerous *modules* that carry out specific tasks. For example, energy supply is universally provided in all forms of life by ATP (adenosine tri-phosphate) via a proton-driven molecular motor. In eukaryotes, these are especially associated with mitochondria as the 'powerhouses' of cells. Therefore, there will be modules that contain the information to make ATP motors and modules that contain the information to make mitochondrial powerhouses. These components are then activated or repressed by the signaling network in a GO/STOP manner to provide energy for and synchronize the daily round of metabolic activity.

THE LAW OF SIGNALS

The universal rule governing cell signaling is that they are *permissive* and not *instructive*. They are GO/STOP messages, and do not contain any instructions as to what is supposed to be done or not done, Figure 8.26. The modules in the cell must therefore already possess the information for *what to do* and *how to do it*, and further possess the information required to interpret what the *GO signal* is and what the *STOP signal is.* All that the signal network needs to do is to send the right sequence of GO and STOP signals for development to proceed from zygote to adult.

This *law of signals* is that signaling networks are *permissive* and not *instructive.*

A crucial consequence of this law is that modules must contain, and maintain, certain basic properties, which are the subject of the *law of modules.*

The law of modules

Fig.8.26: *The crucial element in cell signaling is a GO or STOP signal that does not contain any instructions on what to do or cease doing. The receiver of the signal needs to already know what to do and how to do it. We can liken it to a swimming race in which the starting pistol signals GO. The swimmers must already know what to do and how to do it when the signal is heard.*

A module in engineering is a unit that has a stable internal structure and function such that it can be connected to a number of different other systems and can interact with them, but without the interaction interfering with its own internal structure or function, Figure 8.26. Kirschner and Gerhart attribute life's built-in ability to vary continually to the modularity of its *regulatory system.* Modularity in the regulatory domain is concerned with *information content.*

META-INFORMATION

In neo-Darwinian theory, the only relevant information in DNA was thought to be those which codes for proteins (genes). The rest was thought to be 'junk' leftovers from the past, or useless duplicates of currently functional genes. But when we look to see the other four dimensions of meaning, we come across an entirely different kind of information that is called *meta-information.* Meta-information is *information about information; akin to metadata in the computer world.*

Meta-information in biology is information about how to use the information passed on to the zygote from the parents. A living organism needs to have the following kinds of *meta-information:*

- How to read the information inscribed on, in and around the DNA molecule.
- When to read what bits of information inscribed on, in and around the DNA molecule.
- What to do with the information once it is read from the DNA.
- How to regulate the mechanical structures that implement the information.

- How to repair and maintain the information store and the mechanisms that use it.
- How to pass on the information store and its support systems to the next generation.

There are two crucial features of meta-information that confound all naturalistic explanations for the origin of new biological information. First, it cannot come into existence by a spontaneous random process. A random event is, by definition, one that occurs *independently* of other such events. But meta-information is, by definition, entirely *dependent* upon the information that it relates to. That is, it has no meaning or purpose apart from the information that it relates to. It therefore cannot come into existence by any kind of independent process.

Second, without it, the basic information is of no use. For example, a zygote could have all the genes required to turn it into a human being, but without the necessary meta-information to instruct the cell in how and when to use which genes, the genes themselves would be useless. Information and meta-information are mutually inter-dependent—each is useless without the other.

Meta-information is the *information you need to have* in order to use *the information you want to have* to provide you with the capacity for your survival and the survival of your descendants and your species.

We now have two more universals that constitute laws of life. First, conserved core functional machinery must be coded in two different kinds of functional information—the *primary functional information* (mostly genes), and the *meta-information* needed to implement the primary information (mostly regulatory information). Second, in order to maintain the functional integrity of life, the conserved core information must be kept together in modules that are difficult to break apart, but whose signal circuit connections can be pulled apart and put together again in different ways to produce a built-in system of variation.

The *law of modules* is that the basic module of information has to contain *functionally integrated primary information* plus the necessary *meta-information* to implement the primary information. This information must consist of *what to do*, *how to do it*, what the *GO signal* is and what the *STOP signal* is. This information is not necessarily in coded form—it may be designed into a mechanism that has the required properties.

MUTATIONS

We can now define a *mutation* more precisely as an accident that disrupts the meaning contained in a module or in the signal that activates or represses it. Rarely, a mutation can have survival value—sickle-cell trait in human populations exposed to malaria is a classic case. But too many accidents will lead to more certain extinction, as disruptions are far, far more often deleterious than beneficial and in no cases lead to an increase in information.

There is a huge difference between mutations that cause random accidental damage and the random change capacity that is built-in to the *facilitated variation* system. Independent segregation and random re-combinations of alleles that occur during sexual reproduction produce useable phenotypic variation and do not degrade the machinery of life or its genome. Other modular rearrangements such as insertions, deletions, transpositions and retro-transpositions can also provide potentially useful new combinations. This illustrates why Kirschner and Gerhart used the analogy with Lego® blocks—random rearrangements of blocks does not degrade the integrity of the blocks themselves.

THE LAW OF DEGENERATION

Genomes degenerate over time for two main reasons. First, natural selection is a process of individual extinction—organisms that die without reproducing take their unique store of built-in variation with them to the grave, making it no longer available to future generations. Second, like all organic molecules, DNA is inherently unstable and it accumulates molecular damage with time.

A typical human cell would undergo 2,000 to 10,000 spontaneous DNA hydrolysis damage events every day just because it is an aqueous environment. In addition, DNA is constantly consulted for information during daily metabolism and this causes transcription stress fractures, fork collapse and polymerase fidelity errors. Reactive oxygen species (ROS) produced during normal metabolism causes biochemical disruptions. Environmental toxins and ionizing radiation cause direct physical/chemical damage and indirect injuries through ROS production. Further physical damage arises when cell division pulls chromosomes apart. And at meiosis, during nuclease-mediated genome rearrangements, and during various kinds of recombination events, errors and breaks in the nucleotide sequence regularly occur.

The only reason that DNA functions as well as it does is that cells come equipped with an amazing array of cooperative DNA repair mechanisms. For example, polymerase replication during cell division might produce 6 million errors per cell, but then proofreading machinery can reduce this to 10,000 and then mismatch repair machinery could reduce this to 100. It appears to be impossible, however, to replicate the 6 billion nucleotides in a human cell in a completely error-free manner.

All of life's machinery operates at extremely high cost in molecular damage. Molecule turnover rates have high-wear surfaces in the order of 30 seconds, and each cell in our body typically experiences in the order of ten DNA damage-and-repair events every second. Another evidence is that of the approximately 100 trillion cells in an adult human body, about 70 to 90 billion are dismantled and recycled each day because of irreparable damage, with greatest turnover in high usage areas like blood, intestinal tract and skin. The cell's built-in repair and maintenance systems also suffer molecular damage and so they are unable to attain 100% efficiency, hence inevitable degeneration.

It is not only *genetic* degeneration that is inexorable, but also *epigenetic* degeneration. Epigenetic mechanisms are increasingly being seen to be crucial to how DNA functions. A multitude of epigenetic mechanisms are coming to light and more may yet be anticipated. They include the side chains that are attached to the DNA and control the expression of the associated nucleotides, histone-based nucleosome patterning, chromatin structure, positioning of DNA segments within the nucleus, and transcription and post-transcription errors and modifications.

A large study of identical twins showed that their gene expression patterns were very similar while young, but differences accumulated with age, and this can be largely and perhaps entirely attributed to the accumulation of *epigenetic defects*. Their DNA remains identical, but their epigenetic patterns of DNA expression change in different ways during life. Environment plays a big role in this; different lifestyles produced greater epigenetic differences than identical lifestyles. The authors concluded that accumulation of epigenetic defects would probably occur at a *faster rate* than genetic mutations because the consequences for survival are probably less dramatic and cells do not have repair mechanisms to correct them.

We therefore come to a *law of degeneration* that *the rate of both genetic and epigenetic damage exceeds the capacity of the cell to repair it*, and *most of the damage is too phenotypically insignificant for natural selection to detect and remove from the population*. The net result is inexorable degeneration of organisms and their genomes. Estimates based upon a number of different numerical models indicate a time to extinction of tens to hundreds of thousands of years.

Life is intelligently designed to survive via a built-in system of *facilitated variation* that enables organisms and their offspring to adapt to changing conditions. This leads over time, and across different environments, to diversification of species. Charles Darwin was correct in proposing that the species we see around us today have arisen via the mechanism of *natural selection of natural variation*, but he was wrong in extrapolating it to *all* life. Organisms designed with the capacity for *facilitated variation* would have had enormous initial capacity for rapid diversification from one generation to the next, and vast numbers of new species could have rapidly filled the early earth and rapidly re-colonized the destroyed earth within a few generations after the Flood. However, natural variation is limited by what was built-in to begin with—regulatory signals cannot switch ON features that don't exist in the organism's genome. Despite life's functional beauty, selection depletes gene pools and mutations degrade genomes, and extinction is coming on a time scale of only thousands, not billions, of years. Intelligent design plus rapid extinction points clearly to recent Creation and Fall, as the Bible tells us.

LIFE IS IN THE BLOOD

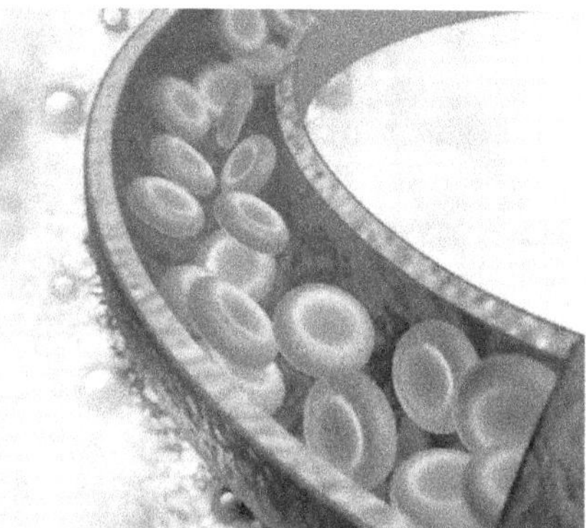

Fig.8.27: "For the life of the flesh is in the blood." Leviticus 17:11

So, says *Leviticus 17:11*, "For the life of the flesh is in the blood.". Everyone recognizes that we must have enough blood flowing around our body or else our bodily functions decline and we die, Figure 8.27. Hitherto, for a long time the exact function of blood was little understood. In what ways has modern science shown *Leviticus 17:11* to be true?

Blood is fundamental to the function of every cell of every component in our bodies. Cells require food to survive, grow, repair themselves and to fulfill their explicit functions, and, to reproduce. Cellular food is transported in blood to provide energy for all the cells' needs. As humans are multicellular organisms, having separate specialized organs with highly sophisticated functions, transport and communication between these structures is essential.

COORDINATION

Do the cells of the body tell the blood how it should work? No. Does the blood carry around everything possible just in case? No. The cells and the blood work together to provide optimum conditions for correct functioning of all the cells—with their different requirements—in all the tissues and organs of the whole body, including the cells of the blood itself.

Blood provides this coordinated environment by regulating acidity/alkalinity (pH), providing oxygen (and removing carbon dioxide and other waste products), and carrying essential vitamins and minerals. Also, blood has to be in the right places at the right times, at the right temperature and pressure, and it carries regulatory messages between organs via blood 'messengers' called hormones. All this is organized within very specific limits—straying outside these (through injury, disease, toxins, etc.) rapidly reduces functionality.

HORMONAL FEEDBACK

Hormones, those important chemical messengers in the blood, are involved in self-regulating feedback systems. These systems stimulate hormone production in times of lack, and suppress it in times of plenty. For example, when we eat, the sugars in the intestine are digested and absorbed into the local bloodstream. This blood then passes through the pancreas and its higher sugar level stimulates production of the hormone insulin. As insulin is distributed in the bloodstream, it reduces the blood sugar to normal levels again by increasing the amount of sugar that all cells take in. In fact, the brain relies almost entirely on sugar (specifically glucose) for its energy supply; hence this feedback system is absolutely critical for proper brain activity. If the blood glucose ever drops too much, we lose consciousness.

The body's systems tend to be wisely over-engineered, so that one might predict that there is also a system to cope with low sugar levels, for example when we exercise and use sugar up. This system uses the hormone glucagon (also from the pancreas) and it works by releasing glucose into the blood from stores located mostly in the liver.

There are about fifteen organs classed as hormone-producing (endocrine) glands, and their products, carried by the blood, affect either every cell in general or specifically target certain cells. Widely known examples are the male and female hormones testosterone and estrogen, adrenaline (epinephrine in the US), the thyroid hormone thyroxine, and many more.

TARGETS

For example, thyroxine regulates the speed of metabolism in every cell, and having the correct amount (within narrow limits) allows normal cellular activity. Too much and we become 'hyper', too little and we are slow and lethargic.

Another example is gastrin. The target organ for gastrin is that part of the inner lining of the stomach which produces hydrochloric acid for digestion. Food in the last part of the stomach stimulates the production of gastrin, which is carried back by the blood to stimulate acid production. This is a positive feedback mechanism in which blood is the essential communicating link.

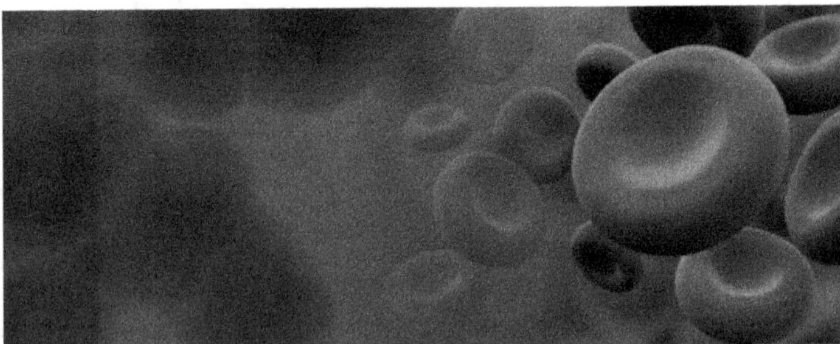

***Fig.8.28**: Blood Platelets*

RED BLOOD

The red color of blood reflects the color of the hemoglobin inside the red blood cells, Figure 8.28. This is because the hemoglobin contains iron. The 'heme' of the hemoglobin molecule in vertebrates (creatures with a backbone) is a porphyrin ring which surrounds ferrous iron atoms. It is the spatial relationship between heme, iron and globin which makes it possible to bind oxygen molecules reversibly—one to each iron—and which makes the system so efficient.

Anticipation-Clotting Cascade

Blood also has a major role in body protection in that it is an integral part of the immune or infection-fighting system, involving antibodies and white blood cells. It also possesses a highly complex mechanism to prevent its own loss from the body (clotting) and to prevent clotting inside the body (thrombosis). The capacity to quickly initiate clotting outside and to limit—even reverse—clotting on the inside is provided by 'cascades'—cumulative processes in which each step of the process is dependent on the one before it. The cascades are of such complexity that new factors, cofactors and regulators are being constantly added to our body of knowledge. It is now known that there are more than a hundred factors or steps that make up the clotting cascade. Such details add to our appreciation of how finely balanced, effective and versatile the system is. Nonetheless, a greater marvel is that such a system, which is there in anticipation of blood loss, internal injury or disease, should be there at all.

Unique Red Blood Cells

Red blood cells (RBCs or erythrocytes) form the majority of the cells in the blood—and a quarter of all cells in the human body. They are unique among all others—in mammals, they have no nucleus and none of the usual energy-producing structures in the cell outside the nucleus. This is a design feature of mammals (creatures which, like us, suckle their young). Normally, a cellular nucleus carries the DNA which instructs the cell on how to perform its functions, including repair and reproduction, at the appropriate times. RBCs cannot do this because instead they are especially designed to carry oxygen, and in humans, having a nucleus would hinder this essential function. Therefore, the nucleus is lost after formation, leaving them with their characteristic biconcave shape.

Two reasons have been suggested for this. First, the relative size of RBCs (6–8 µm diameter and just 2 µm thick) and capillaries (tiny blood vessels) is such that red blood cells often have to deform in order to squeeze through. A nucleus (about 6 µm on average) could prevent passage of the cell and make it get stuck, blocking the circulation.

Second, the shape and deformability of the red blood cell is optimized for the carrying and delivery of oxygen, and it maximizes the amount of hemoglobin that can be packed into the cell. Nevertheless birds, which have a very high oxygen requirement, do fine with nucleated RBCs, so there are other design features in birds that compensate for this.

The system of the red blood cells giving oxygen to the cells of the tissues is reversed when the red blood cell reaches the lungs, where it gives up its carbon dioxide (though this is mostly carried by plasma) and takes on a new load of oxygen. At rest, all the blood (5 liters in an adult) completes a circuit within a minute (spending 1 to 3 seconds in the capillaries). With exercise, circulation is as quick as every 10 seconds. Having a molecule such as hemoglobin which can handle oxygen so quickly and reversibly, when required, is amazing.

BLOOD BYTES

There are about 4–6 million red blood cells (RBCs) in every cubic millimeter of blood; 20–30 trillion of them in each person.

Every day about 1% of these are changed. New RBCs take about 7 days to form in the bone marrow, and are produced at the staggering rate of about 2 to 3 million every second.

Each RBC lasts about 120 days before its components are recycled to form new RBCs.

During its 4-month lifetime, each red cell travels some 500 km (300 miles) around the body, passing through the heart about 14,000 times per day.

Most of our blood vessels are the microscopic capillaries. If the blood vessels in one person were laid end to end, they would be about 150,000 km (100,000 miles) in length—enough to circle the earth at the equator about four times!

WORDS MATTER

FLESH (as used in many English translations of Leviticus 17:11): Hebrew בשר *basar*, the tissues that make up the body, and (by extension) also the body, the living creature.

TISSUE: a collection of cells (not necessarily the same type) grouped for a specific function. E.g., connective tissue, muscle tissue. Blood itself is, technically speaking, also a tissue.

ORGAN: several types of tissue functionally grouped together, e.g., liver, lung.

So, is the life of the flesh in the blood? Although not confirmed by science until modern times, this statement from *Leviticus 17:11* has always been true. Blood actively maintains life by providing a vital function for all cells, tissues and organs, and thus, the life of the whole body. The more we find out about the astounding functional design and complexity of blood, the more marvelous it becomes to us, and the more honor and praise is due its infinite omniscient Creator.

APPENDIX

- 1 µm (micrometre) is a millionth of a metre, or 1/25,400 inch.
- Sugars have linear forms that contain carbonyls—The cyclic forms that occur in nucleic acids also predominate in solution form, but in equilibrium with the linear form. When something reacts strongly with the aldehyde, then more of the linear form is regenerated to replace that which is reacted, so all the sugar molecules will be consumed.
- Bergman, J., Why the Miller-Urey research argues against abiogenesis.
- Truman, R., What biology textbooks never told you about evolution.
- Sarfati, J., Origin of life: instability of building blocks.
- The 'right' and 'left' in terms of chirality refer to the position of the amino group (NH_2) as displayed on a standardized diagram (Fischer projection) of an amino acid.
- With four nucleotides 'letters' comprising DNA and with three read at a time (a 'codon') by the reading machinery, this gives 4x4x4=64 different possibilities (3-letter 'codons').
- Newly discovered DNA repair mechanism, *Science News*, sciencedaily.com, 5 October 2010.
- *Ibid.* p. 336; see Quotable quote: Primeval soup—failed paradigm.
- *Expelled: no intelligence allowed*, Premise Films, 2008.
- Most of the oxygen in the bloodstream (98%) is carried on the hemoglobin in the RBCs; a little is dissolved in the plasma. Most of the carbon dioxide returning to the lungs is carried dissolved in the plasma, with a small amount in the RBCs.

REFERENCES

1. http://evolution.berkeley.edu/evosite/evo101/IIE2aOriginoflife.shtml (accessed 17 October 2013).
2. Myers, P.Z., 15 misconceptions about evolution, 20 February 2008, scienceblogs.com; Matzke, N., What critics of neo-creationists get wrong: a reply to Gordy Slack, pandasthumb.org. Dawkins tries to deal with the origin of life in his book *The Greatest Show on Earth*, where he claims to 'prove evolution'. See Sarfati, J., *The Greatest Hoax on Earth?* ch. 13, 2010, Creation Book Publishers.

3. Kerkut, G.A., *Implications of Evolution*, Pergamon, Oxford, UK, p. 157, 1960 (available online at ia600409.us.archive.org/23/items/implicationsofev00kerk/implicationsofev00kerk.pdf); creation.com/evolution-definition-kerkut.

4. Sarfati, J., World record enzymes, *Journal of Creation* 19(2):13–14, 2005; creation.com/world-record-enzymes-richard-wolfenden. Return to text.

5. Chadwick, A.V., Abiogenic Origin of Life: A Theory in Crisis, 2005; origins.swau.edu/papers/life/chadwick/default.html. Return to text.

6. See for example Potassium ion channel, hydrated ionic radii, creation.com/ionic-error, 21 August 2010.

7. Krogh, A. *et al.*, Predicting transmembrane protein topology with a hidden Markov model: application to complete genomes, *Journal of Molecular Biology* 305(3):567–580, 2001; dx.doi.org/10.1006/jmbi.2000.4315.

8. Transporter Proteins in *Mycoplasma genitalium* G37-; membranetransport.org/index.html (accessed 11 Oct. 2013).

9. Sarfati, J., Origin of life: the chirality problem; creation.com/origin-of-life-the-chirality-problem (updated 2010).

10. Cairns-Smith, A.G., Evolutionist criticisms of the RNA World conjecture, from *Genetic Takeover and the Origin of Life*, 1982; creation.com/cairns-smith-detailed-criticisms-of-the-rna-world-hypothesis.

11. Stanford researchers produce first complete computer model of an organism; news.stanford.edu, 19 July 2012.

12. Sarfati, J., How simple can life be? https://creation.com/how-simple-can-life-be. Research on a synthetic cell, JCVI-syn3.0, showed that 473 genes were essential, 65 of which had no known function: C.A. Hutchison III *et al.*, Design and synthesis of a minimal bacterial genome, *Science* 351:1414, March 25, 2016; doi: 10.1126/science.aad6253.

13. Davies, P., The secret of life won't be cooked up in a chemistry lab: Life's origins may only be explained through a study of its unique management of information, *The Guardian*, Sunday 13 January 2013; guardian.co.uk/commentisfree/2013/jan/13/secret-life-unveiled-chemistry-lab.

14. Davies, P., Life force, *New Scientist* 163(2204):27–30, September 18, 1999.

15. Popper, K.R., "Scientific reduction and the essential incompleteness of all science"; in Ayala, F. and Dobzhansky, T., (Eds.)., *Studies in the Philosophy of Biology*, University of California Press, Berkeley, p. 270, 1974.

16. Smith, C., Lost in translation: The genetic information code points to an intelligent source, 6 May 2010; creation.com/genetic-code-intelligence.

17. Freeland, S.J., *et al.*, Early fixation of an optimal genetic code, *Molecular Biology and Evolution* 17(4):511–18, 2000; mbe.oxfordjournals.org/content/17/4/511.full.

18. Three are usually used as 'stop' codes to mark the end of a protein coding sequence, so 61 are normally used for amino acid coding.

19. Novoa, E.M. and de Pouplana, L.R., Speeding with control: codon usage, tRNAs, and ribosomes, *Trends in Genetics* 28(11):574–581, November 2012; ww2.biol.sc.edu/~elygen/biol655/translation%20speed.pdf.

20. Sarfati, J., New DNA repair enzyme discovered, 13 January 2010; creation.com/DNA-repair-enzyme. Tomas Lindahl, Paul Modrich, and Aziz Sancar won the 2015 Nobel Prize for Chemistry for discovering three different DNA repair mechanisms: Batten, D., DNA repair mechanisms 'shout' creation, *Creation* **38**(2):56, April 2016. Return to text.

21. Cox, M.M., Keck, J.L. and Battista, J.R., Rising from the Ashes: DNA Repair in *Deinococcus radiodurans*, *PLoS Genetics* 6(1): e1000815, 2010; doi:10.1371/journal.pgen.1000815.

22. Catchpoole, D., Life at the extremes, *Creation* 24(1):40–44, 2001; creation.com/extreme and Sarfati, J., Hydrothermal origin of life? *Journal of Creation* 13(2):5–6, 1999; creation.com/hydrothermal.

23. Morelle, R., Darwin's warm pond idea is tested, 13 Feb. 2006; news.bbc.co.uk/2/hi/science/nature/4702336.stm.

24. Prions are sometimes proposed as replicating proteins, but prions cause *existing* proteins to become misshapen; they don't replicate themselves by causing amino acids to line up in the correct sequence to make a copy of the prion (prions are thought to cause 'mad cow disease').

25. Yockey, H., *Information Theory, Evolution and the Origin of Life*, Cambridge University Press, 2005, pp. 118–119.

26. Evolutionist criticisms of the RNA World conjecture; Quotable Quote by Cairns-Smith; creation.com/cairns-smith-detailed-criticisms-of-the-rna-world-hypothesis. See also, Mills, G.C. and Kenyon, D., The RNA World: A Critique, *Origins & Design* 17(1); arn.org/docs/odesign/od171/rnaworld171.htm.

27. Bates, G., *Designed* by aliens? *Creation* 25(4):54–55, 2003; creation.com/aliens.

28. Sarfati, J., Panspermia theory burned to a crisp: bacteria couldn't survive on meteorite, 10 Oct 2008; creation.com/panspermia-theory-burned-to-a-crisp-bacteria-couldnt-survive-on-meteorite.

29. Hoyle, Fred, The Big Bang in Astronomy, *New Scientist* 92:521–527, 1981.

30. For example, Royal Truman has researched the protein ubiquitin, present in eukaryotes, to show that little variation in the sequence is permitted for functionality, so that the chance (naturalistic) origin of such a protein is ruled out; see Truman, R., The ubiquitin protein: chance or design? *Journal of Creation* 19(3):116–127, 2005; creation.com/the-ubiquitin-protein-chance-or-design.

31. Sir Fred Hoyle, as quoted by Lee Elliot Major, "Big enough to bury Darwin". *Guardian* (UK) education supplement, Thursday August 2001 ,23; education.guardian.co.uk/higher/physicalscience/story/0,9836,541468,00.html.

32. Lloyd, Seth, Computational capacity of the universe, *Physics Review Letters* 88:237901, 2002; http://arxiv.org/abs/quant-ph/0110141v1.

33. Morowitz, H., *Energy Flow in Biology*, Academic Press, NY, 1968.

34. Nagel, T., *Mind and Cosmos: Why the Materialist Neo-Darwinian Conception of Nature Is Almost Certainly False*, Oxford University Press, 2012.

35. Yockey, H., *Information Theory and Molecular Biology*, Cambridge University Press, 1992, p. 257.

36. Stuart Kauffman, *At Home in the Universe: The Search for the Laws of Self Organization and Complexity*, Oxford University Press, p. 31, 1995.

37. Harold, F.M., *The way of the cell: molecules, organisms and the order of life*, Oxford Uni. Press, New York, p. 205, 2001.

38. Davies, Paul, The Cosmos Might Be Mostly Devoid of Life: We still have no idea how easy it is for life to arise—and it may be incredibly difficult, *Scientific American*, 1 September 2016; www.scientificamerican.com/article/the-cosmos-might-be-mostly-devoid-of-life.

39. Kirschner, M.W. and Gerhart, J.C., *The plausibility of life: Resolving Darwin's Dilemma*, Yale University Press, New Haven and London, p. 256, 2005.

40. Watchershauser, G., Origin of life: RNA world versus autocatalytic anabolism, *The Prokaryotes*, Vol. 1, 3rd edition, chapter 1.11, pp. 283–275, p. 2006 ,282.

41. Quoted in, Evolution's final frontiers, *New Scientist* 201(2693):42, 2009.

42. Lazcano, Antonio, Historical Development of Origins Research, *Cold Spring Harbor Perspectives in Biology* 2(11): a002089, November 2010; doi: 10.1101/cshperspect.a002089. Return to text.

43. For example, a hybrid between an American paddlefish (family Polyodontidae) and a Russian sturgeon (family Acipenseridae). Kály, J. and 12 others, Hybridization of Russian sturgeon (*Acipenser gueldenstaedtii*, Brandt and Ratzeberg, 1833) and American paddlefish (*Polyodon spathula*, Walbaum 1792) and evaluation of their progeny, *Genes* 11(7):753, 6 Jul 2020; Sarfati, J., Startling sturddlefish, *Creation* 43(1):17, 2020. Return to text.

44. Pennisi, E., Shaking up the tree of life. *Science* 354(6314):817–821. Return to text.

45. Pennisi, E., Hybrids spawned Lake Victoria's rich fish diversity. *Science* 361(6402):539. Return to text.

46. Lamichhaney, S. *et al.*, Rapid hybrid speciation in Darwin's finches, *Science* 359(6372):224–228, 2018.

47. Jeanson, N.T., Mitochondrial DNA clocks imply linear speciation rates within "kinds", *Answers Research Journal* 8:273–304, 2015.

48. Wise, K.P., Mammal Kinds—how many were on the Ark? in: Wood, T.C. and Garner, P.A. (Eds.), *Genesis Kinds: Creationism and the Origin of Species*, Wipf & Stock, Eugene, Oregon, p. 143, 2009.

49. Darwin, C., *On the Origin of Species*, 6th edition, John Murray, London, 1871.

50. Veron, J.E.N. *Coral in Space and Time: the biogeography & evolution of the Scleractinia*, Cornell University Press (Ithaca, NY), 1995. ISBN: 0-801-48263-1.

51. Manica A. and Carter, R.W., Morphological and fluorescence analysis of the *Montastraea annularis* species complex in Florida, *Marine Biology* 137:899–906, 2000 | doi:10.1007/s002270000422.

52. Precht, W.F., Vollmer, S.V., Modys, A.B., and Kaufman, L. Fossil *Acropora prolifera* (lamarck, 1816) reveals coral hybridization is not only a recent phenomenon, *Proc Biological Society of Washington* 132(1):40–55, 2019; | doi: 10.2988/18-D-18-00011.

53. Sanford, J., Carter, R., Brewer, W., Baumgardner, J., Potter, B., and Potter, J., Adam, Eve, designed diversity, and allele frequencies; in: Whitmore, J.H. (Ed.), *Proceedings of the Eighth International Conference on Creationism*, Creation Science Fellowship, Pittsburgh, PA, pp. 200–216, 2018.

54. Paterson, J.R., The trouble with trilobites: classification, phylogeny and the cryptogenesis problem. *Geological Magazine* 157 (special issue 2019 ,46–35:(1.

55. Adrain JM (2011) Class Trilobita Walch, 1771. In Animal Biodiversity: An Outline of Higher-Level Classification and Survey of Taxonomic Richness (ed. Z-Q Zhang). Zootaxa 3148, 104–9.

56. Darwin, C.R., *On the Origin of Species by Means of Natural Selection, or the Preservation of Favoured Races in the Struggle for Life*, 1st ed., John Murray, London, p. 109, 1859; darwin-online.org.uk.

67. Williams, A.R., Molecular limits to natural variation, *Journal of Creation* 22(2):97–104, 2008.

68. Williams, A.R., Mutations: evolution's engine becomes evolution's end, *Journal of Creation* 22(2):60–66, 2008.

69. Polanyi, M., Life's irreducible structure, *Science* 160:1308–1312, 1968.

60. Williams, A.R., Life's irreducible structure, part I: autopoiesis, *Journal of Creation* 21(2):109–115; Part II: naturalistic objections, *Journal of Creation* 21(3):77–83, 2007.

61. Kirschner, M.W. and Gerhart, J.C., *The Plausibility of Life: Resolving Darwin's Dilemma,* Yale University Press, New Haven, CT, 2005. See also, Williams, A.R., Facilitated variation: a new paradigm emerges in biology, *Journal of Creation* 22(1):85–92, 2008.

62. Crotty, S., Cameron, C.E. and Andino, R., RNA virus error catastrophe: direct molecular test by using ribavirin, *PNAS* **98**(12):6895–6900, 2001.

63. Ruse, M., *Darwin and Design: Does Evolution have a Purpose?* Harvard University Press, Cambridge, MA, 2003.

64. Shapiro, J.A., Bacteria are small but not stupid: cognition, natural genetic engineering and socio-bacteriology, *Studies in History and Philosophy of Biological and Biomedical Sciences* 38:807–819, 2007.

65. Nowacki, M., Vijayan, V., Zhou, Y., Schotanus, K., Doak, T.G. and Landweber, L.F., RNA-mediated epigenetic programming of a genome-rearrangement pathway, *Nature* 451:153–158, 2008.

66. Cesari, F., *Cell* polarity: heads or tails? *Nature Reviews Molecular Cell Biology* 9:94–95, 2008.

67. ATP is the most common of a class of similar energy-packaging molecular machines.

68. Sauro, H.M., Modularity defined, *Molecular Systems Biology* 4:166, 2008; Del Vecchio1, D., Ninfa, A.J. and Sontag, E.D., Modular cell biology: retroactivity and insulation, *Molecular Systems Biology* 4:161, 2008.

69. Williams, A.R., Inheritance of biological information, Part I: The nature of inheritance and of information, *Journal of Creation* 19(2):29–35, 2005.

70. Lindahl, T., Instability and decay of the primary structure of DNA, *Nature* 362:709–715, 1993.

71. McCulloch, S.D. and Kunkel, T.A, The fidelity of DNA synthesis by eukaryotic replicative and translesion synthesis polymerases, *Cell Research* 18:148–161, 2008.

72. Fraga, M.F. *et.al.*, Epigenetic differences arise during the lifetime of monozygotic twins, *PNAS* 102(30):10604–10609, 2005.

73. Guyton, Arthur C., *Textbook of Medical Physiology*, Eds Arthur C. Guyton, John E. Hall, p. 838, W.B. Saunders Co., Philadelphia PA 19106, 10th Edition 2000.

74. E.g. there is an international scientific journal dedicated solely to Thrombosis and Haemostasis (Schattauer, ISSN 0340-6245, 12 issues/yr).

75. 'Cell Nucleus', *Encyclopaedia Brittannica: Ultimate Reference Suite 2005.*

76. Sanford, J., Carter, R., Brewer, W., Baumgardner, J., Potter, B., and Potter, J., Adam and Eve, designed diversity, and allele frequencies. In *Proc. 8th Eighth International Conference on Creationism*, ed. J.H. Whitmore, pp. 200–216, 2018; digitalcommons.cedarville.edu/icc_proceedings/vol8/iss1/8.

77. Carter, R.W. and Powell, M., The genetic effects of the population bottleneck associated with the Genesis Flood, *J. Creation* 30(2):102–111, 2018.

78. Cereseto, A. and Giacca, M. Integration site selection by retroviruses. *AIDS Rev.* 6(1):13–21, 2004.

79. Carter, R.W., More evidence for the reality of genetic entropy, *J. Creation* 28(1):16–17, 2014.

80. Cooper, D.N., *et al.*, On the sequence-directed nature of human gene mutation: the role of genomic architecture and the local DNA sequence environment in mediating gene mutations underlying human inherited disease. *Hum. Mutat.* 32(10):1075–1099, 2011. See also Evolution's well-kept secret: Mutations are not random!

9

THE BLACKHOLE

Black holes developed by the collapse of individual stars are relatively small but extremely dense. One of these objects' packs more than three times the mass of the sun into the diameter of a city. This yields a bizarre amount of gravitational force dragging on objects around. Stellar blackholes then devour the dust and gas from their surrounding galaxies, which keeps them growing in size, Figure 9.13.

BLACKHOLES

Black holes are some of the outlandish and most captivating objects in space. They're exceptionally dense, with such strong gravitational attraction that not even light can escape their grip, Figure 9.1.

***Fig.9.1:** A black hole is a place in space where gravity pulls so much that even light cannot get out - Curtsy - Chandra X-ray Observatory NASA/CXC*

The Milky Way could encompass over 100 million black holes, though detecting this voracious entity is very problematic. At the center of the Milky Way lies a supermassive black hole — Sagittarius A*. The immense structure is about 4 million times the mass of the sun and lies roughly 26,000 light-years away from Earth, according to NASA.

The initial image of a black hole was taken in 2019 by the Event Horizon (EHT) collaboration. The outstanding photo of the black hole at the center of the M87 galaxy 55 million light-years from Earth electrified scientists around the world, Figure 9.2.

***Fig.9.2:** supermassive black hole — Sagittarius A* in the Milky Way Galaxy- Curtsy NASA*

BLACKHOLE DISCOVERY

Albert Einstein first foretold the existence of black holes in 1916, with his general theory of relativity. The term "black hole" was devised many years later in 1967 by American astronomer John Wheeler. After decades of black holes being recognized only as theoretical objects.

The initial black hole ever revealed was Cygnus X-1, situated within the Milky Way in the constellation of Cygnus, the Swan. Astronomers observed the first signs of the black hole in 1964 when a sounding rocket detected celestial sources of X-rays according to NASA. In 1971, astronomers stated that the X-rays were coming from a bright blue star orbiting an outlandish dark object. It was proposed that the detected X-rays were an effect of stellar material being stripped away from the bright star and "gobbled" up by the dark object — an all-consuming black hole.

How Many Blackholes are There?

Fig.9.3: *At the center of the Milky Way lies a Supermassive Black hole Sagittarius A* (Sgr A*).*
Curtsy NASA/UMass

The Space Telescope Science Institute stated that roughly one out of every thousand stars is massive enough to become a black hole, Figure 9.3. Subsequently, the Milky Way contains over 100 billion stats, our home galaxy must harbor some 100 million black holes.

However, finding black holes is a difficult task and estimates from NASA suggest there could be as many as 10 million to a billion stellar black holes in the Milky Way.

The nearest black hole to Earth is called "The Unicorn" and is positioned about 1,500 light-years away. The nickname has a dual meaning. Not only does the black hole candidate reside in the constellation Monoceros ("the unicorn"), it is extremely low mass — about three times that of the sun — makes it almost one of a kind.

BLACKHOLE IMAGES

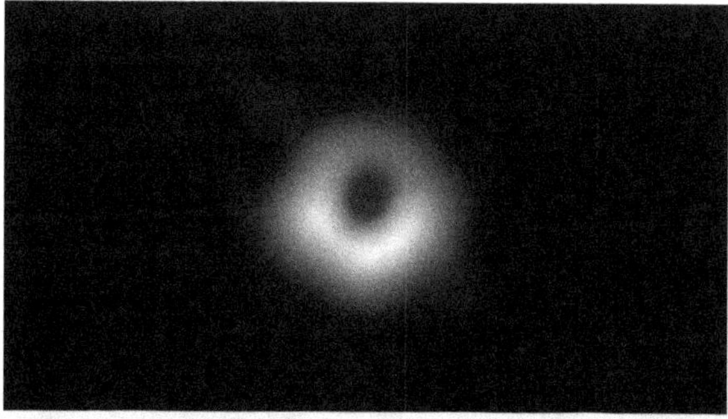

Fig.9.4: *The Event Horizon Telescope (EHT), a planet-scale array of eight ground-based radio telescopes forged through international collaboration, captured this image of the supermassive black hole in the center of the galaxy M87 and its shadow. Curtsy - EHT Collaboration*

In 2019 the Event Horizon Telescope (EHT) collaboration showed the first image ever recorded of a black hole. The EHT captured the black hole in the center of galaxy M87 whereas the telescope was investigating the event horizon or the area past which nothing can escape from a black hole. The image details the sudden loss of photons (particles of light). It also provided a completely new domain of investigation in black holes, now that astrophysicists know what a black hole looks like, Figure 9.4.

In 2021, astronomers exposed a novel view of the enormous black hole at the center of M87, illustrating what the immense structure appears in polarized light. As polarized light waves have a diverse orientation and brightness with respect to unpolarized light, the novel image illustrates the black hole in even more features. Polarization is a unique of magnetic fields and the image allows clarity that the black hole's ring is magnetized, Figure 9.5.

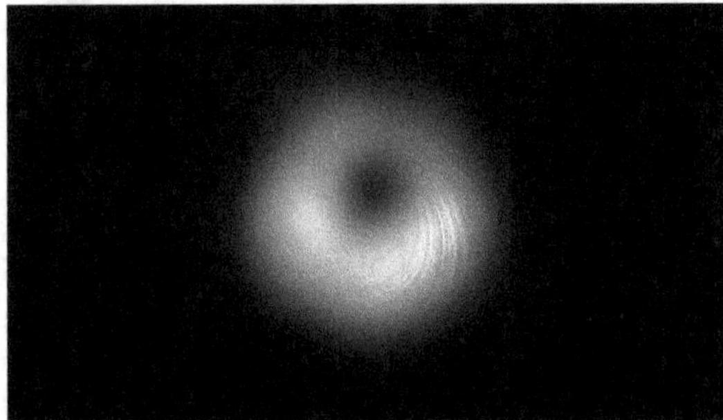

Fig.9.5: Upon the release of the Initial Image of a blackhole in 2019, astrophysicists acquired a new polarized image of the black hole. Curtsy EHT Collaboration

What Do Blackhole Look Like?

Black holes contain three "layers":
- The outer Event Horizon layer and
- The inner Event Horizon layer, and
- The singularity.

The **event horizon** of a black hole is the edge around the mouth of the black hole, once past which light cannot escape. Once a particle reaches the event horizon, it cannot leave. **Gravity** is constant across the event horizon.

The inner domain of a black hole, where the object's mass lies, is recognized as **its singularity**. It is the single point in space-time, where the mass of the black hole is concentrated, Figure 9.6.

Fig.9.6: Black holes are literally a place where the fabric of spacetime is bent to the point of breaking, and they also break our models of how the universe works.

Astrophysicists can't directly see black holes in a similar way they can see stars and other planets and galaxies in space. Instead, they must depend on detecting the radiation black holes emit as dust and gas are drawn into the dense entities. But supermassive black holes, residing in the center of a galaxy, may become masked by the thick dust and gas around them, which can block the revealing emissions.

Fig.9.7: Black Hole are the Absolute Boundary between We Know and what We Don't – Curtsy Shutterstock

Often, as matter is attracted toward a black hole, it rebounds off the event horizon and is thrown outward, rather than being pulled into the jaws. Intense bright jets of material moving at near-relativistic speeds are generated. While the black hole remains unseen, these powerful jets can be observed from great distances, Figure 9.7.

The Event Horizon Telescope EHT's image of a black hole in M87, which was released in 2019, was an exceptional effort, necessitating two years of investigation even after the images were captured. That's because the association of telescopes, which extends across many observatories worldwide, yields an amazing volume of data that is too great to transfer via the internet.

With time, astrophysicist assume to capture other blackholes and construct up a source of what the entities look like. The following target is probably "Sagittarius A*," which is the black hole in the center of our own Milky Way galaxy. "Sagittarius A*" is fascinating as it is calmer than predicted, which may be contributed to magnetic fields smothering its activity, 2019 research stated. Another investigation that year illustrated that a *cool gas halo surrounds Sagittarius A**, which provides unparalleled vision into what the atmosphere round a black hole looks like, Figure 9.8.

Fig.9.8: Dancing with a Black Hole. Astronomers described the strange orbit of a star that loops the monster in the Milky Way, offering more evidence for one of Einstein's ideas – Curtsy The New York Times

CENTER OF THE BLACKHOLE

The singularity at the center of a black hole is the decisive no man's land: a domain where matter is compressed down to an infinitely tiny point, and all understanding of time/space entirely break down. And it doesn't actually exist. Somewhat has to substitute the singularity, but no one knows not exactly sure what.

It could be that unfathomable inside a black hole, matter doesn't get crushed down to an infinitely tiny point. As an alternative, there could be a slightest conceivable formation of matter, the smallest possible concise of volume, Figure 9.9.

Fig.9.9: *A Black Hole Sucking Matter from a Blue Giant Companion Star. The Singularities at the Centers of Astrophysical Black Holes – Curtsy – X-ray: NASA and CXC; Optical: Digitized Sky Survey*

This is entitled a Planck star, and it's a hypothetical possibility intended by loop quantum gravity, which is itself also a highly hypothetical suggestion for generating a quantum version of gravity. In the universe of loop quantum gravity, space and time are quantized — the universe around us is formed of tiny discrete lumps, but at such an unbelievably tiny scale that our movements appear smooth and continuous.

This theoretical entity of space-time offers two benefits;

1.   it takes the imagination of quantum mechanics to its ultimate conclusion, amplifying gravity in a natural way, and

2.   it makes it impossible for singularities to form inside black holes.

As matter compresses down under the enormous gravitational weight of a collapsing star, it faces resistance. The distinctness of space-time stops matter from reaching anything smaller than the "Planck length" (around 1.68 times 10^{-35} meters). All the material that has ever fallen into the blackhole faces compression into a ball not much bigger than this. It is imaginatively less than microscopic, but certainly not infinitely tiny.

This confrontation to continued-compression ultimately forces the material to un-collapse (i.e., explode), making blackholes solely temporary objects. However, due to the extreme time dilation effects around black holes, from our standpoint in the outside universe it takes billions, even trillions, of years before they explode.

There are attempts to get rid of the singularity — one that doesn't rely on untested theories of quantum gravity — is known as the "gravastar." It's once again such a hypothetical thought with a new word.

The difference between a blackhole and a "gravastar" is that, instead of a singularity, the "gravastar" is filled with "dark energy." Dark energy is recognized as a substance that infuses space-time, triggering it to stretch outward. Dark energy is presently active in the greater cosmos, producing our entire universe to accelerate in its expansion.

Fig.9.10: The Gravastar. An Alternative to Black Holes? – Curtsy – Artwork by dnbnOise on deviant art

As matter falls onto a "gravastar," it isn't capable to essentially infiltrate the event horizon (due to all that dark energy on the inside) and consequently just suspends out on the surface, Figure 9,10. However, outside that surface, "gravastars" appear and behave as normal blackholes. (A black hole's event horizon is its point of no return — the edge beyond which nothing, not even light, can escape.)

Nonetheless, current observations of merging blackholes with gravitational wave detectors have theoretically excluded out the existence of "gravastars," as merging "gravastars" will provide a diverse signal than merging blackholes, and outfits like LIGO (the Laser Interferometer Gravitational-Wave Observatory) and Virgo are receiving more and more samples by the day. Though "gravastars" are not merely a no-go in our universe, they are certainly unconvincing.

"Planck stars" and "gravastars" may have overwhelming names, but the realism of their being is in doubt. Therefore, maybe there is a more ordinary clarification for singularities, one that's based on a more subtle — and representative.

The concept of a single point of infinite density comes from our comprehension of stationary, non-rotating, uncharged black holes. Real blackholes are much more interesting characters, especially when they spin.

The swirl of a rotating black hole stretches the singularity into a ring. And according to the math of Einstein's theory of general relativity (which is the only correlated math we have comprehended), once one passes through the ring singularity, one enters a wormhole and pop out through a white hole (the polar opposite of a black hole, where nothing can enter and matter rushes out at the speed of light) into a completely novel and exhilarating patch of the universe, Figure 9.11.

The interiors of rotating black holes are disastrously unstable. And this is due to the very same math that leads to the foretelling of the traveling-to-a-new-universe stuff.

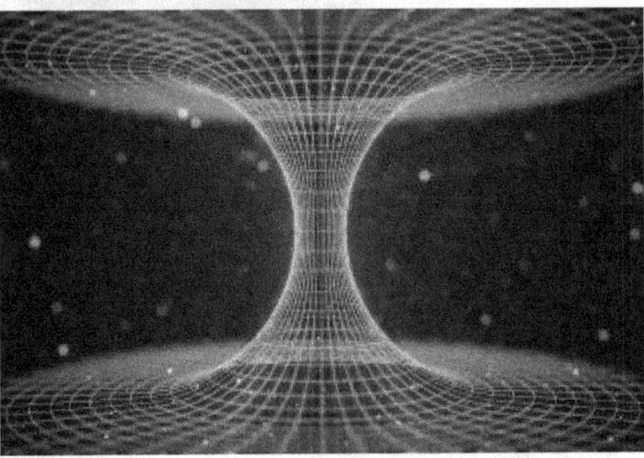

Fig.9.11: Wormholes are Hypothetical Bridges Through Space-Time – Curtsy - Image Getty

The problematic issue with rotating black holes is that they rotate. The singularity, stretched into a ring, is rotating at such a enormous pace that it has farfetched centrifugal force. And in general relativity, strong enough centrifugal forces act like antigravity: they push, not pull, Figure 9.12.

This generates a borderline inside the black hole, called the inner horizon. Outside this region, radiation is falling inward toward the singularity, obligated by the enormous gravitational pull. However, radiation is forced out by the antigravity near the ring singularity, and the turning point is the inner horizon. If one were to experience the inner horizon, one would face a wall of infinitely energetic radiation — the whole past history of the universe, blasted into one's face in less than a blink of an eye.

The development of an inner horizon propagates the seeds for the destruction of the black hole. Nonetheless, rotating black holes surely exist in our universe. This tells us that our math is wrong and something unreal is happening instead.

What is really happening inside a black hole? ***Only the infinite omniscient Creator know.***

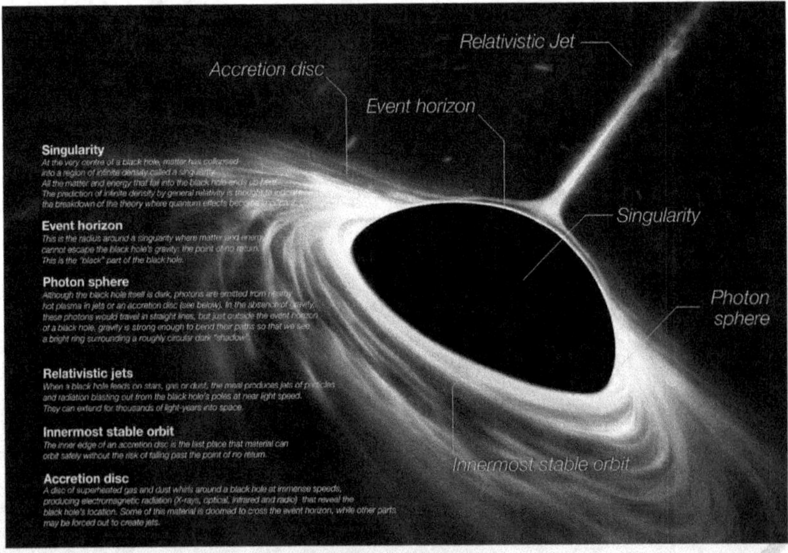

Fig.9.12: Blackhole Anatomy Diagram Illustrates Blackhole Appearance and labeling the different components. Curtsy - ESO

Thus far, astrophysicists have acknowledged three types of black holes Exist:

1. stellar black holes,
2. supermassive black holes and
3. intermediate black holes.

STELLAR BLACKHOLES — SMALL AND DEADLY

As a star burns through the remainder of its fuel, the star may collapse, or implode into itself. For smaller stars (those up to about three times the sun's mass), the new core will develop a neutron star or a white dwarf. However, when a larger star collapses, it endures to compress and generates a astrophysical black hole.

Black holes developed by the collapse of individual stars are relatively small but extremely dense. One of these objects' packs more than three times the mass of the sun into the diameter of a city. This yields a bizarre amount of gravitational force dragging on objects around. Stellar blackholes then devour the dust and gas from their surrounding galaxies, which keeps them growing in size, Figure 9.13.

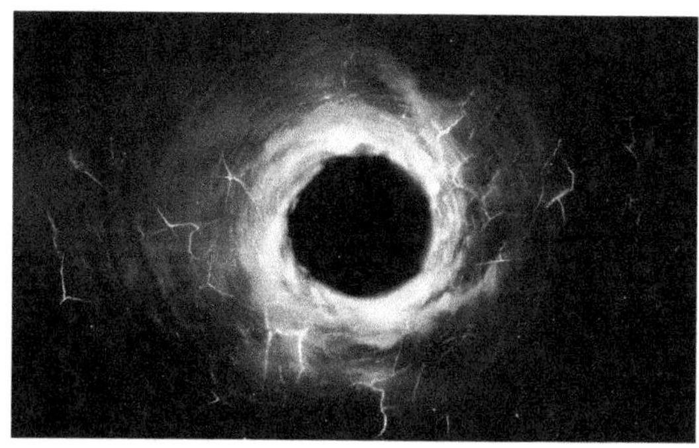

Fig.9.13: Stellar "Ghost" Discovered: Astronomers May Have Detected a "Dark" Free-Floating Black Hole – Curtsy University of California – Berkley

SUPERMASSIVE BLACKHOLES — THE BIRTH OF GIANTS

Small blackholes occupy the universe, Nonetheless, their cousins, supermassive blackholes, dominate the universe. These mammoth black holes are millions or even billions of times as enormous as the sun but are about the similar size in diameter. Such blackholes are supposed to lie at the center of every galaxy, counting the Milky Way.

Astrophysicists are not sure how such huge blackholes spawn. Once these goliaths have shaped, they snatch mass from the dust and gas around them, material that is abundant in the center of galaxies, permitting them to grow to even more massive sizes, 9.14.

Supermassive blackholes may be the product of hundreds or thousands of miniature black holes that combine. Huge gas clouds could also be accountable, imploding together and speedily accreting mass. A third possibility is the collapse of a stellar cluster, a collection of stars all falling together. Fourth, supermassive blackholes could arise from huge clusters of dark matter. This is an entity that we can observe through its gravitational consequence on other objects; though, we don't distinguish what dark matter is composed of as it does not emit light and cannot be directly observed.

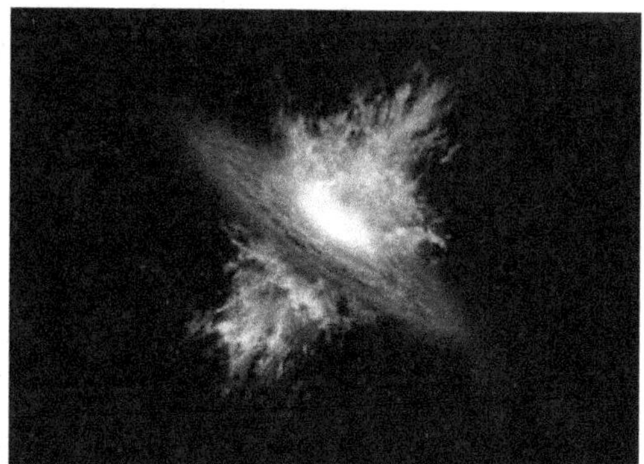

Fig.9.14: Supermassive Black Hole - These Black Hole "Winds" Blew about 13.1 billion Years Ago – Curtsy credit: ALMA (ESO/NAOJ/NRAO)

INTERMEDIATE BLACKHOLES

Astrophysicists once believed that blackholes originated in only small and large sizes, but investigations have revealed the likelihood that midsize, or intermediate, blackholes (IMBHs) could exist. Such forms could form as stars in a cluster collide in a chain reaction. Numerous of these IMBHs forming in the same region could then ultimately fall together in the center of a galaxy and create a supermassive blackhole.

In 2014, astronomers found what seemed to be an intermediate-mass black hole IMBHs in the arm of a spiral galaxy. Also, in 2021 astronomers took advantage of an ancient gamma-ray burst to detect one.

"Astronomers have been searching hard for these medium-sized black holes," investigation co-author "Tim Roberts," of the University of Durham in the United Kingdom, said in a statement. *"There have been hints that they exist, but IMBHs have been acting like a long-lost relative that isn't interested in being found."*

Studies, from 2018, proposed that these IMBHs may exist in the heart of dwarf galaxies (or very small galaxies). Observations of 10 such galaxies (five of which were formerly unidentified to science prior this latest survey) exposed X-ray behavior — common in blackholes — signifying the existence of black holes from 36,000 to 316,000 solar masses. The information came from the "Sloan Digital Sky Survey," which studies about 1 million galaxies and can detect the kind of light frequently detected coming from black holes that are picking up neighboring debris.

BINARY BLACKHOLES: DOUBLE TROUBLE

Fig.9.15: Illustration of a Supermassive Blackhole with a Companion Blackhole Orbiting around it. Curtsy - Caltech–IPAC

In 2015, astronomers using the Laser Interferometer Gravitational-Wave Observatory (LIGO) detected gravitational waves from merging stellar black holes.

"We have further confirmation of the existence of stellar-mass black holes that are larger than 20 solar masses — these are objects we didn't know existed before LIGO detected them," "David Shoemaker," the spokesperson for the LIGO Scientific Collaboration (LSC), said in a statement. LIGO's observations also deliver understandings into the direction a blackhole rotations. As two black holes coiled around one another, they can rotate in the same direction or the opposite direction, Figure 9.15.

There are two philosophies on how two-fold blackholes formulate. The first proposes that the two black holes in a binary formulate at about the same time, from two stars that were born together and died explosively at about the same time. The two-fold stars would have had the same rotational orientation as one another, so the two blackholes left behind would as well.

Under the second model, blackholes in a stellar cluster sink to the center of the cluster and couple up. These companions would have random spin orientations with respect to one another due to LIGO Scientific Collaboration. LIGO's observations of companion blackholes with dissimilar rotation orientations deliver robust evidence for this formation theory.

"*We're starting to gather real statistics on binary black hole systems,*" said LIGO scientist "Keita Kawabe" of Caltech, who is based at the LIGO Hanford Observatory. "*That's stimulating as some models of blackhole two-fold formation are to some extent preferred over the others even now, and in the future, we can further narrow this down.*"

Blackhole Facts

- If you fell into a blackhole, theory has long proposed that gravity would stretch you out like spaghetti, though your death would come before you reached the singularity. However, a 2012 study published in the journal Nature recommended that quantum influence would enable the event horizon to behave much like a wall of fire, which would instantaneously burn you to death.

- Blackholes don't suck. Suction is triggered by pulling something into a vacuum, which the massive blackhole definitely is not. Instead, objects fall into them just as they fall toward anything that exerts gravity, like the Earth.

- The first object well-thought-out to be a blackhole is Cygnus X-1. Cygnus X-1 was the subject of a 1974 friendly wager between Stephen Hawking and fellow physicist Kip Thorne, with Hawking betting that the source was not a blackhole. In 1990, Hawking accepted defeat.

- Miniature blackholes may have generated immediately alleged after the Big Bang. Speedily stretching space may have compressed some regions into tiny, dense blackholes less massive than the sun.

- If a star passes very near to a black hole, the star can be torn apart.

- Astrophysicists approximated that the Milky Way has anywhere from 10 million to 1 billion stellar blackholes, with masses roughly three times that of the sun.

- Blackholes continue enormous feed for science fiction books and movies.

WHERE DO BLACKHOLES LEAD TO?

Where do You Go Travelling Through a Blackhole?

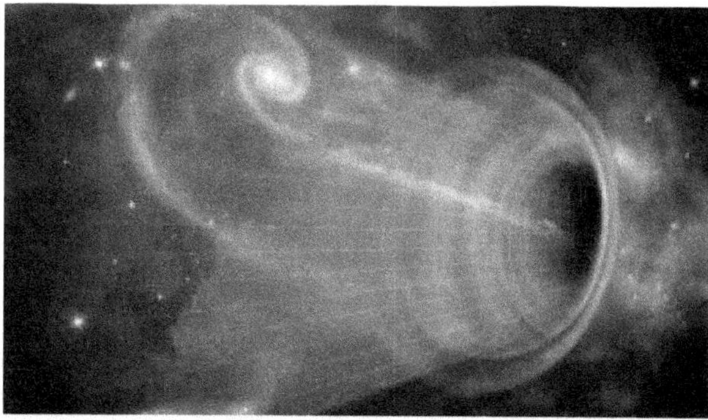

Fig.9.16: Where do black holes lead to? – Curtsy - All About Space Magazine

You are about to leap into a black hole. What could conceivably await should — against all odds — you by some means survive? Where would you end up and what enticing fictions would you be capable to entertain if you succeeded to climb your way back?

The meek reply to all of these questions is, as "Professor Richard Massey" describes, *"Who knows?"* As a Royal Society research fellow at the Institute for Computational Cosmology at Durham University, Massey is completely mindful that the mysteries of blackholes are bottomless, Figure 9.16.

"Falling through an event horizon is literally passing beyond the veil — once someone falls past it, nobody could ever send a message back," he said. *"They'd be ripped to pieces by the enormous gravity, so I doubt anyone falling through would get anywhere."*

If that sounds like an unsatisfactory — and throbbing — answer, then it is to be predictable. Ever since Albert Einstein's general theory of relativity was well-thought-out to have foretold blackholes by connecting space-time with the action of gravity, it has been recognized that blackholes generated from the death of an enormous star leaving behind a small, dense remainder core. Assuming this core has more than nearly three times the mass of the sun, gravity would overpower to such a degree that it would implode on itself into a single point, or singularity, recognized to be the blockhole's infinitely dense core.

The consequential uninhabitable blackhole would have such an authoritative gravitational pull that not even light could evade it. Therefore, should you then realize yourself at the event horizon — the point at which light and matter can only pass inward, as projected by the German astrophysicist "Karl Schwarzschild" — there is no escape. According to Massey, tidal forces would diminish your body into strands of atoms (or 'spaghettification', as it is also recognized) and the object would ultimately end up crushed at the singularity. The clue that you could pop out wherever — maybe at the other side — seems utterly fanciful.

WORMHOLE

Over the years scientists have observed into the likelihood that blackholes could be wormholes to other galaxies. They may even be, as some have proposed, a path to allegedly another universe, Figure 9.17.

Fig.9.17: A New 'Einstein' Equation Suggests Wormholes Hold Key to Quantum Gravity

Such a concept has been floating around for some time: Einstein teamed up with "Nathan Rosen" to hypothesize bridges that connect two different points in space-time in 1935. However, it grew some fresh ground in the 1980s when astrophysicist "Kip Thorne" — one of the world's leading authorities on the astrophysical consequences of Einstein's general theory of relativity — elevated a conversation about whether objects could physically travel through them.

Reading Kip Thorne's popular book about wormholes is what first got scientists excited of the prospects. "Massey" said: *"But it doesn't seem likely that wormholes exist."*

Indeed, Thorne, who lent his skilled advice to the production team for the Hollywood movie Interstellar, wrote: *"We see no objects in our universe that could become wormholes as they age,"* in his book *"The Science of Interstellar"* (W.W. Norton and Company, 2014). Thorne told Space.com that *"journeys through these theoretical tunnels would most likely remain science fiction, and there is certainly no firm evidence that a black hole could allow for such a passage."*

However, the difficulty is that we can't get up close enough to see for ourselves. We can't even take photographs of anything that takes place inside a blackhole — if light cannot escape their immense gravity, then absolutely nothing can be captured by any camera. As it stands, theory recommends that whatever goes beyond the event horizon is merely added to the blackhole and, furthermore, as time distorts close to this boundary, this will appear to take place incredibly slowly, therefore, an answer would not be quickly forthcoming, Figure 9.19.

"I think the standard story is that they lead to the end of time," said "Douglas Finkbeiner," professor of astronomy and physics at Harvard University. *"An observer far away will not see their astronaut friend fall into the black hole. They'll just get redder and fainter as they approach the event horizon [as a result of gravitational red shift]. But the friend falls right in, to a place beyond 'forever.' Whatever that means."*

Fig.9.18: Artist's concept of a wormhole. If wormholes exist, they might lead to another universe. Nonetheless, there's no evidence that wormholes are real or that a black hole would act like one. –
Curtsy Shutterstock

BLOCKHOLE LEADS TO WHITE WHOLE

Certainly, if blackholes open the passage to another part of a galaxy or another universe, there would require to be something opposite to them on the other side. Could this be a white hole — a hypothesis put forward by Russian cosmologist "Igor Novikov" in 1964! Novikov suggested that a blackhole links to a white hole that exists in the past. Dissimilar a blackhole, a white hole will permit light and matter to depart, however, light and matter will not be able to enter, Figure 9.18.

Astrophysicists have continued to investigate the possible link between black and white holes. In their 2014 research published in the journal Physical Review, physicists "Carlo Rovelli" and "Hal M. Haggard" stated that *"there is a classic metric satisfying the Einstein equations outside a finite space-time region where matter collapses into a black hole and then emerges from a while hole."* Needless to say, all of the material black holes have gulped could be vomited out, and blackholes may become white holes as they die.

Far from abolishing the information that it absorbs; the collapse of a black hole would be stopped. It would in its place, experience a quantum bounce, permitting information to escape. Should this be the circumstance, it would shed some light on a suggestion by former Cambridge University cosmologist and theoretical physicist Stephen Hawking who, in the 1970s, investigated the prospect that blackholes emit particles and radiation — thermal heat — as a result of quantum fluctuations.

"*Hawking said a black hole doesn't last forever*," "Finkbeiner" said. Hawking computed that the radiation would cause a blackhole to lose energy, shrink and disappear, as described in his 1976 paper published in Physical Review. Assumed his claims that the radiation emitted would be random and encompass no information about what had tumbled in, the blackhole, upon its explosion, would erase heaps of information.

This indicated Hawking's idea was in conflict with quantum theory, which articulates information can't be demolished. Physics states information just becomes more problematic to find because, should it become lost, it becomes impossible to distinguish the past or the future. Hawking's idea led to the 'black hole information paradox' and it has long mystified scientists. Some have said Hawking was merely wrong, and the man himself even professed he had made an error during a scientific conference in Dublin in 2004.

Therefore, do we go back to the concept of blackholes emitting preserved information and throwing it back out via a white hole? Maybe, Figure 9.10. In their 2013 study published in Physical Review Letters, "Jorge Pullin" at Louisiana State University and Rodolfo Gambini at the University of the Republic in Montevideo, Uruguay, applied loop quantum gravity to a black hole and discovered that gravity upsurged towards the core but decreased and dumped whatever was entering into another region of the universe. The outcome gave extra credibility to the idea of blackholes serving as a gateway. In this research, **singularity does not exist**, and so it doesn't formulate an impenetrable barrier that ends up crushing whatever it encounters. It also means that information doesn't disappear.

Blackhole Goes No Where

Nonetheless, physicists "Ahmed Almheiri," "Donald Marolf," "Joseph Polchinski" and "James Sully" still convinced Hawking could have been on to something. They worked on a theory that became known as the "AMPS firewall," or the black hole firewall hypothesis. By their calculations, quantum mechanics could practicably turn the event horizon into an enormous wall of fire and anything coming into contact would burn in an instant. In that logic, blackholes lead nowhere because nothing could ever get inside.

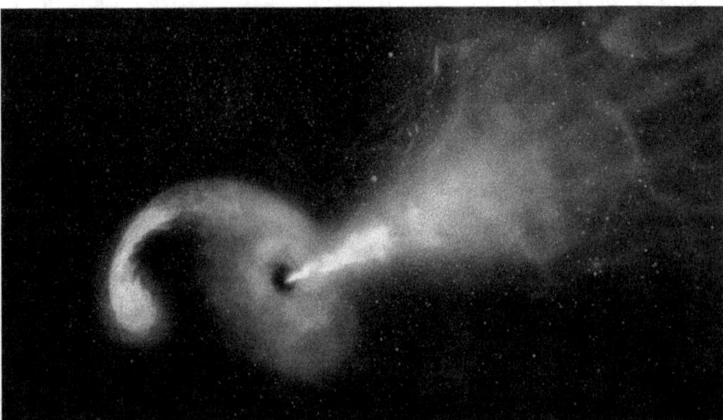

Fig.9.19: The Concept of Blackhole emitting a White Hole

This, nonetheless, disrupts Einstein's general theory of relativity. Somebody crossing the event horizon shouldn't essentially feel any great adversity because an object would be in free fall and, according to the equivalence principle, that object — or person — would not feel the extreme influences of gravity. It could obey the laws of physics present elsewhere in the universe, but even if it didn't go against Einstein's principle it would weaken quantum field theory or suggest information can be lost.

In 2014, Hawking published a study in which he avoided the presence of an event horizon — hoping there is nothing there to burn — saying gravitational collapse would result in an 'apparent horizon' instead.

This horizon would hang light rays attempting to move away from the core of the blackhole, and would keep it up for a "period of time." In his rethinking, ostensible horizons for the time being hold matter and energy before dissolving and releasing them later down the line.

This description best fits with quantum theory — which says information can't be demolished — and, if it was ever established, it proposes that anything could discharge from a blackhole, Figure 9.19.

Hawking went as far as propagating **black holes may not even exist**. "Blackholes should be redefined as metastable bound states of the gravitational field," he wrote. ***There would be no singularity***, *and while the apparent field would move inwards due to gravity, it would never reach the center and be consolidated within a dense mass.*

And yet anything which is emitted will not be in the form of the information swallowed. It would be impossible to understand what went in by looking at what is coming out, which creates problems of its own — not least for, say, a anthropological being who found themselves in such an frightening position. They'd never feel the same again!

It is certain, this specific mystery is going to gulp up many more scientific hours for a long time to come. "Rovelli and Francesca Vidotto" lately proposed that a component of dark matter could be shaped by remnants of evaporated blackholes, while Hawking's paper on blackholes and 'soft hair' was released in 2018, and defines how zero-energy particles are left around the point of no return, the event horizon — an idea that proposes information is not lost but arrested.

This flew in the face of the no-hair proposition which was articulated by physicist "John Archibald Wheeler" and operated on the basis that two blackholes would be vague to an observer because none of the distinct particle physics pseudo-charges would be well-maintained. It's an idea that has got scientists chatting, but there is long-way to go before it's verified as the answer for where blackholes lead. If only we could find a way to hop into one.

SUPERMASSIVE BLACKHOLE

The Event Horizon Telescope (EHT)

Fig.9.20: *Event Horizon Telescope*

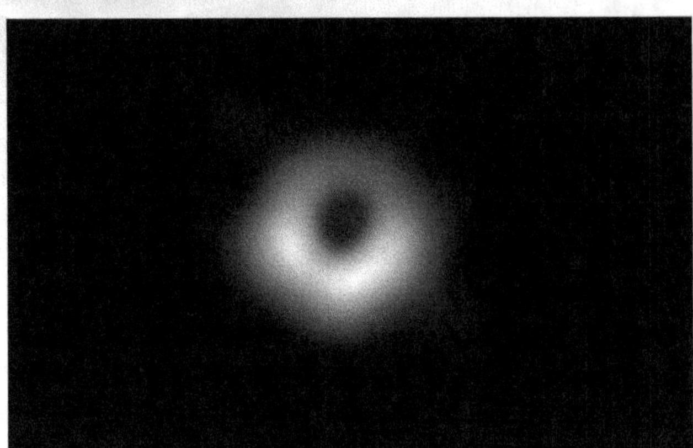

Fig.9.21: Using the Event Horizon Telescope (EHT), scientists obtained an image of the black hole at the center of galaxy M87, outlined by emission from hot gas swirling around it under the influence of strong gravity near its event horizon. Curtsy – Credit: Event Horizon Telescope Collaboration et al.

On 10 April, 2017 the internationally synchronized announcement was made of the first ever image of the event horizon (EH) of the supermassive blackhole at the center of the distant galaxy Messier 87 (M87), Figure 9.21. The galaxy is 55 million light-years away and the supermassive black hole was established to have a mass of 6.5 billion suns. The sun weight is 1.989×10^{30} kg.

The Event Horizon Telescope (EHT), Figure 9.20 — It is a planet-scale array of eight ground-based radio telescopes collaborated through international teamwork—was developed to capture images of black holes at the center of galaxies.

This is the research of many astronomers employing millimeter wave radio-telescopes" across the planet. By stitching together, the power of 8 state-of-the-art mm Wave radio-telescopes they fundamentally turned the planet into one giant radio-telescope, Figure 9.22. By employing such a large telescope and millimeter wavelengths they added resolution never gained before that permitted them to capture the synchronized stitched images of the event horizon, which is about the diameter of our solar system.

The consequences this far are reliable with all foretold by Einstein's General Relativity theory.

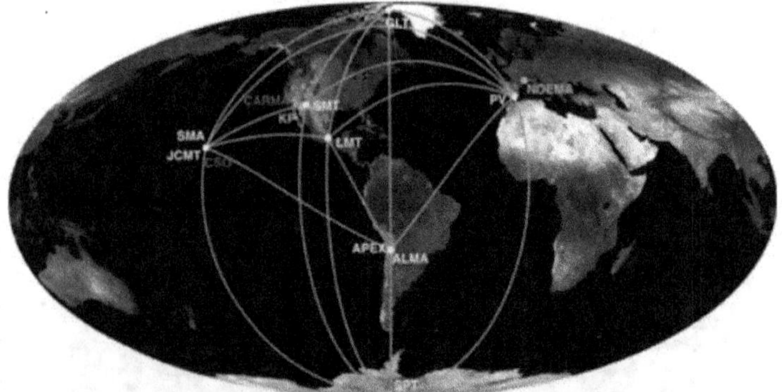

Fig.9.22: Map of the EHT. Stations active in 2017 and 2018 are shown with connecting lines and labelled in yellow, sites in commission are labelled in green, and legacy sites are labelled in red. Nearly redundant baselines are overlaying each other, i.e., to ALMA/APEX and SMA/JCMT. Such redundancy allows improvement in determining the amplitude calibration of the array. Curtsy – Event Horizon Telescope collaboration et al.

From the biblical faithful believers' perspective this is, yet again, excellent operational science. There is nothing novel here that disproves the biblical timeline of about 6,000 years because that is dependent on historical science contemplations. It is not an operational science question. The data was captured from the different telescopes and was accumulated and processed over a period of about a year, however, those initial observations were documented over a period of *7 days* in April of 2017. During those days the supermassive black hole was 'observed'. In the same manner over the 24-hour period Day 4 of Creation Week about 6,000 years ago all the stars and galaxies, with supermassive black holes, were 'observed' at the earth as God created them *(Genesis 1:16–19)*. **God spoke and "it was so."**

Creationist cosmologists have proposed numerous ways in which God could have done this with starlight from distant stars visible at the earth in the biblical timeframe. However, eventually, Creation Week was a miraculous sequence of events, and God might well have done things in a manner that is not accessible to us.

COLLAPSING STAR

The evidences for black holes is very robust. They are unquestionably a theoretical likelihood from Einstein's theory of General Relativity. Moreover, once a star depleted of nuclear fuel, then the outer pressure would no longer match gravity. Accordingly, the star must implode, Figure 9.23.

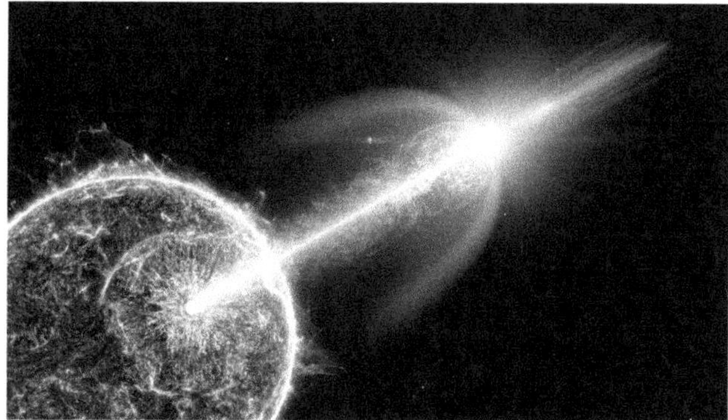

Fig.9.23: Collapsing star produces one of the most epic cosmic explosions ever seen – Curtsy - Science Communication Lab

There are other recognized forces that avert a comprehensive implosion into a black hole. Electron degeneracy pressure halts the collapse and leaves the star as a white dwarf but gravity overwhelms this if what is left of the burnt-out star is greater than 1.4 solar masses (the Chandrasekhar limit).

Then the next barricade is neutron degeneracy pressure, which stops the implosion as a neutron star, but gravity will probably overcome this at about 3 solar masses (the Tolman–Oppenheimer–Volkoff limit). Unless a hypothetic quark degeneracy pressure kicks in, there would be nothing to stop the star collapsing completely into a black hole.

There are *observations* consistent with black holes, quite independent of any theories of their origin from stars. They tend to attract matter that forms an accretion disc around the equator. This matter falls inward and releases much gravitational potential energy. In fact, this is an extremely efficient mass-to-energy conversion process, turning %40 of the mass into energy, compared to only about %1 with thermonuclear fusion. This can be observed as a strong X-ray source. The process also results in powerful relativistic jets from the poles. This would explain X-ray binaries: a black hole sucks matter away from a companion star.

There are also observations consistent with supermassive black holes at the center of galaxies, including our own. A certain star (S2) orbits around something in the galactic center at a distance of 17 light hours (about three times that of Pluto), and period of only 15 years (Pluto's is 248 years). This is consistent with the gravitational pull of 4.1 million solar masses. The solar mass is 1.989×10^{30} kg.

Furthermore, if this object were not smaller still, the star would collide with it; it's likely only to be 6.25 light-hours, not much more than Pluto's orbit. But the only object known to theory that could compact over 4 million solar masses into such a small volume is a black hole.

How can We Observe Distant Stars in a Young Universe?

- If the universe is young and it takes millions of years for light to get to us from many stars, how can we see them?
- Did God create light in transit?
- Was the speed of light faster in the past?
- Does this have anything to do with the big bang?
- What about Relativity

Numerous galaxies are billions of light-years away. Since a light-year is the distance light would travel over the time period of one year, and we can see such galaxies, does this mean that the universe1 is very old?

Despite all the biblical and scientific evidence for a young earth/universe, this has long been a seemingly intractable problem. However, any scientific understanding of origins will always have opportunities for research—problems that need to be solved. We can never have complete knowledge and so there will always be much to learn.

BIG BANG LIGHT TRAVEL PROBLEM

It's vital to learn that the most widely held cosmology, the standard secular big bang theory, has a problem of its own with time and light travel, called the "horizon problem." According to the big bang, the universe allegedly began in a fireball from which all matter in the universe is ultimately derived, Figure 9.24. For galaxies to have any hope of forming at all during the expansion process, the fireball must have begun with an uneven distribution of temperatures. However, we see radiation coming from the cosmos, in all directions of the sky that has a very *uniform temperature*. This is the cosmic microwave background (CMB) radiation and its temperature has been measured to be uniform to one part in 100,000.

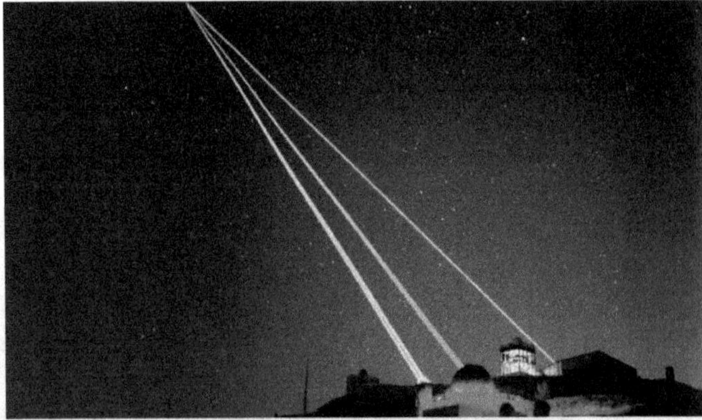

Fig.9.24: Big Bang Light Travel Problem

If the regions started at uneven temperatures, and are now almost at the same temperature, then energy must have been transferred from hot regions to cooler ones. The fastest way that energy can be transferred is by radiation, at the speed of light. Consider, then, a region of space **10 billion light years** (a light year is the distance light travels in one year = 6,000,000,000,000 miles) away from earth in the north sky, and the other 10 billion light years in the south. They are 20 billion light years apart. However, since the big bang was allegedly only 13.7 billion years ago, this is not enough time for light to have travelled from one region to the other. Yet the background temperature is almost identical.

However, the problem for the big bang is even **more severe** than this. The CMB radiation is alleged to be the radiation that appeared when the temperature of the initial fireball cooled enough for it to become transparent to radiation. This is alleged to have happened about **300,000 years** after the initial fireball appearance. Consequently, only those regions within about 300,000 light-years of each other could have become uniform in temperature during this time. Yet we have regions separated by at least **20 billion light-years that are at essentially the same temperature**.

This horizon problem gave rise to hypothetical **fudge factors** such as faster-than-light '*inflation*' of space being added to the big bang— expanding by a factor of 1050 (186,000 x 1050 miles/s) in 10^{-33} seconds. However, there is no known mechanism to start or stop the process in a smooth fashion—it is effectively a naturalistic 'miracle.' Even "New Scientist" asked whether inflation was "just wishful thinking". "Dr Paul Steinhardt," winner of the 2002 Dirac Medal for his contributions to inflation theory, wrote an article, featured on the cover of Scientific American as "*Quantum Gaps in the big bang: Why our best explanation of how the universe evolved must be fixed—or replaced.*" Steinhardt identified four ways in which **inflationary theory fails**.

Other big bang cosmologists have even suggested that the speed of light (radiation) may have been much faster in the past. Therefore, no-one can correctly claim this issue as a reason not to believe the Bible, because the standard secular big bang cosmology has a similar problem.

At this point we could just say, '*The big bang has miracles without any miracle worker, so, surely we Christians can have miracles with a miracle worker!*' Creation Week was, after all, a miraculous event.

LET THERE BE LIGHT

A few decades ago, perhaps the most common explanation from biblical creationists was that God said "Let There Be Light," Figure 9.25: therefore, Adam could view the stars immediately without having to wait years for the light from even the closest ones to reach the earth. While we should not limit the power of God, this has some immense difficulties.

Fig.9.25: Let There Be Light – Genesis 1:3

It would mean that whenever we look at a very distant object, what we apparently observe happening never really happened at all. For instance, say we see an object a million light-years away that appears to be rotating; that is, the light we receive in our telescopes carries this information, 'recording' this behavior. However, according to the 'created in transit' explanation, the light we are now receiving did not come from the star, but was created 'enroute.'

This would mean, for 10,000-year-old universe, that anything we see happening beyond about 10,000 light-years is actually part of a gigantic picture-show of things that have not actually happened, showing us objects that may not even exist.

To explain this problem further, consider an exploding star (supernova) at, consider, an accurately measured distance of 100,000 light-years. Remember, we are using this explanation in a 10,000-year-old universe. As the astronomer on Earth watches this exploding star, astronomer is not just receiving a beam of light. If that were all, then it would be no problem at all to say that God could have created a whole chain of photons (light particles) already on their way. However, what the astronomer receives is also a particular, very specific pattern of variation within the light, showing the changes that one would expect to accompany such an explosion—a predictable sequence of events involving neutrinos, visible light, X rays and gamma-rays. For example, because most neutrinos pass through solid matter as if it were not there, while light is slowed down, we can detect a massive neutrino burst before the light reaches us.

The light and neutrino burst carry information recording an apparently real event. The astronomer is perfectly justified in interpreting this 'message' as representing actual reality—that there really was such an object, which exploded according to the laws of physics, brightened, emitted X-rays, dimmed, and so on, all in accord with the expected outcomes of known physical laws.

Everything the astronomer observes is consistent with this, including the spectral patterns in the light from the star, giving us a chemical signature of the elements contained in it. However, the 'light created enroute' explanation would mean that this recorded message of events, transmitted through space, had to be contained within the light beam from the moment of its creation, or planted into the light beam at a later date, without ever having originated from that distant point. (If it had started from the star—assuming that there really was such a star—the light beam would still be 90,000 light-years away from Earth, if the universe was 10,000 years old and the speed of light constant.

To create such a detailed series of signals in light beams reaching Earth, signals which seem to have come from a series of real events but in fact did not, has no conceivable purpose. Worse, it is like saying that God created fossils in rocks to fool us, or even test our faith, and that they don't represent anything real (a real animal or plant that lived and died in the past). This would be a strange deception for a holy God to engage in.

Did Light Always Travel at the Same Speed?

An obvious solution would seem to be a higher speed of light in the past, allowing the light to cover the same distance in less time. This seems at first glance a too-convenient ad hoc explanation. Some years ago, "Barry Setterfield" raised such a possibility to a high profile by showing that there seemed to be a decreasing trend in the historical observations of the speed of light (c) over the past 300 years or so. Setterfield (and his later co-author, Trevor Norman) produced evidence in favor of their 'cdk' (speed of light "c" decay) theory (speed of light decay). They believed

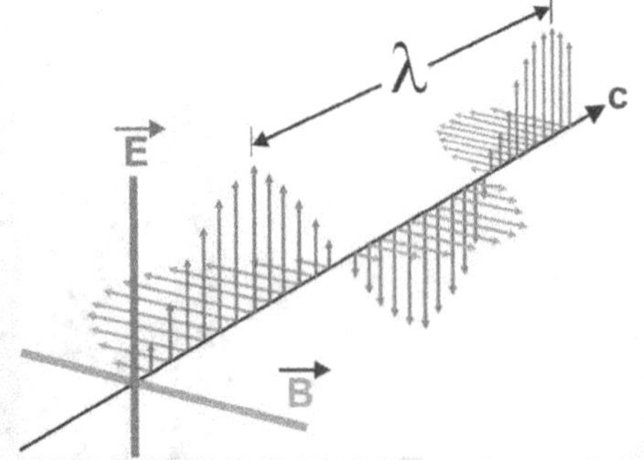

Fig.9.26: Light Travels at Constant Speed 186,282 mi/s

that it would have affected radiometric dating results, and even have caused the red-shifting of light from distant galaxies, although this idea was later overturned, and other modifications were made also.

Many attacked the idea on the fallacious grounds that Einstein's Special Relativity said that the speed of light could not change. It actually just says that the speed of light measured by observers will be invariant regardless of the speed of the source or observer.

Much debate raged to and fro among capable people within creationist circles about whether the statistical evidence really supported cdk or not. The biggest difficulty, however, is with certain physical consequences of the theory. If c had declined the way Setterfield proposed, these consequences should still be discernible in the light from distant galaxies, but they are apparently not. High-precision tests of Einstein's Theory of General Relativity, in our galaxy, using co-orbiting pairs of neutron stars, where at least one is a pulsar, within thousands of light-years distance, indicate the same value for c as we measure locally.

In short, none of the theory's defenders have been able to answer all the problems raised. Interestingly, big bang defenders treated the idea of cdk with contempt, but then one of their own, "João Magueijo," proposed a similar idea to rescue the big bang from its own light-travel (horizon) problem!

NEW CREATIONIST COSMOLOGIES

Nevertheless, the cdk theory stimulated much thinking about the issues. For example, creationist physicist Dr Russell Humphreys says that he spent a year, on and off, trying to get the cdk theory to work consistently, but without success. However, the thinking inspired him to develop ideas for a new creationist cosmology as an alternative to big bang theory.

This sort of development, in which one creationist theory, cdk, is overtaken by another, is a healthy aspect of science. The basic biblical framework, because it comes from the Creator, is non-negotiable, as opposed to the changing views and models of fallible people seeking to understand the data within that framework (evolutionists also often change their ideas on exactly how things have made themselves, but never whether they did; that materialistic framework remains non-negotiable).

Speed of Light – No Change

Consider that the time taken for something to travel a given distance is the distance divided by the speed it is travelling. That is, Time = Distance (divided by) Speed.

When this is applied to light from distant stars, the time calculates out to be billions of years. Some have sought to challenge the distances, but they are very unlikely to be substantially wrong.

Astronomers use many different methods to measure the distances, and no informed creationist astronomer would claim that errors would be so vast that billions of light-years could be reduced to several thousand, for example. Even our own Milky Way Galaxy is about 100,000 light years across.

If the speed of light (c) has not changed, the only thing left in the equation is time itself. In fact, Einstein's Relativity Theory has been telling the world for a hundred years that time is not an absolute. Scientists may not know what time is but they do know how to measure it. Nowadays very precise and exact atomic clocks measure the rate or flow of time and it has been measured to vary from place to place.

In fact, two things have been observed to distort the flow of time— one is speed and the other is gravity. Einstein's general theory, the best theory of gravity we have at present, indicates that gravity distorts time.

This effect has been measured experimentally, many times. Clocks at the top of tall buildings, where gravity is slightly less, run slightly faster than those at the bottom, just as predicted by the equations of General Relativity (GR).

Gravity distorts time so that a clock on the top of a mountain will run faster than a clock on the plains (Picture)

GRAVITY DISTORTS TIME

Most people think of the universe as having a center and an edge. This means that if you were to travel into space, you would eventually come to a place beyond which there was no more matter. In this understanding, Earth is near the center, as it appears to be as we look out into space. This might sound like common sense, as indeed it is, but all modern secular cosmologies deny this. That is, they make the assumption that the universe has no boundary—no edge and no center—dubbed the 'cosmological principle'. In this assumed universe, every galaxy would be surrounded by galaxies spread evenly in all directions, Figure 9.27. In such a universe, all net gravitational forces cancel out and there is no preferred direction, so there are also no net effects of movement of astronomical objects.

Fig.9.27: *Scientists have created a detailed new map of the matter in the universe –*
Curtsy - Depositphotos

This is a philosophical assumption; that is, religious. And it is made to remove Earth from its apparently privileged position near the center of the universe, because that's what the Bible implies—***that Earth is the focus of God's attention in creating the universe***. The views of respected cosmologist "George Ellis," once a colleague of the famous Stephen Hawking; as reported by Scientific American:

"People need to be aware that there is a range of models that could explain the observations" Ellis argues. "For instance, I can construct you a spherically symmetrical universe with Earth at its center, and you cannot disprove it based on observations." Ellis has published a paper on this. "You can only exclude it on philosophical grounds. In my view there is absolutely nothing wrong in that. What I want to bring into the open is the fact that we are using philosophical criteria in choosing our models. A lot of cosmology tries to hide that."

Not only can you have such an understanding of the universe, but it actually fits the evidence better than the no-center, boundless universe assumed by secularists. There is now observational evidence that the universe has a center. For example, galaxies appear to have a large-scale structure centered near our galaxy. These observations do not fit the materialists' no-center, unbounded, randomly generated universe, but are consistent with a universe designed by a creator.

The big bang has many other problems, so much so that even many secularists are calling for a radical rethink:

"Big bang theory relies on a growing number of hypothetical entities—things that we have never observed. Inflation, dark matter and dark energy are the most prominent. Without them there would be fatal contradictions between the observations made by astronomers and the predictions of the big bang theory."

According Einstein's to General Relativity, if the universe has a boundary and center, then there can be net gravitational effects on a cosmological scale and these can affect the flow of time during its history. Depending on the how the universe was created, clocks could have run at different rates on Earth compared to other parts of the universe. In other words, it is no longer enough to say God made the universe in six days. He certainly did (*Exodus 20:11* and *Genesis 1*), but six days as measured by which clocks? If we say 'God's time' we miss the point that He created the flow of time as we now experience it; He is outside of time, seeing the end from the beginning. Equally seriously, God inspired Scripture to instruct us (*2 Timothy 2:15– 17*). This entails that words and logical inferences must be the same for God and man, otherwise Scripture would not be able to equip us with truth He reveals.

Figure 9.27. A 3-D spherical ball of space and matter has a center and thus a net gravitational force. In the big bang model, the matter of our universe is imagined to be spread over the surface of a 4-dimensional or higher dimensional space, which has no center (balloon analogy).

New Approaches

We now have two creationist cosmologies that could explain how God created everything in six earth days and Adam and Eve could see distant starlight. Both these concepts are rather mind-stretching, but we should not be surprised that when we are trying to get a glimpse of the miracle of creation it is not easy to understand God's ways are higher than our ways!

Dr Russ Humphreys

Dr Humphreys had an earlier model, as explained in the book, Starlight and Time, but it failed to account for observations in relation to nearby galaxies. He has developed a new explanation of light-transit-times, to explain how light travelled from the distant cosmos and reached Earth, all during one ordinary-length day on Earth, the fourth day of creation week. This understanding depends on the effect of gravity on time, gravitational time dilation.

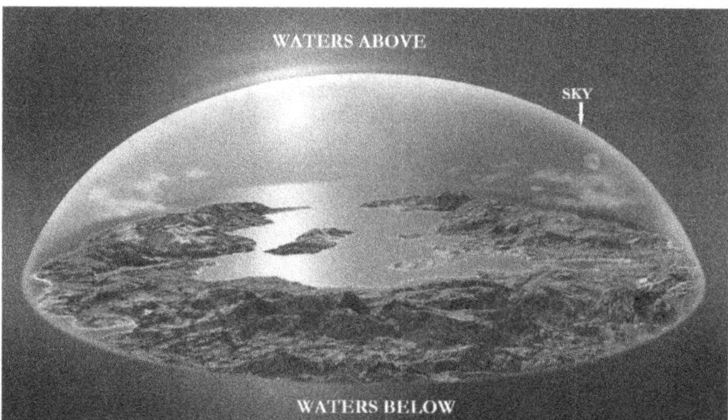

Fig.9.28: According to Dr Humphreys, the waters above the heavens (Psalm 148:4) today are possibly a thin veil of ice particles, or scattered planet-sized spheres of water covered with thick crusts of ice, at the edge of (surrounding) the universe.

Humphreys takes the "waters that are above the heavens" (*Psalm 148:4 cf. Genesis 1:6–10*), to mean that God created the universe with a massive layer of water that encircles the universe, Figure 9.28. If the mass of this water were very large, it would have a large effect on the flow of time throughout the universe. And then there is the effect of God's creating the stars during the fourth day of creation week as well (*Isaiah 40:26*). He also takes it that God 'stretching out the heavens', mentioned in various places in Scripture, refers to the expansion of the universe, especially during the fourth day. This expansion could have started on Day 2, when God created the 'expanse' (Hebrew raqia, KJV "firmament", *Genesis 1:7*).

The model indicates that early on the fourth day, Earth plunged into a zone of timelessness. In this zone all physical processes, including clocks, come to a complete stop. The spherical zone of timelessness expands out from the earth at the speed of light, engulfing the newly-created stars and galaxies. After reaching the most distant galaxies, the timeless zone reverses direction and begins shrinking back toward the earth at the speed of light. As it does so, it uncovers the new galaxies, so that the light can be seen on Earth. Dr Humphreys: "When the sphere reaches zero radius and disappears, Earth emerges, and immediately the light that has been following the sphere will reach Earth, even light that started billions of light-years away. On the fourth day, an observer on the night side of the earth would see a black sky one instant, and a sky filled with stars the next instant."

A universe with a center and an edge, plus Humphreys' concept of the waters above, provided an explanation for the 'Pioneer anomaly', which is a small but strange deceleration of four outgoing spacecraft: Galileo, Ulysses, and Pioneers 10 and 11.

1. Dr John Hartnett

Dr John Hartnett has taken a different approach, which uses a different aspect of Einstein's relativity theory. His cosmology applies a concept developed by Israeli cosmologist "Dr Moshe Carmeli" (1933–2007) called 'cosmological relativity'. Carmeli argued that to adequately describe the large-scale structure of the universe, in addition to length, breadth, depth, time (four dimensions), another measure, or dimension, was needed: the velocity of the expansion of space. This dimension has an effect on gravity and time—hence 'cosmological general relativity'. Carmeli's ideas have been successful in explaining long-standing astronomical puzzles, such as high redshift supernovas, galactic rotation observations, spheroidal galaxy anomalous dispersion, and expansion of the large-scale universe. A great strength of "Carmelian" relativity is that it does away with hypothetical unobserved entities such as dark matter and dark energy, both of which are needed for big bang cosmology.

Carmeli developed his cosmology with the assumption of the cosmological principle, no center and no edge to the universe, but Hartnett realized that these ideas also worked with a universe with a center and an edge. Furthermore, with this approach, an acceleration (increasing velocity) of the expansion of space, such as could be expected on the fourth day of the creation week, would have profound implications for time during that period. Time dilation results, but not due to a net gravitational effect—it is due to the enormous ***accelerated stretching of the fabric of space***, Figure 9.29. This means that on Day 4, the clocks in the outer reaches of the expanding universe were running very fast compared to clocks on Earth. This allows time for distant starlight from the galaxies being created on the fourth day to travel to Earth and be visible to Adam and Eve. Again, it's the fourth day as measured by Earth clocks, the clocks the Bible uses.

Fig.9.29: Space-Time - Accelerated Stretching of the Fabric of Space

What if no-one had ever thought of the possibility of time dilation? Many might have felt forced to agree with those scientists (including some Christians) who have asserted that there was no possible solution— vast ages for Earth are a fact because we can see distant stars, and the Bible must be 'reinterpreted' (massaged) or rejected. Many have urged Christians to abandon the Bible's clear teaching of a recent creation because of these 'undeniable facts.'

However, this reinterpretation of Scripture would also mean that Earth is old and the rocks containing fossils under our feet are old. Therefore, this also entails, if it is logically thought through, accepting that there were billions of years of death, disease, and bloodshed before Adam, thus, eroding the Creation/Fall/ Restoration historical framework presented in the Bible—the framework in which the Gospel makes sense, and upon which western civilization has been built, with all its many benefits.

However, even without the new ideas that seem to solve the problem, such an approach would still have been wrong-headed. ***The authority of the Bible should never be compromised*** by mankind's 'scientific' proposals. One little previously unknown fact, or one change in a starting assumption, can drastically alter the whole picture so that what was 'fact' is no longer so.

This is worth remembering when dealing with other areas of difficulty which, despite the substantial evidence for Genesis creation, still remain. As shown, this particular area of difficulty is shared by the big bang theory, and creationists should point this out. Only God possesses infinite knowledge. By basing our scientific research on the assumption that His Word is true, instead of the assumption that it is wrong or irrelevant at points where today's 'science' cannot explain it, our scientific theories are much more likely, in the long run, to come to represent reality accurately. However, creation was a miraculous process and we must recognize that God is able to do things that we, in our human limitations, will struggle to understand. And big bangers invoke secular (God-less) 'miracles' to try to solve the same problems.

BLACK HOLES AND GALAXIES

The educational and media system often contrasts 'creationists' and 'scientists.' However, in every issue of *Creation* magazine we prove them wrong with an interview with a highly qualified scientist who is also a creationist. Not too many of them have been astronomers or astrophysicists like Dr Blietz.

His interest in science started at the tender age of five. Markus's father took him kite flying, and the boy was very interested in why the kite could fly and stay intact. This was the beginning of the interest in science, Figure 9.30.

Fig.9.30*: Giant Black Holes and Their Galaxies – Curtsy NASA*

Black hole is predicted by Einstein's General Theory of Relativity if an unimaginable amount of mass accumulates in a very small space. E.g., the sun (mass 2×10^{30} kg) would need to be compressed into only 6 km (4 miles) diameter. While the centers of the galaxies are proposed to be *supermassive* black holes. These would be up to 100 million times the sun's mass, concentrated in a volume with a diameter smaller than the distance from Earth to the moon (384,000 km / 239,000 miles).

Black hole is 'black,' because its mass is concentrated, which enables it to generate an extremely strong gravitational force. The gravitational force bends the four-dimensional space fabric, including time. As a result, even light, which has no rest mass and moves at 300,000 kilometers (186,000 miles) per second, cannot escape. Since no light should reach us, physicist "John Wheeler" called them 'black holes.' Nonetheless, why do galactic nuclei emit so much light? Shouldn't they be pitch black as well?

"The strong gravitational force causes nearby gas clouds, which surround the black hole, to spiral into it. Due to the spiraling, the clouds are forming a disc, a so-called accretion disc. While this occurs, the gas clouds accelerate and emit highly active X-rays. This radiation is then supposed to hit other gas clouds, further away from the center, which then heat up and emit longer wavelength radiation, which can be observed as visible light. Using special instruments, one can observe the spectral distribution of the light emitted by these further out gas clouds, which permits us to discover the chemical composition of the gas as well as the gas velocities. This again permit us—under certain assumptions—to calculate the central mass of the black hole."

MIRACLE OF FAITH

Remarkably, an astrophysicist name "Dr Blietz" became a believer in Jesus Christ late in life. When he lost his father to cancer, and had a mental 'burnout,' "so extreme that he could not even read one single word anymore! Even the smallest decisions were too much for him. Basically, his brain, the instrument which he trained all his life and which he was proud of, all of abruptly went out of operation."

While he was in such dilemma, his secular science could not explain what has happened to him. He wondered regarding the clear reality of "***good vs evil***," which occupied his thought constantly. However, the difference between good and evil made sense to him. If there is indeed a Creator, then "Dr. Blietz" decided to read the Bible, which was owned by his wife, who was not yet a reborn Christian, he said:

"I read the Gospel according to Matthew. Almost immediately I understood that Jesus was a real, historical person; that He came to fulfill a mission; and that I needed him urgently."

Then a Christian friend lent him a small booklet that "explained the full plan of God, from the beginning of the creation, to the coming of Jesus on this earth, His crucifixion and resurrection, His second coming, the final judgment and the creation of a new heaven and earth. He knew this was the truth. In the booklet there was also a prayer, where one could confess his sins and give his life to Jesus. he did not hesitate a moment; he fell on his knees and delivered his life to Jesus."

VITAL CREATION

That was an interesting point: this booklet started from creation. Yet many evangelists discourage talk about creation, and say, "*Just preach the Gospel.*" But Dr Blietz responds:

"If the Bible is not reliable in its historical statements, how can it be true in other statements? If Jesus didn't speak the truth about Genesis, how could we trust what He was saying about sin, the cross, resurrection and everlasting life? Jesus Christ *is* the truth; and if He affirmed the literal creation of the world in six normal-length days, we Christians should do the same. If, however we compromise and try to marry millions and billions of years with the creation account of the Bible, we may easily pull folk away from the truth of the Bible and the Gospel."

But what about science, which would be important to a published scientist? Dr. Blietz's friends pointed out:

*"**Science is a human endeavor to find the truth about the world of matter.** It is not fully reliable, because hypothesis, theories and models change over time. Additionally, science cannot say anything about the big questions of '**where do we come from, why are we here, and where do we go?**' Science is limited in its abilities and should not step over the clear limits which have been set by God."*

ASTRONOMY SUPPORTS THE BIBLE

Also, despite the astronomical limitations, science offers good provision for the Bible. Dr Blietz listed a number of areas in geology and biology, in addition to his own field of astronomy. For example, the existence of comets in our solar system, as he explains:

Fig.9.31: Webb Telescope's Pillars of Creation Shows Things Hubble Couldn't – Curtsy - NASA / ESA / CSA / STScI

"Comets are like dirty snowballs circling around the sun on highly elliptical orbits. Every orbit, they lose material, because they start to melt when they come close to the sun. After less than 10,000 years they should have disappeared completely. Evolutionist scientists have therefore 'invented' the so-called Oort Cloud, which, according to their theory, should act as a source for replenishing the comets, and which they say is located in the most distant parts of our solar system. However, despite intensive search in the last century, up to now there is not the smallest piece of evidence of this hypothesis. If the Bible is true, one would of course expect many comets to still be 'alive', because 6,000 years is just not enough for most of the comets to have melted."

Dr Blietz was also highly critical of the big bang dogma. For example, he said:

"The big bang model assumes the existence of so-called dark matter and dark energy. Neither of these have ever been observed in the laboratory. However, they are desperately needed to uphold the model and to avoid contradiction with the observational data."

"The big bang also has no reasonable explanation for the virtual non-existence of antimatter. This is an enemy of normal matter: when they meet, they annihilate each other with intense release of energy. But according to the model, an equal amount of antimatter and matter should have been generated."

DISTANT STARLIGHT

Dr Blietz has published papers in astrophysics journals on Seyfert galaxies, e.g., those classified as NGC 1068 and NGC 7469. Nonetheless, they are 47 million and 200 million light years away respectively, Figure 9.31. Accordingly, *"How could one study light from these galaxies if the universe is only 6,000 years old?"*

The answer is found in the pages of the Bible, where the Lord God, the Infinite Creator had formed Time/Space as inseparable one entity. If you somehow pull time, you are actually pulling the fiber of space, Figure 9.32. The Lord God in His perfect wisdom explained how He allowed light to travel to earth through stretching space/time as stated in the following Biblical verses:

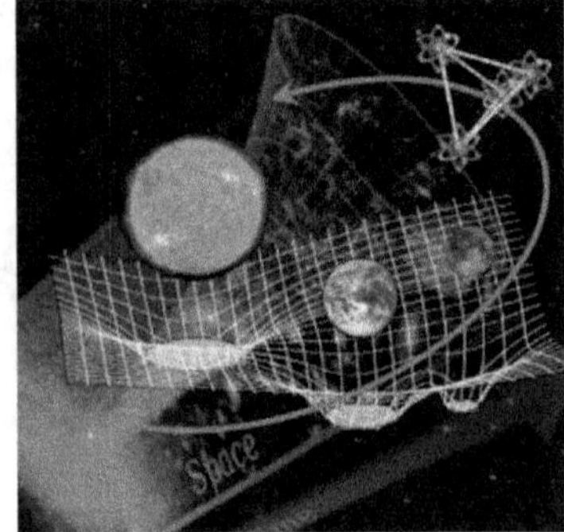

Fig.9.32: *It is I who made the earth and created mankind on it. My own hands stretched out the heavens; I marshaled their starry hosts. Isaiah 45:12*

Isaiah 42:5

This is what God the LORD says the Creator of the heavens, *who stretches them out*, _ who spreads out the earth with all that springs from it, who gives breath to its people, and life to those who walk on it.

Isaiah 44:24

Ò This is what the LORD says your Redeemer, who formed you in the womb: I am the LORD, _the Maker of all things, *who stretches out the heavens*, who spreads out the earth by myself,

Isaiah 45:12

It is I who made the earth and created mankind on it. My own hands *stretched out the heavens; I marshaled their starry hosts*.

Isaiah 48:13

My own hand laid the foundations of the earth, and my right hand *spread out the heavens; when I summon them, they all stand up together*.

Isaiah 51:13

That you forget the LORD your Maker, *who stretches out the heavens* and who lays the foundations of the earth, that you live in constant terror every day because of the wrath of the oppressor, who is bent on destruction?

Jeremiah 10:12

But God made the earth by his power; he founded the world by his wisdom and *stretched out the heavens by his understanding*.

Jeremiah 51:15

Ò He made the earth by his power; he founded the world by his wisdom *and stretched out the heavens by his understanding*.

Zechariah 12:1

The LORD, *who stretches out the heavens*, who lays the foundation of the earth, and who forms the human spirit within a person, declares: Ò I am going to make Jerusalem a cup that sends all the surrounding peoples reeling. Judah will be besieged as well as Jerusalem.

CHRISTIANS STUDY SCIENCE

It is sufficient to believe in Jesus Christ, who executed the act of creation, indeed can open our eyes and give us the correct view of our world. Before a scientist becomes a Christian, he/she never felt truly content with the evolutionary world view, which many had adopted.

Darwinian evolution produced too many contradictions and left open too many questions. Only the truth in the Word of God is able to give a full, comprehensive answer to our basic questions of death and life.

However, believing scientists warn that numerous in the scientific establishment will ridicule and persecute dissenters. All the same, scientists may "study science withing an unadulterated pure and honest perception to be able to better serve the infinite creator God. It is evident God reaches any scientists attempting to reach Him. Knowing God through reaching Him is precisely what God wants every one of us to do.

STAR TREK WARP SPEED

Gene Roddenberry's classic sci-fi drama, Star Trek, made famous the warp drive, a theoretical concept whereby a spacecraft travels Faster Than Light (FTL).

Fig.9.33: *In the Star Trek TV series and movies the starship USS Enterprise could travel faster than light by engaging its warp drive.*

A 'trekky' enthusiast once told me that the warp speeds described on the television shows and in the movies may be calculated as follows.

Warp factor *w,*

from the original *Star Trek* series, means that the spacecraft travels at

w^3 times the canonical speed of light ($c \cong 300{,}000$ km/s or $186{,}000$ miles/s).

Therefore:

warp factor $w = 7$ means the space craft travels at $7^3 = 343\ c$.

It would be unusual to hear that the starship the *USS Enterprise,* Figure 9.33, had exceeded warp factor 9, which is about $= 9^3 = (9 \times 9 \times 9 = 729)$ times the speed of light.

To travel even around the local neighborhood of our galaxy warp factor 9 (from the original TV series) just won't do it. The nearest star to our solar system is about four light-years away. Accordingly, travelling at warp 9, you would take two days to get there. Not too bad. However, what about going to other star systems?

To travel 50 light-years, which is a very small distance in the galaxy and which includes very few stars— only 64 sun-like stars—would take you 25 days at this speed. Within a distance of 100 light-years from Earth there are known to be only 512 stars of the same spectral class as our sun and very few of those might be candidates for a solar system that could potentially support life. Therefore, it would be much better to be able to travel 100 light-years quite quickly; but that would take you 50 days. However, in the TV shows they often arrive in just a matter of hours.

Years later, the next TV series—*Star Trek: The Next Generation*—solved this problem by introducing a new formula that increased the warp speed in a highly non-linear fashion, such that a warp factor of 10 meant infinite speed across the galaxy. Thus, distances were no longer a roadblock to get around the universe in the 40 minutes or less available in a TV episode. For example, warp factor $w = 9.9999$ is equal to nearly $200{,}000\ c$. At that speed you could hypothetically travel right across the galaxy in only six months.

MIGUEL ALCUBIERRE WARP DRIVE

Nonetheless, that is all science-fiction. What about a *real warp drive*? In 1994, a Mexican physicist by the name of Miguel Alcubierre came up with a proposal for stretching the fabric of space-time in a way which would, theoretically, permit (Faster Than Light FTL) travel.

Miguel Alcubierre discovered a theoretical solution of Einstein's field equations producing what has been called *Alcubierre Warp Drive*. Needless to say, it is a highly speculative mathematical model, which specifies how space, time and energy interact.

Fig.9.34: Warp Bubble

To put it simply, this method of space travel involves stretching the fabric of space-time in a wave which would (in theory) cause the space ahead of an object to contract while the space behind it would expand. An object inside this wave (i.e., a spaceship) would then be able to ride this region, known as a "warp bubble" of flat space, Figure 9.34.

The spacecraft hypothetically generates the warp bubble in its bow wake and collapses it behind the craft as it travels through space-time, Figure 34. The speed of light within the bubble remains the same speed of light, c. The spacecraft undergoes no local acceleration—no huge g-forces are experienced by the craft. Thus Captain James T. Kirk, or anyone else on board, would not spill his/her coffee even as they go to warp 9.9999.

Nonetheless, the idea of warp drive presents a few problems.

1. There are no known methods to create such a warp bubble in a region of space that would not already contain one.

2. Assuming there was a way to create such a bubble, there is not yet any known way of leaving once inside it.

That's a problem. Once you enter hyperspace, or "subspace" as they called it in the original *Star Trek* series, you are stuck there, never to leave it again. That would have been a good story for an episode in the series.

In 2012, NASA's Advanced Propulsion Physics Laboratory (*aka* Eagleworks) even announced that they had begun conducting experiments to see if a "warp drive" was possible. The team lead scientist, "Dr Harold Sonny White," described their work in a NASA paper titled *Warp Field Mechanics 101*. In 2013, White and members of Eagleworks published the results of their 19.6-second warp field test under vacuum conditions. These results, which were deemed to be inconclusive, were presented at the 2013 Icarus

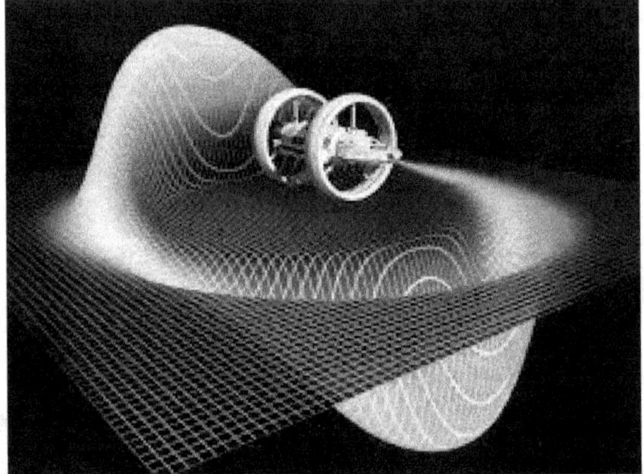

Fig.9.35: Visualization of a warp field, according to the Alcubierre Drive.

Interstellar Starship Congress held in Dallas, Texas. I cannot find any report of any progress after that time.

Real physical evidence also suggests that FTL space travel is impossible. It is a problem of energy, which is well understood by physicists. Alcubierre's theory requires the use of large amounts of a never observed type of "exotic matter" that violates known physical laws. It is needed to produce a negative type of energy to fold space into the warp bubble. This hypothetical negative energy involves either tachyons (which travel faster than light speed) or a naked singularity (which is a black hole without the event horizon).

Still, one of the most dubious is "Dr. Kiguel Alcubierre" himself. He listed a number of concerns, starting with the vast amounts of exotic matter that would be needed. "The warp drive on this ground alone is impossible," he said. "*At speeds larger than the speed of light, the front of the warp bubble cannot be reached by any signal from within the ship,*" he said. "*This does not just mean we can't turn it off; it is much worse. It means we can't even turn it on in the first place.*"

Nevertheless, even if such a drive could be developed it has some somber problems to overcome. When the spaceship reaches its destination, it has to stop, and that's when the problems begin. Metaphorically hell breaks loose!

Researchers from the University of Sydney have done some advanced crunching of numbers regarding the effects of FTL space travel via Alcubierre drive, taking into consideration the many types of cosmic particles that would be encountered along the way. Space is not just an empty void between point A and point B. It's rather full of particles that have mass (as well as some that do not.) What the research team—led by Brendan McMonigal, Geraint Lewis, and Philip O'Byrne—has found is that these particles can develop "swept up" into the warp bubble and focused into regions before and behind the ship, as well as within the warp bubble itself.

When the Alcubierre-driven ship decelerates from superluminal speed, the particles, where the bubble has gathered are released in energetic outbursts. In the case of forward-facing particles, the outburst can be *very* energetic—enough to destroy anyone at the destination directly in front of the ship, Figure 9.35.

Anyone waiting at the destination would be blasted into oblivion by a beam of gamma rays resulting from the extreme blue-shifted particles released from the forward region of the warp bubble as the spaceship comes out of warp drive. But this is all theoretical anyway.

PURE FANTASY – CONSIDERING FASTER THAN LIGHT TRAVEL

The galaxy is enormous in size (about 100,000 light-years in diameter) and those who believe that life evolved on Earth see no distinction between our home planet and any other planet out there that might support life. Like many others, "Gene Roddenberry" imagined a universe full of creatures that have evolved along different evolutionary paths, producing all sorts of strange creatures not seen on Earth. Among them are many different forms of "*alien intelligent life.*"

There are those who imagine either that mankind will one day travel out there and meet intelligent aliens, like "*Klingons or Vulcans,*" or, that those alien civilizations have evolved to such an advanced state that they may have already invented warp drives and the ability to travel across the vastness of space to Earth.

This is pure fiction, which is evidenced by the plethora of sci-fi TV shows and movies that incorporate warp-drive-powered spacecraft. The alternative in the movies for sub-light-speed spacecraft is *sleep chambers* where the travelers are put into "hyper-sleep"—a form of suspended animation—for 100 years or more to make the enormous journey, even to just one of those nearby stars.

Recently astrophysicist "Paul Davies" acknowledged the rarity of life in the universe. He admitted that Earth may be the only place where life exists. Certainly, Earth is the only planet in the universe known to support life. This admission makes nonsense of belief in alien intelligent life on other planets outside our solar system. The question here for the Christian believer is: Did God create life on other planets? And if not, why did He create such a large universe?

Warp drives are really just about a type of faith—a ***blind faith***. That type of faith posits that intelligent life is 'out there' (contrary to all evidence to the contrary) and that mankind will eventually break the seemingly impossible barrier of the vastness of space and break out into the galaxy.

The alien races are so much more advanced than we are, and therefore they must have conquered the problem of warp drive, is *pure storytelling based on evolution*. It is a substitute for the Creator God.

Darwinian evolution is *assumed to have increased the brain power* of some putative alien race to the point where they teach tensor calculus in nursery school. Consequently, their adults are more *godlike* than we are and find manipulation of space, time and energy a trifling matter. With their machines they can create space and time and manipulate energy and gravity to make safe intra-galactic, even inter-galactic, travel possible. Hence, they can visit distant places in the galaxy at will.

This section of this book "The Origin of the Universe" started out with science fiction and now it ends with science fiction. Trust me; it is more fiction than science because the starting premise—"*Darwinian evolution*"—is pure fiction. Professor "Richard Dawkins" once famously said:

"Evolution has been observed. It's just that it hasn't been observed while it's happening."

It has never been observed but that is what science requires—observation of evidence. What Dawkins is really referring to is historical science, not operational science. The latter can be observed while it is happening. The former cannot. As the famous evolutionary biologist "Ernst Mayr" admitted, "*For example, Darwin introduced historicity into science. Evolutionary biology, in contrast with physics and chemistry, is a historical science—the evolutionist attempts inappropriate techniques for the explication of such events and processes. Instead, one constructs a historical narrative, consisting of a tentative reconstruction of the particular scenario that led to the events one is trying to explain.*"

Perhaps the main reason NASA even considered conducting "warp drive" experiments is the entrenched belief in *Darwinian evolution* and the implied existence of *aliens* from other star systems. Hence the need to go out there and meet them.

Some even believe they have already made the trip there (Roswell, New Mexico), while others say only *interdimensional* travel is possible. The latter folk communicate with '*aliens from other stars systems*' using meditation and other New Age psychic practices. They have bypassed the impossibility of FTL travel and claim thought is not so limited. Hence, some say that they are in daily communication with '*little green men.*' It makes you wonder where this will lead!

THE BLACK HOLE AND GOD

The query of heavenly omnipresence in light of the current visual evidences of a black hole, is worthy of intensive investigations. The subject sheds light on the belief of God's omnipresence by contemplating on recent world interpretations, scientific theory, and the philosophies of personification and incorporeal nature of God's being. The object of this study proposes an understanding of Godly omnipresence against the setting of Psalm 139, Figure 9.36.

Fig.9.36: *35 New Black Hole Collisions Captured, Including One Created Merger – Curtsy IFLScience*

are the set texts to demonstrate God's omnipresence. I merely quote these texts as examples of an presumed comprehension of God's omnipresence that, in my opinion, caused by a product of the contextual cosmological considerations of those in Scripture who provided witness to their faith. I would like to resist that the idea of divine omnipresence was both a feasible and reliable doctrine, seeing the cosmological understanding of the time.

Throughout biblical times, explicitly in the development of the creation account of Genesis 1, it was a shared belief that the cosmos contained of a three-tier universe. Van Dyk (1987:10) delivers the following clarification of the manner the cosmos was professed (paraphrased).

First, it was held that the earth was a flat disc, suspended on pillars. The earth offered the first-level stage, on which life could be existed and experienced. The sky disconnected the water above the earth from the earth (and the water below the earth). Above the sky and its water, one discovers the heavens, the residence place of God, and the space in which the spiritual functions.

The heavens (or Heaven) is the layer above the earth, the canvas of the spiritual, encompassing the elements that point to purpose and ultimate divine realism. The third level established in the levels below the earth. In Judaism, *Sheol,* or the realm of the dead, gave space for those who did not find presence in the realm of earth, or yet existing with God in the heavens.

From this standpoint, it surely makes sense that, if God were above the sky, with a full sight of the whole earth and all that is in and below it, the existence of God was unavoidable. No wonder the psalmist, especially, uttered that God's reach was everywhere! The omnipresence of God, in this biblical world interpretation and in philosophical customs (as we will explore later), was carefully secured to the knowledge of God's omnipotence and omniscience. "Meanwhile God is everywhere, he is causally active all over creation and able to distinguish all things instantly" (McGuire & Slowik 2012:280).

The initial Jewish dogma of creation "declares that God is existing equally in the entirety of creation" (McGuire & Slowik 2012:280). Not only was God comprehended to be existing everywhere, recognizing all things, and able to do all things, but in this cosmos, God was comprehended to be the causal primer of all things. The involvement of life itself was a exhibition of God's existence in the world. Not only did life provide testament to the existence of God, but so did the exhibition of the notion of divine judgement (justice), where it was held that divine blessing or curse would exhibit itself as a judgement on the expression of life lived.

The righteous would flourish, while the wicked would discover God's wrath - a formula that is believable only with a comprehending that God is concurrently omnipresent and omniscient.

The questioned belief (in early Judaism) of the absenteeism of an *afterlife* stated to this fact. God's justice would illustrate in a blessed life for the righteous, whereas a life of anguish was in store for the wicked - justice occurred in this life, not in the next. Even in the inquisitorial of divine justice in, for example, the Book of Job, divine justice is clarified in the human incapability to understand the existence of God throughout the universe.

When Job questions God why he, a righteous man, should undergo so much anguish, God replies with a series of oratorical interrogations, starting with: "Where were you ...?" (Job 38:4). Through the questions, Job is made to realize that humanity has a limited perception, for human beings live within the limitations of space/time. God's justice is perfect, since God is omniscient, and God is all-knowing, since God is omnipresent. Humankind is cognizant of locality, of immanency, but God is able to be perfect and to adjudicate justly, for God is indeed omnipresent, omniscient and omnipotent.

It was only much later, with the mounting comprehending that we do not live in a fixed three-tier universe, that the conception of God's omnipresence turn out to be more complex. As the interpretation of the cosmos transformed, so did the interpretation of divine omnipresence. To fast-track the dialog past

Ptolemy, Copernicus and Galileo, it is predominantly in Isaac Newton that we discover a innovative shift in interpretation divine omnipresence.

"McGuire," an enthusiastic scholar of Newton, illuminated how Newton himself toiled with the conception of omnipresence, bearing in mind his increasing understanding of the universe. To Newton, the predicament of God's omnipresence demonstrated in the greatness of space/time. While Newton, to a large extent, interpreted space to be infinite and some heavenly bodies to be static in space, it was obvious that God was not *above* creation, as designated by biblical world views.

Newton made sense of God's omnipresence in this infinite cosmos by signifying that the mere fact of its existence, the expanse of the universe attested to the presence of God, even the ideas of the cosmos that were unobservable and inaccessible to human observation and influence.

McGuire (1978:119) translates Newton, stating the following:

By purpose of its eternity and infinity, space will neither be God nor wise nor powerful nor alive, but will simply be enlarged in duration and magnitude; while God by purpose of the eternity and infinity of space (that is, by reason of his eternal omnipresence) will be rendered the most perfect being.

A fixed star, whether it has come into being as the first of all stars or after a series of previous stars, whether the number of the stars be finite or infinite, will not by this means be either more perfect or more imperfect: God, nevertheless, will be established to be more powerful, wiser, better, and in every manner more perfect from the eternal sequence and infinite number of his works, that He would be from works merely infinite.

To Newton, God was still the **external craftsman**, whose handiwork can be observed in all of creation. The mere vastness of creation carries demonstration to the presence of God, for nothing in the cosmos existed deprived of God. Because space is infinite, God's presence is infinite, and since time is eternal in duration, God's presence is also eternal (McGuire 1987:125).

This does not mean that God's omnipresence is sealed in the created order. To Newton, God's omnipresence is not a consequence of material space/time locality; God's omnipresence pre-empts space and eternal duration (McGuire 1987:126), being the contributory mover of bringing all things in time and space into existence.

This has consequences for the relationship between God and God's creation. If we take a linear view of time/space, if God is *outside* time and space, in essence, pre-stating time and space, then God can be understood as *causa prima efficiens* - and free will may well be negated.

Is there something besides God that keeps the whole of the cosmos coherent?

To Newton, the shared denominator in all of eternal space is time; everything moves in the same time frame in a linear direction. Qu (2014:436449) compares this conception of time and space to the views of Einstein and Barth.

To Einstein, time is not a shared denominator and is conditional on the changes in space as established by gravity, spatial speed, mass, and the like. God's omnipresence in this instance turn out to be even more problematic, as there is no precise viewing platform located in time, no point where God can split eternal past and future into a fixed moment in time (and in space). God's omnipresence would, consequently, cause "time-disturbances" and incongruity in the being of God, where, if God is in space and time, some parts of God will be moving quicker, while other parts will move slower. We will get to the non-divisive nature of God later.

"Barth," in turn, attempted to make logic of the eternal and inherent natures of God, unfolding God's presence as eternal immanence, piercing the concepts of our space and time - *historie*. Consequently, "God's eternity is both transcendent and immanent in human time" (Qu 2014:436). To Barth (2010:611, 46-48, 91-97), the personification must be observed as the essential point around which God is concurrently immanent and transcendent whereas being completely and comprehensively present in both states. The next section discusses the notion of incarnation.

We can accomplish that the conception of God's omnipresence has experienced increasing encounters as our interpretation of the universe has spread-out. Omnipresence was first comprehended in a motionless and limited cosmos, where God's presence was related to God's capability to observe all things and do all things. This cosmos, as well as Newton's cosmos, shadowed a time-linear course, where God moved alongside the whole cosmos, being completely present in each moment (Shults 2007:48). God's omnipresence was recognized to be causal and determinative of the cosmos' existence. With time, the accepted sciences have progressively rejected any linear interpretation of causality and temporality (Shults 2007:48) and, henceforth, probed the nature of a theological (and philosophical) conception of divine omnipresence.

IS GOD THEREFORE IN A BLACK HOLE?

To the biblical writers, black holes were not recognized and this would, consequently, have been a ridiculous question. If they distinguished black holes, then God would still be above the firmament, above the black holes, knowing, seeing and being capable to impact black holes, according to God's divine will.

To Newton, the presence of black holes would have been an pointer that God was the causal mover of the existence of black holes. Since God is, black holes could exist, but God would move parallel in conjunction with black holes in the linear continuum of time.

To Einstein, a divine existence in a black hole would have been problematic, as God's being would have had to involve the concurrent collapse of time/space and matter as God would continue outside the black hole in the rest of the cosmos. It would not be a query of *Is God in a black hole?* But rather *How/ when/where would God be in a black hole?*

To Barth, the personified Christ affirms to the unchangeable nature of a transcendent God within the experience of earthly time and space. God is unchangeable; consequently, God would be the same in a black hole as God would be in the person of Jesus Christ.

OMNIPRESENCE AND THE INCORPOREAL NATURE OF GOD

Another feature of the question of omnipresence is of *how* God is present -whether God is existing in body or whether God is a present *force* without a body. We refer to the latter as the incorporeal nature of God.

To Dyck (1977:85), there is a certain relationship between God's omnipresence and incorporeality. If God is all over the world, then it would not be of any logic for God to be restricted or limited to the boundaries of an embodied form. Dyck (1977:85) then asks the question:

Is it a contradiction to speak about an embodied omnipresent being?

Here, the misunderstanding centers around the notion of "body", which suggests form. This, in turn, indicates spatial limitation and may even deduce that God is a form of matter. If this were factual, and if we were to adopt that Newton is correct in signifying that the infinite nature of the universe in space and time is symptomatic of the presence of God, we could settle that God is embodied in and through the universe.

Dyck (1977:86) challenges this thought by drawing a discrepancy between God and the universe; God and the universe are not the same, consequently, antithetical any conception of pantheism. To be fair to Newton, he did not propose a form of pantheism, but he was fairly obstinate that both space and God have an incorporeal extension; they are both infinite (in duration and in spatial infinity), neither God nor the cosmos is limited, but they are not the same (McGuire & Slowik 2012:290). Space and time are "physical characteristics that stand as external affections of divine being" (McGuire & Slowik 2012:306). Newton, consequently, inclines more towards an incorporeal omnipresence than the omnipresence of an personified being.

This idea did not begin with Newton, but is already expressed in the writings of, for instance, "Thomas Aquinas."

To Aquinas, the incorporeal nature of God is not about the physical (embodied) presence of God; it denotes to the contact with divine power as experienced through the cosmos (Aquinas 1964:283). God is the causal mover, allowing all things into to come toexistence, deprived of whom, nothing can exist or accomplish its divine resolve in the greater scheme of the universe. There are clear lines between Aquinas and Newton on the topic of the incorporeal nature of God's presence.

Another problem with pantheism, or any method of doctrine of an personified God, would be the proposition that the presence of God will be superior in the greater things and less in the smaller. God is, then, equivalently divisible according to space, time, or any other dimension of our choosing in the cosmos!

DAHL (2014:76) POINTS OUT THIS QUANDARY, STATING

Since God fills all things, they must encompass only portion of God. Nonetheless, if God is not dividable, it cannot be more of Him in bigger than smaller parts of the world. God must consequently, fill things with the entire of himself, nonetheless, still nothing can encompass God comprehensively.

On this point, we would like to draw from three theologians, whose present work emphases on theology and emergence (complexity theory).

The first is the Danish theologian, "Niels Gregersen." To Gregersen (2010:173-176), God is part of all of creation's procedures - the universe in its infinite condition functions, upholds itself, and carries on notwithstanding any unambiguous *external* force. However, we speak of *someone* like God, signifying that God has made Godself recognized in a *language* that we comprehend.

To "Gregersen," God's incorporeal presence is part of "the entire pliable matrix of materiality" (Gregersen 2010:176), but what makes it distinct from pantheism is the personal method in which the incorporeal presence turn out to be personified in the conveyance of self-revelation. We can only speak of incarnation *sub specie anthropos* (Bentley 2016:2), where our account of the personified nature of God's presence is locked within the boundaries of human existence, experience and knowledge, permitting us to understand the Incarnation, utilizing exclusively our frame of reference (Bentley 2016:2).

"Gregersen's" notion of "deep incarnation", therefore, proposes that God's incorporeal nature and embodied self-revelation are not contraries, but that the embodied self-revelation is a separate form of communication with a level of complexity that operates in, and comprehends the language of embodiment.

The second is "Klaus Nürnberger." Nürnberger defines God as the definitive source and destiny of realism. Compatible with Gregersen, Nürnberger's (2016:15) comprehending of God pivots on the knowledge that God is not a force or power outside of the dominion of physics, but that God is personally involved in the cosmos as both its source and destiny. It is in God's transcendence that God's presence turn out to be engrained in the language of incarnation.

Likewise, "Van Huyssteen" (2006:10) proposes that the difference between immanency and transcendence is built realities, expressed by our own epistemologies and ontologies. There exists one realism in which both the transcendent and the immanent, the incorporeal and the personified natures of God are equally exact.

It is stimulating that theological language proposes that this apparently opposing nature of God is the collective lived experience of many. Take, for instance, African perceptions of God. "Byaruhanga-Akiiki" (1980:360) sums it up stunningly: In African imageries of God, God is believed to be in all things, henceforth all medicines can work to address issues, as the Creator's power is there. The transcendent turn out to be evident, not in the restrictions of bodily form, but rather by becoming embodied in the expression of power through nature, which God permeates.

From a more Western standpoint, "Oord" (2019) proposes that God's touch is originated in a human partnership with nature and with other people. Though the presence of God is probed, particularly during times of suffering and sorrow (Dicken 2013:132-151), it is correspondingly factual that the presence of God is experienced through the participating presence and action of those around us. There is thus, a connection between *presence* and affect.

Is God in a black hole?

When take into account the concepts of "embodied omnipresence" and the "incorporeal nature of God", we would be accurate to state that, if God were present in a black hole in embodied form, the laws of physics would most definitely act on the being of God; God, with all other matter, time and space, would implode into Godself. The incorporeal nature of God's omnipresence, however, can be present in a black hole without God's being unfavorably affected. However, the same incorporeal presence of God is the presence of God manifest in the Incarnation. This leads us to the next point of discussion.

ANTHROPOMORPHISM, THEODICY AND THE DYNAMIC NATURE OF GOD

The incorporeal nature of God undoubtedly makes sense, and it refutes countless of the philosophical stumbling blocks surrounding the fluctuating manifestations of time and space, predominantly as it relates to black holes. "Gregersen, Van Huyssteen" and "Nürnberger" offer some form of resolution between the immanence and transcendence of God, but incorporeal realism, as specified earlier, absences in the personal dimension of a god-figure. If God were only an incorporeal presence, our experience of God would be very alike to "tapping into the *Force* of Star Wars." We require something more - we need a physical presence of God that turn out to be like us, speaks like us and, more than this, speaks our language. We need a God who comprehends, not simply a God who is an unseen force, hovering throughout eternal time and space without *persona*. It is significant to find God somewhere, even in symbolism. In Scripture and tradition, this attempt to locate God has found expression in different sacred metaphors: the presence of God in the Ark of the Covenant, the Temple, the people of God, the Church, the Sacraments - these are metaphors that create a sense of God's realism, intimate and physical presence among us.

A metaphysical presence becomes ubiquitous, "in which we are all absorbed" (Dicken 2103:135). Such an omnipresent force does not appear to address adequately the experience of suffering and pain, nor of a personal interest that shows justice, empathy, and intimate presence.

If God is incorporeal, how do we account for unjust pain, destruction and suffering?

The "Problem of Divine Hiddenness" proposes that there is too much suffering to warrant the existence, particularly the incorporeal and omnipresent existence of God (Oakes 2008:115). In MacLeish's play *J.B.: A play in verse,* based on the Book of Job, the character Nickles expresses the question regarding a God who permits suffering in the following words:

If God is God, He is not good, if God is good, He is not God; take the even, take the odd (MacLeish 1989:14).

Translated in a different way: If God is omnipresent, then is God good? If God is good, can God be omnipresent, and if so, then how?

The problem with theodicy is that it undertakes some form of stasis in the created order, that reality (the reality of lived experience) is standard, predisposing an intended, universal notion of good and prosperity. Kauffman (2016:74) suggests that this idea is not a true reflection of reality. The universe, each moment, can be divided between actuals and potentialities. The right conditions and actions transform potentialities into realities. Suffering, pain and the like are, therefore, the actuals of a particular set of potentials that materialized, and have absolutely nothing to do with divine predeterminism, will, or influence. The universe itself is dynamic, giving rise to life, death, suffering, and prosperity, as it turns out. He and Suchocki (2010) further propose that a dynamic universe needs a dynamic God. If God were static and the universe dynamic, there would be a growing gap between the existence of the universe and the presence of God.

God is dynamic, along with the universe - God is not static in the sense that God is locked into a being or in a body, which will limit the possibilities of *who* God can be, *where* God can be - along with the dynamism of the universe, the dynamic nature of God makes for endless possibilities of God's being and God's locality (Suchocki 2010:39-58). This makes God a partner in the experience of life and, hence, open to responding and inspiring responses in a dynamic universe.

God primordially and everlastingly enjoys a assuredness of satisfaction in the everlasting enfoldment of the world, and this satisfaction is appetitive, everlastingly generating a superjective nature that induces the becoming of finite occasions (Suchocki 2010:51).

Taking these points into consideration, God is present in a black hole in the sense that God becomes the fulfilment of all potentialities and actuals, even the actual of black holes that seem to contradict the intuitive notion of a universe unfolding and expanding.

GOD'S OMNIPRESENCE

One of the go-to passages in Scripture regarding God's omnipresence is undoubtedly Psalm 139. The psalmist asks the question: "Where can I go from your Spirit? Where can I flee from your presence?". By exploring the heavens, the depths, and the far sides of the seas, the psalmist exclaims that God's presence is to be found everywhere.

Undoubtedly, as we explored earlier in this writing, the psalmist's understandings were shaped by his consideration of the universe. Is there anything that we can learn from the psalmist in trying to answer our research question?

We would like to argue that there is indeed an implied epistemology in this Psalm that helps us gain a different standpoint on the question, not an epistemology that necessarily answers the question, but one that helps us place the question in perspective. It certainly appears that Psalm 139 is not so much about the omnipresence of God as it is about God's presence being inescapable (Oakes 2008:157).

Is there a difference?

The difference is that the psalmist does not begin from the standpoint of transcendence, breaking into immanence. The Psalm starts from the perspective of an imminent, personal relationship between the psalmist and God and then works outwards. The predicament we face in asking whether God is in a black hole is that we first have to contend with metaphysics and then try to reconcile the subsequent presuppositions with a notion of a person(al) God. This is virtually impossible. Even if we try to make sense of it through theological language such as "incarnation" or "transcendent immanence", it still remains a mystery that makes God either impersonal, confined, limitless something that is simply not within our frame of reference.

To start, like the psalmist from a viewpoint of personal experience, leading to bigger and greater circles, I suggest that the psalmist has a much better chance of making sense of the great mystery of God that, he discovers, transcends his notion of experienced reality. Let me illustrate this by breaking down the Psalm in its various stages of unfolding.

In verses 1 to 6, the psalmist does not ask the complicated, mysterious laden questions of God's existence. He simply states the known reality he experiences, namely that he, the psalmist, feels that God knows him personally. The extent of God's omnipresence and omniscience is located within the psalmist's lived experience - this is where God's omnipresence and omniscience make sense. God's omnipresence is located in the expression: "You know me better than I know myself". One would assume that, to the psalmist, this would be enough, but he does not stop here. The psalmist then moves from his own person to a wider context.

In verses 7-12, the psalmist extends the omnipresence of God to space outside his lived experience. Is there a place where the psalmist can escape God's presence? The psalmist argues and is in awe that the same presence that is experienced in person is the presence that will be experienced irrespective of the psalmist's movements and searching! The personal, intimate God is consistently encountered wherever the psalmist may find himself.

One can already note a question of immanence and transcendence in this shift. The psalmist makes sense of the consistent presence of God by describing God as the causal mover, the one in whom all things (and all beings) find their identity (vv. 13-16). It is only because God is the same primal mover of all things that it is possible to experience the personal God in the impersonal spaces of that which exists outside ourselves. The psalmist then makes a profound statement:

In verses 17 and 18. Although God is experienced in the personal, intimate spaces of being, God is beyond our comprehension and not embodied in our limited experiences of reality. Despite this God who confounds our thinking, the psalmist still draws back to the personal God who is known and who makes Godself known in experienced reality.

True to his world view and to the notion of God's omnipresence, omniscience and omnipotence, this personal God who is incomprehensibly equally and similarly present in and through all things brings balance to all of experienced reality. God's justice (vv. 19-20) brings equilibrium in an inconsistent world. Without God, the balance of creation and all that is in it would not exist. Hence, without God's omnipresent justice, the world as we know it could not exist or continue to exist.

In verses 21 and 22, the psalmist pledges his allegiance to this personal and transcendent God. To live life, to experience the reality of self and this created order, is to become part of the divine movement (and divine wisdom) in the realm of experienced reality.

In Verses 23 and 24, the concluding verses, the Psalmist draws back to the personal. God, who is personal, who is consistent outside the psalmist as God is within, who is beyond understanding, yet the one who brings order in this creation, the one who allows participation from God's created beings in order to experience life, is the God whom the psalmist asks again to speak to him in a personal and intimate language, and so, to become the source of the psalmist's inner conviction and the great motivator, drawing the psalmist to Godself.

If we were to answer the question: "If God is omnipresent, then is God in a black hole?", then we could follow in the psalmist's footsteps. By starting with black holes, with the "out there", with the mystery of transcendence and the complicated permutations of space and time, it would be difficult to bring God back to a personal being with personal interest in us. Conceivably, we should start with God-talk in the space of the personal. From a human, created perspective, perhaps we should start with what we can be "certain" of, that we know God as a personal and immanent God. Nonetheless, , God is not locked in our personal experience. The same presence experienced in the mystery of worship is the presence to be experienced throughout the universe, irrespective of where we may be looking. The personal becomes transcendent. The transcendent God is the personal God. The *person* of God is the incorporeal presence of God.

If the psalmist were to be asked the question:

THE FIRST IMAGE OF A BLACK HOLE

On 10 April 2019, the science community was in a state of euphoria. The first image of a black hole had been taken. Until then, black holes had been matters of scientific hypothesis, nonetheless, presently, with this image, the unseen became seen. Undeniably, seizing an image of a black hole is no mean feat, since that the gravitational field in a black hole is exceptionally poweful that no light is capable to escape it. Henceforth, what light might be obtainable to demonstrate what the black hole essentially appears like? Scientists clarified that the *image* of the black hole is not a solitary image collected by one telescope; it is a prudently constructed incorporation of data (nealy 5 petabytes over-all) offered by a grid of telescopes crossways the world. This information was then interpreted into an image by utilizing composite algorithms to harmonize and sort through the information collected from the EHT (Event Horizon Telescope) (Lutz 2019).

Black holes are captivating constructions; their presence amazes our thinking of space/time. They are the unpreventable breakdown of matter, and space/time into itself - the most stunningly enthralling and damaging cosmological construction that we distinguish.

Not only scientists were fascinated by this achievement; as a scientist with an interest in astrophysics, my ears gained vigor and my mind began reconnoitering the doctrinal questions raised up by this event. What is the denotation of this? Did God create this? Is God present in the black hole? If a space traveler somehow accomplished to travel to the black hole, could they have a divine experience, sensing close to God? What ensues to God on the event horizon? Is God dissimilar outside the black hole to what God is in the black hole? I distinguish that these questions are no more irrelevant to the existed experience of people everywhere around the world than enquiring: "How many angels can you fit on the head of a pin?" Nevertheless, these questions are vital as they direct our comprehending of how Christian faith and science formulate part of our knowledge of life and our thoughtfulness of our own meaning in view of the universe.

The very thought of black holes summons theological belief and the reconsidering of theological doctrines that we typically take for granted.

This writing undertakes into reconsidering the doctrine of omnipresence in the settings of these super cosmological structures. If we consider that God is *everywhere,* then what do we really mean? Do we embrace in our considerations that God would even be in to some degree like a black hole? This writing investigates the following themes:

1. The conception of omnipresence in tandem with contextual cosmological considerations.
2. Omnipresence and the intangible nature of God.
3. Anthropomorphism, theodicy and the self-motivated nature of God.
4. A ultimate revisiting of Psalm 139.

OMNIPRESENCE AND COSMOLOGY

What do we actually think when we utter that God is everywhere all over the universe (omnipresent)? First, we require to consider that the concept of divine omnipresence is not novel, particularly to the Judeo-Christian custom. While it is not openly stated in Scripture, the omnipresence of God surely appears to be an implied concept in numerous texts. In *Genesis 1:2*, we read that the **Ruach** *(Spirit)* of God hovered over the waters, signifying that the existence of God was not limited to the heavenly realm, but was already current within the realm of nature. In Psalm 139, the psalmist adopt this understanding further, by declaring that, when he explored the heavens, the depths, and the far sides of the seas, he found that God's scope was all over the world. In the book of Job, serious questions are asked about justice and fairness in light of God's all-encompassing presence. I will not offer a full exegetical explanation of these texts, nor do I suggest that these

"Is God in a black hole?"

Then perhaps his answer would be:

"Where can I go from your Spirit?

If I live life on earth, you are here. If I get drawn into a black hole, you are there".

Is this not enough to hold together the seemingly irreconcilable differences in our understanding of immanence and transcendence, embodiment and incorporeal nature, and infinity and the limits of our reality of space and time? The only place where we can speak of, is here, whether here is here, or here is in a black hole.

APPENDIX

- Operational science involves repeatable measurements that yield consistently reliable results.
- G stars within 100 light-years, solstation.com, accessed January 2017.
- See Warp Speed Calculators, anycalculator.com, accessed January 2017.
- Without the warp bubble around the spacecraft the occupants would experience enormous acceleration or deceleration while moving up to or down from speeds near the speed of light.
- In principle, FTL warp drive experiments (not sci-fi stories) fall into the category of operational science. The repeated failure of such experiments to produce any warp effect exemplifies the power of operational science.

REFERENCES

1. Hartnett, J.G., Life on Earth 2.0—Really? August 2015; creation.com.
2. Williams, M., What is the Alcubierre "Warp" Drive? universetoday.com, January 2017.
3. One 2014 paper (Lee, J.S., and Cleaver, G.B., The Inability of the White-Juday Warp Field Interferometer to SpectrallyResolve Spacetime Distortions, arvix.org, accessed January 2017) claims that the White-Juday Warp Field Interferometer is incapable of detecting any spacetime distortions from the NASA group's electrically charged plate experiments.
4. NASA—"Is It On the Verge of Discovering 'Warp Bubbles' Enabling Dreams of Interstellar Travel?", dailygalaxy.com, accessed January 2017.
5. Major, J., Warp drives may come with a killer downside, universetoday.com, accessed January 2017.
6. Hartnett, J.G., Aliens are all around us, 20 December 2016; creation.com.
7. Bates, G., Did God create life on other planets? *Creation* 29(2):12–15, March 2007.
8. SETI (Search for extra-Terrestrial Intelligence) searches must now have been going on for 50 years with absolutely no success. See Hartnett, J.G., SETI really? BibleScienceForum.com, 29 September 2016.
9. Hartnett, J.G., Wow! Communications from little green men? April 2016; creation.com.
10. 'Battle over evolution' Bill Moyers interviews Richard Dawkins, *Now*, pbs.org, accessed January 2017.
11. Mayr, Ernst (1904–2005), Darwin's Influence on Modern Thought, based on a lecture that Mayr delivered in Stockholm on receiving the Crafoord Prize from the Royal Swedish Academy of Science, 23 September 1999; published on scientificamerican.com, accessed January 2017.
12. Bates, G., *Alien Intrusion*, Creation Book Publishers, pp.66–69, 2015.

OUR MILKY WAY GALAXY

OLD GALAXIES IN A YOUNG UNIVERSE

Rethink Big Bang Ideas

This year stretched the imaginations of many astronomers and cosmologists. They have discovered amazing features at the outer reaches of the universe. And they cause nuisances for those with blind faith in naturalistic origin theories—including a big bang about 14 billion years ago..

Fig.10.1: Spiral Galaxy NGC 4414 as pictured by the Hubble Space Telescope. This extremely large galaxy is rich in clouds of interstellar dust. This is seen as the dark streaks silhouetted against the arms. The measured distance from Earth is about 60 million light years. Curtsy - NASA

Last January, 2022 a team of astronomers announced the discovery of a massive and distant string of galaxies, Figure 10.1. By their own dating methods, they were looking at a structure within only 2 billion years of the universe's inception. This was much too early for such a complex structure to have evolved naturally.

Later this year, astronomers announced another anomalous discovery. This time, they found individual galaxies at allegedly advanced stages of galactic 'evolution' in a part of the sky named the 'redshift desert.' They used the Gemini North Telescope, with an 8-metre mirror, on the summit of Mauna Kea on the big island of Hawaii.

This area of the sky is supposed to be so old and so close to the beginning of everything that it was believed nothing as complex as a galaxy should, or could, exist there.

Under big bang assumptions, astronomers looking into the redshift desert are seeing the universe as it was 8 to 11 billion years ago, at a time when it was 'only' 3 to 6 billion years old. This part of the sky had not previously been widely explored. Astronomers believed it contained objects too faint and dim to study properly. However, recent advances in telescope optics have allowed astronomers to make a systematic study of the redshift desert, the Gemini Deep-Deep Survey (GDDS), Figure 10.2.

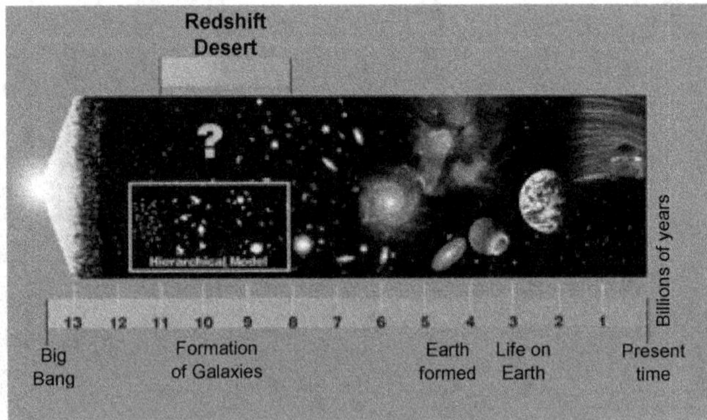

Fig.10.2: Gemini Observatory – Curtsy NASA

What the GDDS astronomers found was totally unexpected. Where they had expected to see young, small, still-developing galaxies, they found more than 300 fully mature galaxies, just like those seen near our own galaxy, the Milky Way.

Team member Dr "Karl Glazebrook" from Johns Hopkins University says the find presents a huge challenge because their 'star-forming youth is in fact long gone.' He explained:

'We expected to find basically zero massive galaxies beyond about 9 billion years ago, because theoretical models [based on the big bang] predict that massive galaxies form last. Instead we found highly developed galaxies that just shouldn't have been there, but are.'

This is a story that is sounding more and more familiar.

A CHRISTIAN BELIEVER VIEW

Thanks to new developments in Earth-based optical technology and orbiting telescopes such as the James Webb Telescope and the Hubble Space Telescope, astronomers have been able to detect fainter light from more distant objects. Therefore, they can probe the most distant reaches of space and detect objects so faint that astronomers 10 years ago did not even know they existed.

These new discoveries have shaken current theories of star and galaxy formation:

1. Elements thought to 'evolve' within the incinerators of ancient stars over many billions of years have been found 'only' 2.5 billion years after the big bang, under their own dating system.

2. Very complex strings of galaxies, claimed to be hundreds of light-years in size, have been found at a time when only small, isolated proto-galaxies should exist, Figure 10.3.

3. And now, massively complex galaxies and supermassive black holes have also been found too early in the evolutionary life of the universe to be explained by conventional theories. Consequently, what is the Christian believers' response to these latest, amazing discoveries?

In *Genesis 1:14–19*, God tells us when He created the heavenly bodies—the planets, stars and galaxies that make up our amazing universe. The passage teaches that God commanded, **'Let there be lights'**, and the command was fulfilled with rapid formation of these objects— 'and it was so'—all within **Day 4**. This is further reinforced in *Exodus 31:17*, **'for in six days the Lord made the heavens and the earth, and on the seventh day he abstained from work and rested'**. Also, *Psalm 33:6* declares, **'by the word of the Lord were the heavens made, the starry host by the breath of his mouth'.**

In the big bang model of the origin of the universe, galaxies started small. These small galaxies then began to collide. Eventually, after many billions of years, large, mature galaxies, like our own, were formed. This is called the hierarchical model of galaxy formation.

If the original heavenly bodies were created mature, then we would expect to see fully formed galaxies everywhere, even in the most distant parts of the universe. We should not be surprised to see massive strings of galaxies or to find supermassive black holes in all regions of space. In a nutshell, mature galactic structures are not a problem for creationist astronomers.

Australian physicist and Christian cosmologist, Dr "John Hartnett," says that these recent discoveries are very significant for a Christian understanding of the universe. 'This has enormous significance because [the big bang astronomers] are saying they don't see how such a structure could form so quickly according to the big bang model.'

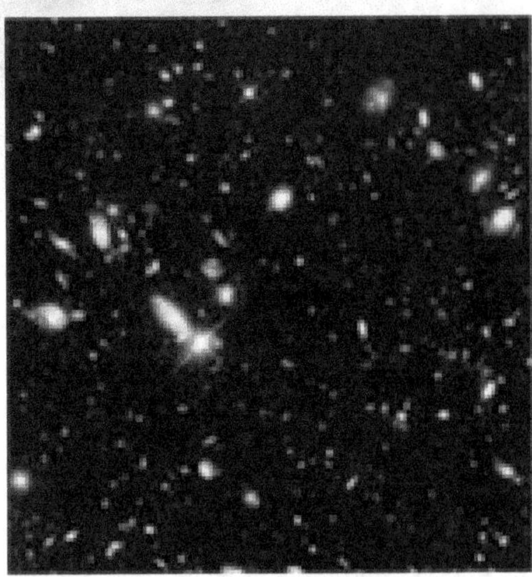

Fig.10.3: Very Complex Strings of Galaxies – Curtsy NASA

IMAGES CHALLENGE BIG BANG

Weighing over 11,000 kg at launch in 1990, this remarkable instrument has since been in low Earth orbit 600 km above the ground. Among the telescope's most amazing images is this 'deep-field' view, Figure 10.3, into the farthest reaches of the universe. The picture covers an area of the sky approximately equal to that covered by a small coin viewed from 23 m (75 ft) away. Over 1,500 galaxies have identified in this image, which took the camera 10 days of constant time exposures to capture, Figure 10.4. The galaxies are **four billion** times fainter than that which can be seen by the human eye.

Fig.11.4: Low Earth Orbit Telescope 600km above Earth – Curtsy NASA

Dr Hartnett believes that the redshift methods used to measure the distances to these objects are flawed. A growing list of evolutionary astronomers and cosmologists, such as Dr "Halton Arp," agrees that the big bang interpretations of the redshifts are flawed. "Arp" documented many pairs of objects that have greatly different redshifts, supposedly showing that they are vast distances apart and receding at hugely different speeds. Yet there is also connecting material between them, meaning that they must be the same distance away.

If the distances are wrong, then an object may appear small and dim not because it is incredibly distant, but because it really *is* small and dim. And faulty distances mean that any theory based on them—such as the big bang—is faulty too!

As telescope technology continues to improve and astronomers are able to probe more easily the darkest depths of the universe, it is likely more and more of these big-bang–challenging discoveries will arise.

These mature galaxies present a major problem for evolutionary scientists. However, the underlying models have become so flexible that it is only a matter of time until they are modified to explain away such problems. Nevertheless, for the biblical Christian, these discoveries, and others like them, are sound and exciting evidence in support of the biblical creation account. This account, unlike its evolutionary counterparts, is divinely inspired, so does not need any modification and change whenever new discoveries are made.

REDSHIFT ANSWERS

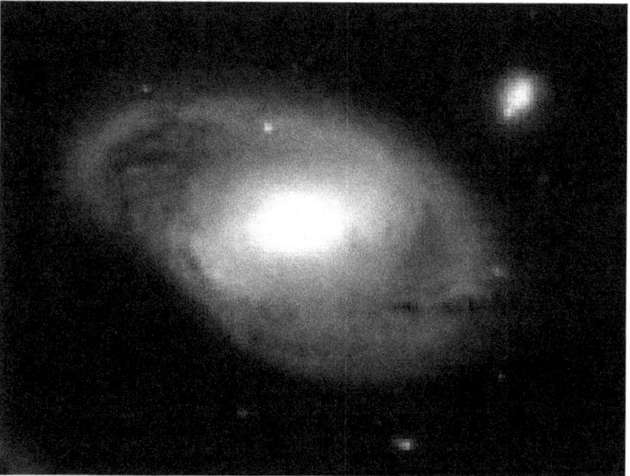

Fig.10.5: *Barred Spiral Galaxy, NGC 4319 – Curtsy NASA*

Barred spiral galaxy, Figure 10.5, NGC 4319, and the much smaller quasar, Markarian 205. Light from objects that are moving away from us is 'stretched' and shifted in color towards red (redshifted).

Fig.10.6: *"Halton Arp" Correctly Processed the Image of Barred spiral galaxy, NGC 4319*

According to the big bang idea, objects with greater redshifts are farther from us. From the redshifts, the quasar in the picture, Figure 10.5, should be much farther away than the galaxy. But the picture by the astronomer "Halton Arp," Figure 10.6, shows matter apparently bridging from the quasar to the galaxy, suggesting that they are close. NASA's recently- incorrectly, published image shows **no bridge** between the galaxy and the quasar. "Arp" and other experts say that when the original NASA image is adjusted appropriately, Figure 10.6, the bridge can be seen. Some other quasar–galaxy pairs show similar bridging. Such evidence would raise huge problems for big bang cosmology.

Image Processing experts said: The abnormality of the Red-Shist in Figure 10.7, prompts us to process this image. The result is in accordance with the claims of "Halton Arp," the bridge is there, and very much so.

Fig.10.7: The Supermassive Black Hole– Curtsy - NASA

SUPER MASSIVE BLACK HOLE – BLAZARS AT THE BEGINNING OF TIME

In another big-bang–defying discovery, astrophysicists from Stanford University claim to have discovered one of the biggest, most distant black holes ever found.

The supermassive black hole is uninspiringly dubbed Q0906+6930 after the coordinates at which it is found, Figure 10.7. Astronomers believe it exists at the center of an extremely distant galaxy in the direction of the northern hemisphere constellation Ursa Major (Great Bear). This galaxy is said to have an 'active nucleus.'

The black hole was detected by narrow jets of high-energy particles being ejected from its poles. Such jets are only visible when they are aimed exactly in the direction of the earth, meaning these types of objects—nicknamed blazars—are only rarely observed.

The black hole, Figure 10.7, is believed to be more than 10 billion times the mass of our Sun and *supposedly* formed 12.7 billion years ago, when the universe was 1 billion years old.

The big problem presented by this *blazar* is its size. In big bang terms, it is just too big to have formed in the 'mere' billion years since the big bang itself. The scientists behind the discovery have been challenged by its implications, 'How do you take something big enough to hold 1,000 solar systems and as heavy as all of the stars in our Milky Way galaxy put together, and quickly crunch-collapse it [in such a short period of time]?'

Of course, size and maturity are not a problem when the Bible, rather than man's fallible ideas, is used as a starting point.

Some Christian cosmologists believe that the type of galaxy supposedly containing this blazar played an important role in the initial creation process. It is possible that **galaxy creation on Day 4 of Creation Week** involved galaxies with active nuclei, i.e., black holes.

STAR WITNESSES YOUNG UNIVERSE

We cannot use science to *prove* the age of the universe because science can only deal with what is observable now. We can measure the rates of all manner of things in the present. However, to use these as 'clocks' to estimate ages, we have to assume a history, which in turn depends upon our *beliefs* about where we came from. The Bible gives us an eyewitness record of what happened, the order, and the timeframe, which 'science' cannot tell us.

Nevertheless, today's widespread belief in a very old universe fails to account for many 'clocks' that indicate a far younger age. Here are two examples.

I. Blue stars in galaxies

Blue stars are the biggest and brightest of all 'main sequence' stars, but this means they burn up their nuclear fuel very fast. Indeed, they burn so fast that the biggest ones could not last more than a million years, and the smallest around 10 million years. Yet blue stars abound in spiral galaxies, including our Milky Way. This suggests that these galaxies cannot be even one million years old. This problem for the belief that the galaxies are billions of years old is 'solved' by assuming the blue stars, Figure 10.8, formed more recently than the rest of the galaxy. However, no one has observed such star formation and there is not even a viable mechanism for it to happen.

Fig.10.8: Blue Stars Evident of Young Universe

II. Neutron Stars in Globular Clusters

Globular clusters are compact, ball-shaped groups of stars that orbit the center of a galaxy. They supposedly contain 'very old' stars. The secular big bang story has great difficulty explaining them. Astronomers have seen numerous fast-moving neutron stars in globular clusters. These are thought to arise from supernovas (exploding stars) within the cluster, where a neutron star is created that is 'thrusted' out at very high speed. With the compact sizes of globular clusters and the high speed of the neutron stars, all neutron stars should be ejected from such clusters in less than two million years. Many globular clusters should have emptied in a few *thousand* years. A major study of this so-called 'retention problem' called it a "long-standing mystery." These observations, too, are consistent with ***a young age of the universe***.

EXPLODING STARS REVEALS YOUNG UNIVERSE

Supernova Remnants

A *Supernova*, or violently exploding star, Figure 10.9, is one of the most brilliant and powerful objects in God's vast cosmos. On average, a galaxy like our own, the *Milky Way*, should produce one Supernova every 25 years.

When a star has exploded in this way, the huge expanding cloud of debris is called a *Super-Nova-Remnant* (SNR). A well-known example is the "Crab Nebula" in the constellation of "Taurus," produced by a supernova so bright that it could be seen during daytime for a few weeks in 1054AD. By applying physical laws, as well as using powerful computers, astronomers can predict what should happen to this cloud.

According to their model, the *Super-Nova-Remnant* (SNR) should reach a diameter of about 300 light years after 120,000 years. Accordingly, if our galaxy was billions of years old, we should be able to observe many SNRs this size. Nonetheless, if **our galaxy is 6,000–10,000 years old**, no SNRs would have had time to reach this size. Therefore, the number of observed SNRs of a particular size is an excellent test of whether the galaxy is old or young. In fact, the results are consistent with a universe thousands of years old, but are a puzzle if the universe has existed for billions of years. The conclusions can be seen from the simple Table 10.1, shown below, However, calculations are stated and detailed in a section below:

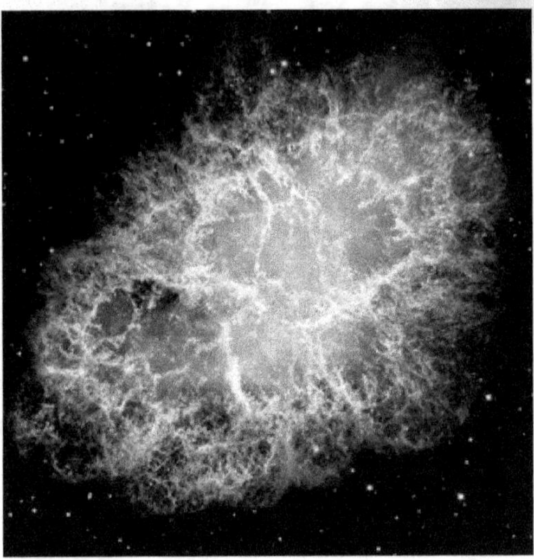

Fig.10.9: *Supernovas – A Mega–Explosion in Space. Some of these have been seen from the earth. The "Crab Nebula" here as it is today, is the remnant of a Supernova which was seen in the year 1054 AD and Remained Visible to the Naked Eye for about a year – Curtsy NASA*

Table 10.1 : Number of Observable Super-Nova-Remnants Predicted

Supernova Remnant Stage	Number of Observable SNRs Predicted If Our Galaxy Were!		Number of SNRs Actually Observed
	Billions of Years Old	7000 years old	
First	2	2	5
Second	2260	125	200
Third	5000	0	0

As can be readily seen above, a young universe model fits the data of the low number of observed SNRs. If the universe was really billions of years old, there are about 7000 missing SNRs in our galaxy.

Not only that, but the predictions for the Milky Way's satellite galaxy, the "Large Magellanic" Cloud are also consistent with a young universe. Theory predicts 340 observable SNRs if the Large Magellanic Cloud (LMC) were billions of years old, and 24 if it were 7000 years old. The number of actually observed SNRs in the LMC is 29.

As the evolutionist astronomers "Clark" and "Caswell" say, '*Why have the large number of expected remnants not been detected*?' and these authors refer to '*The mystery of the missing remnants.*'

There should be no mystery—*Psalm 19:1* says: '*The heavens declare the glory of God; and the firmament sheweth his handiwork.*' Supernovas declare His mighty power, but are still only finite expressions. The low number of their remnants is a pointer to God's *recent* creation of the heavens and earth.

THE DEVELOPMENT OF THE SUPERNOVAS

An ordinary star is a gigantic ball of gas, about a million times more massive than the earth—our sun is a medium-sized star. It is potentially stable for a long time, because the energy produced by the core produces an enormous outward pressure, which balances the inward force of gravity on its huge mass.

However, when the nuclear fuel runs out, there is no longer any force to balance its gravity. If the star is very massive, most of it collapses very fast — in about two seconds. This releases a huge amount of energy—one supernova will out-shine all the billions of stars in its galaxy. The collapse is so violent that the electrons and nuclei are crushed together and produce a core of neutrons. This core is so dense that a *teaspoonful would weigh 50 thousand million* tons on earth. It cannot be compressed any further, so the incoming material from the rest of the star meets a solid wall. This material bounces off the core, rushes outward and shines very brightly. The remaining core, only about 20 km in diameter, is called a *neutron star*. Because it is spinning very fast, and has a strong magnetic field, we observe regular radio pulses, so the object is called a *pulsar*.

The energy produced by a supernova is mind-boggling: 10^{44} joules. It is the same as if each and every gram of the earth's mass was converted to a nuclear bomb 200 times more powerful than the one dropped on Hiroshima. That amount of energy would fuel 80 million sun-like stars for 100 years!

CALCULATIONS DETAIL

A widely-accepted model of supernova expansion predicts three stages:

The First Stage: Starts with debris hurtling outwards at 7000 kilometers per second, Figure 10.10. After the material has expanded for about 300 years, a blast wave forms, ending the first stage. By this time, it reaches a diameter of about 7 light years. This is an immense object—about 25,000 times larger than our solar system, which is 'only' about eight light hours across (about 8600 million km or 5400 million miles).

Since the first stage should last about 300 years and one SNR should occur every 25 years, there should now be 300/25 first stage SNRs in our galaxy, or about 12. We should not expect to see them all—astronomers calculate that only about 19% of SNRs should be visible, that is about two of the 12. It makes no difference whether the universe is thousands of years old as the Bible indicates, or billions of years old as evolutionary theory asserts. Actually, we see five first stage SNRs (this is within the uncertainty range of the calculation).

The Second Stage: Super-Nova-Remnants (SNR), known as the adiabatic or "Sedov" stage, is a very powerful emitter of radio waves, Figure 10.10. This is predicted to expand for about 120,000 years and reach a diameter of about 350 light years. After this, it starts to lose thermal (heat) energy and begin the third stage.

Now, if the universe was billions of years old, we would predict (remember, one supernova every 25 years, and taking into account SNRs in the 300-year first stage) that in our galaxy there would be about (120,000–300)/25 second stage SNRs, or about 4800. However, if the universe has only existed for about 7000 years, then there would be only enough time for (7000–300)/25, or about 270. Astronomers calculate that 47% should be visible, so evolutionary/uniformitarian theory predicts about 2260 second stage SNRs, while the

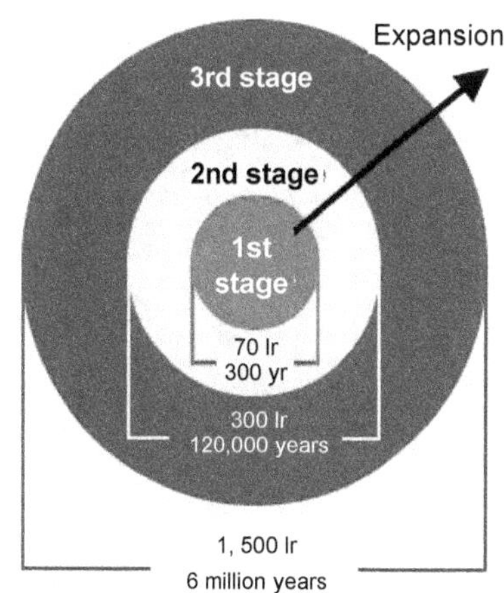

Fig.10.10: The Three Predicted Stages of Supernova Development

Biblical Creation theory predicts about **125**. The actual observed number of second stage SNRs is a good test of which theory best fits the facts.

There are actually only 200 second stage SNRs observed in our galaxy! **This is in the right ball park for Biblical creation, but is totally different from evolutionary predictions.** Evolutionists at present have no answer to the problem of the missing supernova remnants.

The Third Stage: or "Isothermal," stage is theorized to emit mainly heat energy, Figure 10.10. This stage would theoretically only start after 120,000 years, and would last about one million to six million years. The SNR would end its career when it either collided with similar SNRs at a diameter of about 1400 light years, or became so dispersed that it would be indistinguishable from the 'vacuum' of space at a diameter of about 1800 light years.

One calculation makes the generous (to evolutionary theory) assumption that the third stage starts at about 120,000 years and a diameter of about 340 light years, and lasts to an age of one million years and 650 light years. Thus, if the universe was billions of years old, there should be (1,000,000–120,000)/25 third stage SNRs in our galaxy, or about 35,000. Of these, about 14% should be observable, or about 5000. However, if the universe is only about 7000 years old, no SNR should be old enough to have reached the third stage, so there should be absolutely none, under currently accepted models. This is another test of the two theories, an **old** *vs.* **a young** universe. There are actually no third stage SNRs observed in our galaxy!

THE SUN – PRECIOUS STAR

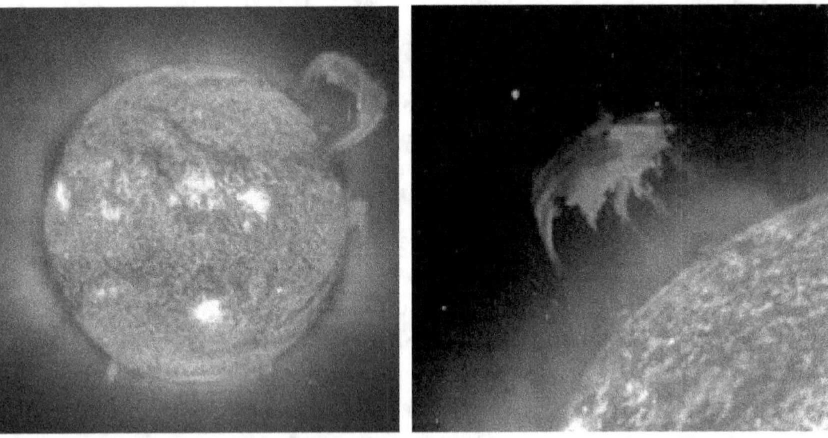

Fig.10.11: The earth is Seen (on right) Dwarfed here in Approximate Relative Size to the Sun. The Massive Fiery Plumes (known as coronal ejections) Seen here Would Encompass the Earth Many Times over – Curtsy - NASA.

The sun, Figure 10.11—this hot, bright ball of plasma dominates the daytime sky, and is by far the most massive object in our solar system. It provides heat and light to earth; it is no ordinary star.

The Origin of the Sun in Our Galaxy

According to God's Word, the Bible, the sun did not always light the earth. It wasn't made till Day 4 of Creation Week, while the earth was created on Day 1. This refutes ideas like '*God used evolution*' and '*God created over billions of years*', because they all assert that the sun arose before the earth.

For the first three days of existence, the earth was lit by the light emanated from the Glowing of God's Glory, "**Let there be Light**, *Genesis 1:3*." This was not created light, as many Christian scholars claim. The Scripture provides clear attributes of God. One of His distinctive attributes is "**God is Light**," 1 John 1:5. God does not create Himself as He is the Essence of the Glory of Light. Accordingly, when God desired to illuminate the world on Day 1, He said "Let there be light," and He did not say let us create light on Day 1. The earth hemisphere that faced the glorious light of God was illuminated. We do not know for certain, if the Lord allowed the Earth to be rotated on the first day or on the fourth day. However, we clearly know that the glorious light of God was illuminating the world from Day 1 until day 4 where the sun and the stars were created. Additionally, the Earth was created on Day 1, while the sun was created on Day 4. The rotation of the of the Earth was referenced to the precise location of the sun. Therefore, it is logical to assume that the Earth was not made to rotate on the first day, nevertheless it was made to rotate on the 4th Day.

We may recall that when the Israelites departed Egypt the Lord guided them by a cloud by Day and pillar of fire by night. The fire and the cloud did not appear prior the necessity of God's guidance. Similarly, it is likely that the Earth rotation took place precisely as the sun was made. We assume the Glory of God's light either to illuminate the world He created for a period of time He called it Day, and cease His light for a period of time representing night.

Alternatively, He may Have allowed the Earth to rotate while His Glorious light never ceased for 4 Days, until the sun and the sars were created on the 4th Day.

On Day 1 (*Genesis 1:3*), while the day/night cycle was probably caused by the earth's rotation relative to this glorious directional light source. Then according to Genesis 1:16-19, "God made two great lights—the larger one to govern the day, and the smaller one to govern the night. He also made the stars. God set these lights in the sky to light the earth, to govern the day and night, and to separate the light from the darkness." And God saw that it was good.

It is possible that the earth was made to rotate on Day 4. And evening passed and morning came, marking the fourth day.

The Glory of Gods Light

Nevertheless, the glorious light of God will return to Earth to directly illuminate the entire world without the need of the sun. It is stated in (*Revelation 21:23*), in New Jerusalem, there will also be no need for the sun, because God will provide the light once again. Meanwhile, we can appreciate the wonder of the Sun-star God has provided for us.

How Precious is the Sun

Anti-theists are fond of dismissing the sun as a run-of-the-mill star in a not-too-special place in a galactic spiral arm. It is true that many stars are far bigger and brighter than the sun. However, saying that bigger stars are more important is as illogical as saying that a 7–foot man is more important than a 5–foot woman.

Recent research has called the sun 'exceptional.' Our sun is among the top 10% (by mass) of stars in its neighborhood. It is actually an ideal size to support life on earth. There would be little point in having a red supergiant star like Betelgeuse, Figure 10.12, because it is so huge that it would engulf all the inner planets! Nor would we want a star like the blue-white supergiant Rigel, 25,000 times as bright as the sun, and emitting too much high-frequency radiation, Figure 10.13.

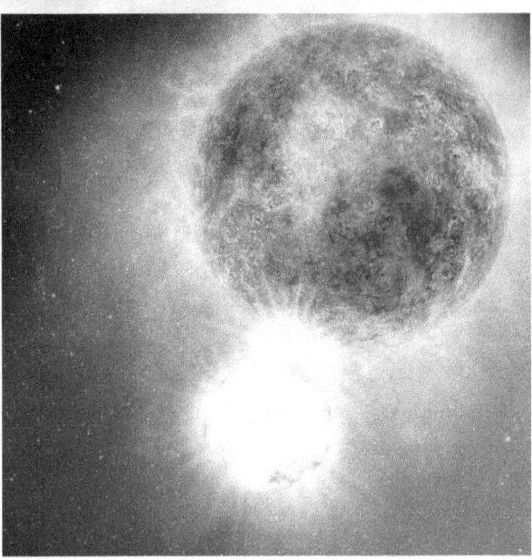

Fig.10.12: Supergiant Star Betelgeuse Blew Its Top in a Violent Explosion, Baffling Scientists

Conversely, a star much smaller than our sun would be too faint to support life, unless the planet were so close to the star that there would be dangerous gravitational tides.

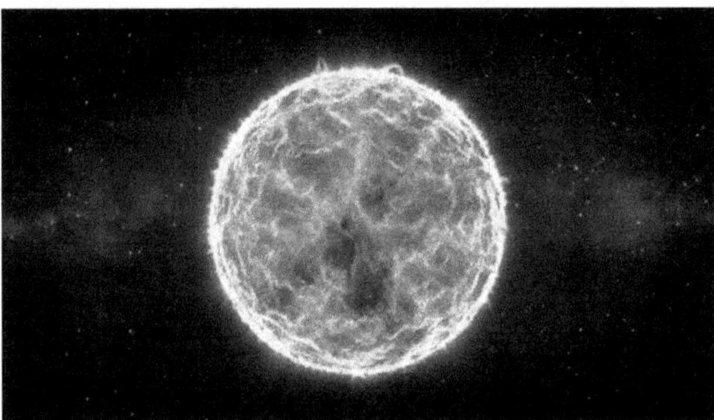

Fig.10.13: Rigel, A Magnificent Bluish-White Supergiant Located 772.9 Light Years Distant

The sun is in an ideal environment. It is a single star—most stars exist in multiple-star systems. A planet in such a system would suffer extreme temperature variations. The sun's position in our spiral Milky Way Galaxy is also ideal. Its orbit is fairly circular, meaning that it won't go too near the inner galaxy where supernovae, extremely energetic star explosions, are more common. It also orbits almost parallel to the galactic plane—otherwise, crossing this plane would be very disruptive. Furthermore, the sun is at an ideal distance from the galactic center, called the *co-rotation radius*. Only here does a star's orbital speed match that of the spiral arms—otherwise the sun would cross the arms too often and be exposed to supernovae.

Our sun is a powerful object, often throwing out flares, and every few years (usually around sunspot maximum—Sunspots, Galileo and heliocentrism), more violent ejections called coronal mass ejections, Figure 10.11. They cause huge electric currents in earth's upper atmosphere and disrupt power grids and satellites.

In 1989, one disabled a power grid in northern Quebec. But the sun turns out to be an 'exceptionally stable' star. Three astronomers recently studied single stars of the same size, brightness and composition of the sun. Almost all of them erupt about once a century in *super-flares* 100 to 100 million times more powerful than the one that blacked out Quebec. If the sun were to erupt in such a super-flare, it would destroy earth's ozone layer, with catastrophic results for life, Table 10.2.

Table 10.2 : Sun Facts

Mean distance from earth	149,600,000 km or 92,937,000 miles (1 astronomical unit (AU))
Diameter	1,392,000 km or 864,950 miles (10^9 × earth)
Mass	1.99×10^{30} kg (330,000 × earth)
Volume	$10^{18} \times 1.412$ (1,300,000 × Earths)
Mean density	1.41 g/cm^3 (1/4 earth)
Temperature	5,470 °C (9,880 °F) surface, 14,000,000 °C (25,000,000 °F) core
Power output	3.86×10^{26} watts
Escape velocity at surface	618 km/sec or 384 miles/sec (55 x earth)
Rotational period (days)	26.9 (equator), 27.3 (sunspot zone, 16°N), 31.1 (pole), all synodical[25]

HOW DOES SUN SHINE

In 1939, "Hans Bethe" proposed that the sun and other stars are powered by *nuclear fusion*—this theory earned him the 1967 Nobel Prize for Physics. In fusion, extremely fast-moving hydrogen nuclei join to form helium—this requires temperatures of millions of degrees. Some mass is lost and converted into a huge amount of energy as per Einstein's famous formula $E = mc^2$

Thus, the sun would be essentially a gigantic hydrogen bomb. If fusion were totally responsible for the sun's huge power output of 3.86×10^{26} watts, four million tons of matter would be converted every second into energy—this is huge, but negligible compared to the sun's enormous total mass.

That fusion is responsible for at least part of the sun's energy output is supported by the sun's huge flux of *neutrinos*; ghostly particles that can usually pass through light-years thicknesses of matter untouched.

However, if nuclear fusion were the *sole* source of power, then we would expect to observe three times more neutrinos than we do. This shortfall has been tentatively explained by the idea that neutrinos alternate between three types. This would require that they have mass, although previously they were universally regarded as massless.

Alternatively, two-thirds of the sun's energy could be provided by *gravitational collapse*, through conversion of gravitational potential energy to heat and light as the sun's gases collapse inwards. This theory was proposed by the great physicist "Hermann von Helmholtz" (1821–1894). It was the chief theory until the prominence of Darwinism, which could not tolerate that it would put an upper limit on the sun's age at 22 million years—far too short for evolution.

Observations suggesting the sun is shrinking at a rate of at least 0.02 seconds of arc per century, give some support to the notion. This would be ample for collapse to be a significant energy source. But the shrinkage is controversial, even among Christian believers. In any case, since nuclear fusion is at least a partial source of energy, Helmholtz's age limit cannot be strictly applied.

Astrophysicists Phillip F. Schewe, Ben Stein, and James Riordon published their findings in *The American Institute of Physics* on 24 April 2002. Their research seems to provide conclusive evidence for neutrino oscillation. Previously, detectors were able to pick up only electron neutrinos. But this new experiment at the Sudbury Neutrino Observatory (SNO) were able to detect the missing neutrino flavors, the mu and tau neutrinos that undergo 'neutral current' reactions.

This is consistent with other lines of evidence that fusion is the primary source of energy, e.g., standard physical models indicate that the core temperature is high enough for fusion. This means that neutrinos must have a very tiny rest mass after all—experimental data must take precedence over the theories of particle physicists that neutrinos have zero rest mass. Therefore, Christian believers should *no longer* invoke the missing neutrino problem to deny that fusion is the primary source of energy for the sun. Therefore, it cannot be used as a young-age indicator—nor an old-age indicator for that matter.

However, the solar astronomer "John Eddy" commented:

'*I suspect … that the sun is 4.5 billion years old. However, given some new and unexpected results to the contrary, and some time for some frantic recalculations and theoretical readjustment, I suspect that we could live with Bishop Ussher's value for the age of the earth and sun [about 6,000 years]. I don't think there is much in the way of observational evidence to conflict with that.*'

EVOLUTIONARY PROBLEMS WITH THE SUN

Evolutionists believe that the solar system formed from a cloud of dust and gas 4.5 billion years ago. This *nebular hypothesis* has many problems. One authority summarized: '*The clouds are too hot, too magnetic, and they rotate too rapidly.*'

One major problem can be shown by accomplished skaters spinning on ice. As skaters pull their arms in, they spin faster, Figure 10.14. This effect is due to what physicists call the *Law of Conservation of Angular Momentum*.

Angular momentum = mass × velocity × distance from the center of mass,

and always stays constant in an isolated system. When the skaters pull their arms in, the distance from the center decreases, so they spin faster or else angular momentum would not stay constant.

In the formation of our sun from a nebula in space, the same effect would have occurred as the

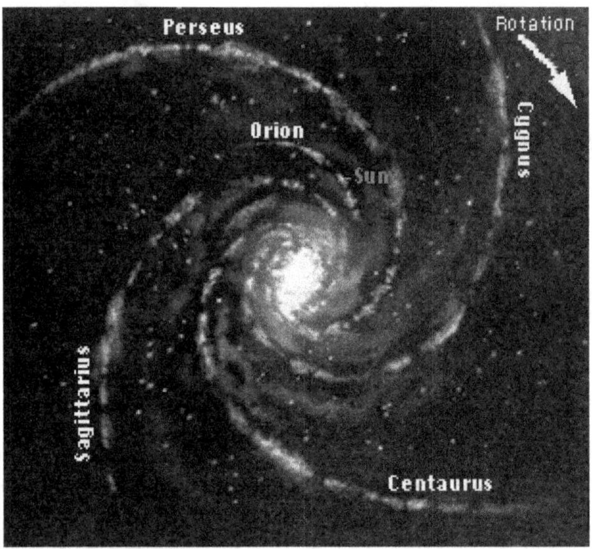

Fig.10.14: The Solar System Orbits around the Center of the Milky Way Galaxy

gases allegedly contracted into the center to form the sun. This would have caused the sun to spin very rapidly. Actually, our sun spins very slowly, while the planets move very rapidly around the sun. In fact, although the sun has over 99% of the mass of the solar system, it has only 2% of the angular momentum. This pattern is directly opposite to the pattern predicted for the nebular hypothesis. Evolutionists have tried to solve this problem, but a well-known solar-system scientist, Dr "Stuart Ross Taylor," has said in a recent book, '*The ultimate origin of the solar system's angular momentum remains obscure.*'

Another problem with the nebular hypothesis is the formation of the gaseous planets. According to this theory, as the gas pulled together into the planets, the young sun would have passed through what is called the *T-Tauri phase*. In this phase, the sun would have given off an intense solar wind, far more intense than at present. This solar wind would have driven excess gas and dust out of the still-forming solar system and thus there would no longer have been enough of the light gases left to form Jupiter and the other three giant gas planets. This would have left these four gas planets smaller than we find them today.

Sunspots, Galileo and Heliocentrism

Sunspots look like dark patches on the sun. They can be seen to move, and analyzing them shows that different parts of the sun rotate at different rates, unlike a solid body. Sunspots come and go in cycles of about 11.2 years. Galileo Galilei (1564–1642), Figure 10.15, systematically studied sunspots in 1611 and realized that they upset the prevailing Aristotelian/Ptolemaic view that the heavenly bodies were 'perfect spheres.'

Today we realize that sunspots are vortices of gas on the sun's surface, and appear dark because they are several thousand degrees cooler. Analysis of their light spectra shows that the sun's magnetic field is especially strong in sunspots.

Galileo supported the theory of Nicolaus Copernicus (1473–1543) that the earth and other planets move around the sun. Anti-Christian propagandists make much of the conflict between Galileo and the Church,

Fig.10.15: Galileo Galilei (1564–1642)

or religion *vs* science. But Galileo thought that the much simpler mathematics of the Copernican system compared to the unwieldy Ptolemaic system would best reflect God's mathematical simplicity (i.e., God is not composed of parts but is Triune). *The New Encyclopædia Britannica* identifies Galileo's main opponents as the scientific establishment:

'The Aristotelian professors, seeing their vested interests threatened, united against him. They strove to cast suspicion on him in the eyes of the ecclesiastical authorities because of [alleged] contradictions between the Copernican theory and Scriptures.'

Both sides should have realized that all movement must be described *in relation to something else*—a *reference frame*—and from a descriptive point of view, *all reference frames are equally valid*. The Bible writers used the *earth* as a convenient reference frame, as do modern astronomers talking about 'sunset'; speed limit signs also depend on the earth as a reference frame. Using the sun (or the center of mass of the solar system) is the most convenient for discussing planetary motions.

Eclipse!

Fig.10.16: A Time-Sequence Composite of the August 21, 2017 Total Solar Eclipse (Photo Curtsy: VW Pics/Universal)

On 11 August 1999, large numbers of people from England to India were fortunate to behold the awesome sight of a total eclipse of the sun, Figure 10.16. This is possible because the moon is almost exactly the same angular size (half a degree) in the sky as the sun—it is both 400 times smaller and 400 times closer than the sun. This looks like design.

The moon is gradually receding from the earth at 4 cm (1½ inches) per year. If this had really been going on for billions of years, and mankind had been around for a tiny fraction of that time, the chance of mankind living at a time so they could observe this precise size match-up would be remote. (Actually, this recession puts an upper limit on the age of the earth/moon system at far less than the assumed 4.5 billion years).

During a total eclipse, the sun's outer atmosphere, the *corona*, is visible. This comprises extremely thin ionized gas, which is extremely hot. At 2 million °C (3.6 million °F), it is about 350 times as hot as the sun's surface. This has been a mystery, because heat normally flows from hot objects to cooler ones. A promising theory (which still needs work) involves the sun's strong magnetic field—reconnection of magnetic flux lines could release large amounts of energy into the corona. This could have applications in fusion power research. Recent photographs show that the coronal loops comprise several finer loops, and that they are heated strongly at the base. A new model has the gas, mainly ionized iron, travelling upwards for 400,000 km at 100 km/sec then cooling as it crashes back down on the sun's surface.

JESUS ON THE AGE OF THE EARTH

People were There from the Beginning

Jesus declared a young world, but leading theistic evolutionists say He is wrong. The standard secular timeline, from an alleged 'big bang' some 15 billion years ago to now, is accepted by most people, even in the evangelical Christian world, even though many would deny evolution. Some would even say that to dispute billions of years is to place an unnecessary stumbling block in the way of any scientifically-minded potential converts.

Sadly, this is in contrast to the teaching of the Lord Jesus Christ, the Creator made flesh, as well as several of the biblical authors, which makes it plain that this is wrong—people were there *from the beginning* of creation, Figure 10.17. But in the evolutionary timeline, people have only been around for one or two million years—this puts them toward the *end* of the timeline. This means that He is most definitely claiming that the world *cannot* be billions of years old.

Year	Person born	World Population	Jews
4000 BC	Adam	7,000,000	0
3000 BC	Noah	14,000,000	0
2000 BC	Abraham	27,000,000	0
1500 BC	Moses	35,000,000	2,000,000
1000 BC	David	50,000,000	2,700,000
500 BC	Nehemiah	100,000,000	2,800,000
—	Jesus	250,000,000	3,000,000
570 AD	Muhammad	230,000,000	3,200,000
1483 AD	Luther	400,000,000	5,000,000
1732 AD	Washington	900,000,000	7,000,000
1882 AD	Chester A. Arthur	1,500,000,000	7,800,000
1942 AD	Roovevelt	2,200,000,000	11,000,000
2021 AD	[Now]	7,900,000,000	15,000,000

Fig.10.17: People were There from the Beginning of Creation.

For example, dealing with the doctrine of marriage, Jesus says in *Mark 10:6*

"But **from the beginning of the creation**, God made them male and female."

In Luke 11:50–51, Jesus also says: "That the blood of all the prophets, which was shed **from the foundation of the world**, may be required of this generation; From the blood of Abel to the blood of Zacharias … ". And in *Romans 1:20*, the Apostle Paul says of God: "For his invisible attributes, namely, his eternal power and divine nature, have been clearly perceived, **ever since the creation of the world**, in the things that have been made. So, they are without excuse."

Paul is plainly saying that people have been able to perceive these attributes of God in His creation ever since the creation of *the world*. Not ever since people were created.

Comparing the appearance of people on the timelines, Figure 10.18 below, which are both to scale, is instructive. Jesus, speaking around 4,000 years after creation, was correct to say that Day 6, when humans were created, was effectively '**the beginning of creation**' as seen from thousands of years later. By contrast, a creation fifteen billion years ago on the secular timescale would put humans at the *end* of the time scale. It shows clearly how the acceptance of the secular timeline starkly contrasts with the statements of Jesus.

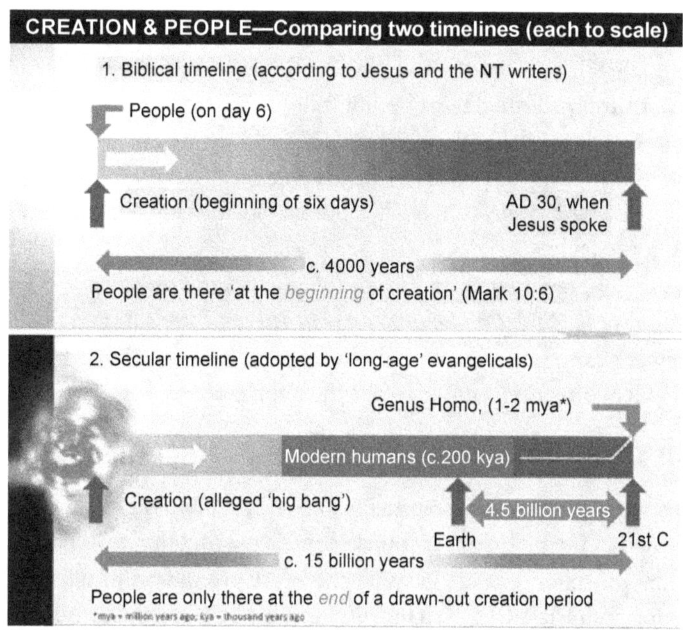

Fig.10.18: *Creation Timeline Vs Big Bang Timeline – Curtsy CMI*

Today, the vast majority of Christians is not only secular academia, but also theological institutions, Bible colleges, etc. believe—and many teach—that the secular 'billions of years' is fact. When one tries to find out how they deal with these repeated references, responses vary. But the 'explaining away' that takes place (whenever the problem is not simply ignored) invariably makes it plain that the authority being deferred to is not the **Word of God**, but rather current **secular opinion**.

The most striking (and sad) example of this switch in authority source I know of comes from a personal experience. In New Jersey, USA, several years ago, I had arranged to sit down over a hot drink with a distinguished university professor, a Christian who was well-known for his active opposition to a straightforward view of Genesis. At that time, he was actually the head of a grouping of Christian academics which had been openly set up to provide opposition to the inroads our ministry was making. Over the years, this group has unfortunately been very effective in persuading most Christian training institutions that compromising on biblical creation in favor of secular thinking (evolution, long ages) is the only 'respectable' position.

This professor himself, in addition to his secular science qualifications, was well regarded in the theological arena as well as being very biblically literate. He had at that time already been a frequent guest lecturer at several leading American evangelical training institutions.

During our courteous exchange, I asked him about the above comments by Jesus in relation to the age of the world. I asked, "Isn't it clear that Jesus taught and believed that the world was young?"

A STUNNING RESPONSE

I expected him to do as other Christian evolutionists have done—to try to find ways to torture the text to escape these obvious implications. Instead, he said that he totally agreed that Jesus believed in a recent creation of all things.

Somewhat taken by surprise, I said, "Well, how do you deal with that, then?" (He would of course have assumed, correctly, that I knew of the long-age position of this prominent organization of theistic evolutionists.) His answer simply stunned me, to put it mildly. He said:

"*Jesus didn't know as much science as we do today.*" His words burned themselves indelibly on my memory, while the recollection of my response has faded somewhat. But I recall saying something about Jesus being the Creator, God made flesh; He was there at creation, He does not lie, that sort of thing. To which his reply was once again unforgettable:

"*Ah, but that's where it gets very complex—it has to do with the theology of the Incarnation, where Jesus deliberately laid aside many of the things that had to do with His pre-incarnate divinity.*"

Our conversation was nearing the end of its allotted period in any case, but I recall being so stunned by this that it took me till well afterwards to fully process the implications.

What does it All Means?

Firstly, and very importantly, the professor's comments were a clear admission that the words of the Lord Jesus Christ Himself, as recorded in the Bible, confirm that He believed that things were recently created.

Remember that this professor was at the time the most prominent of all the professing evangelical academics that were being enthusiastically welcomed into Bible colleges and seminaries—to tell them why it was OK to believe in evolution and long ages. He obviously saw it as hopeless to try to claim other than what the Lord is clearly saying in this Bible text. And this is despite many attempts by others to 'explain away' this huge stumbling block for long-agers.

His way of being able to hold onto his theistic evolutionary view was to claim that Jesus was not lying, it was just that He was poorly informed. This was because when He as God the Son became flesh, laying aside aspects of His divinity included divesting Himself of all knowledge about what really happened when He had created all things.

If I had the presence of mind, an appropriate response might have been to ask something like the following:

"*OK, let's assume for the sake of the argument that firstly, creation was by evolution, over millions of years of death and suffering—and that Jesus did perform some sort of lobotomy on Himself, so that He could no longer recall what really took place. Therefore, He just understood Genesis in the most natural straightforward way, not realizing what the real truth was. What you're claiming in that case amounts to this: That God the Father, knowing the real truth, permitted not just the Apostles, but His beloved Son, while on Earth, to believe and teach things that were utter falsehoods. Furthermore, it means that the Father permitted these false teachings to appear—repeatedly—in His revealed Word. With the result that for some 2,000 years, the vast majority of Christians were seriously misled about such things as not just the time and manner of creation, but gospel-crucial matters such as the origin of sin, and of death and suffering."*

The Lord Jesus repeatedly made it clear that His words and actions were on the Father's authority, in all respects. Some examples are firstly *John 8:28*: So, Jesus said to them, "*When you have lifted up the Son of Man, then you will know that I am he, and that I do nothing on my own authority, but speak just as the Father taught me*". And *John 12:49–50*: "*For I have not spoken on my own authority, but the Father who sent me has himself given me a commandment—what to say and what to speak. And I know that his commandment is eternal life. What I say, therefore, I say as the Father has told me.*"

One thing is very clear from all this. Namely, that the erroneous belief that 'science' insists that evolution and long ages are 'fact' is the most serious challenge to biblical authority, and thus to the faith in general, that Christendom has ever faced. If even Jesus' words in Scripture can't be trusted on some issues, how are we supposed to trust anything in the Bible at all? See also about the 'kenotic heresy'.

Other leading theistic evolutionists have similarly made plain their belief that Jesus was mistaken. For example, on the American theistic evolutionary site BioLogos, led by Francis Collins, there appeared the following:

"If Jesus as a finite human being erred from time to time, there is no reason at all to suppose that Moses, Paul, John wrote Scripture without error. Rather, we are wise to assume that the biblical authors expressed themselves as human beings writing from the perspectives of their own finite, broken horizons."

This is all the more serious because Jesus and the apostles used the history, they taught to back up the theology that they taught. The Resurrection (1 Corinthians 15), marriage (Mark 10:1–12), atonement (Romans 5:12–21), and Heaven (Revelation 21–22:5) are only a few of the areas in which compromising Christians are theologically crippled, because they don't have the same strong stand on Genesis that Jesus and the apostles did when they taught about these areas.

What a tragedy that so many Christian leaders have been bluffed and intimidated into assuming that secular interpretations of the evidence should dictate their understanding of God's Word. And right at a point in history when there are more scientific reasons than ever to confirm the utter rationality of trusting the Bible, not evolutionary conclusions.

THEISTIC EVOLUTION AND THE KENOTIC HERESY

Kenoticism, also known as kenotic theology or kenotic Christology, is an unbiblical view of Christ's nature!

This error from many leading theistic evolutionists is not a new idea. It was rejected by the Church in general as the **kenotic heresy** in the 4th Century already, but has been revived in modern times, and for reasons as shown in the main text.

This asserts that in the Incarnation, Jesus emptied Himself of divine attributes, which is a misunderstanding of *Philippians 2:6–7*:

"[Jesus] Who, being in very nature God, did not consider equality with God a thing to be grasped; rather, he emptied Himself by taking the very nature of a servant, being made in human likeness."

This does indeed talk about 'emptying' (kenosis), but what does it actually *say*? "He emptied Himself by taking … ". That is, He didn't empty anything *out of* Himself, such as divine attributes; rather, His emptying of

Himself was by *taking*. That is, it was a subtraction by means of adding—*adding human nature* to His divine nature, not taking away anything divine.

This is what makes our salvation possible: he "shares our humanity" (*Hebrews 2:14–17*), and is our "kinsman–redeemer" (*Isaiah 59:20*); but He is also fully divine so He can be our Savior (*Isaiah 43:11*) and can bear the infinite wrath of God for our sins (*Isaiah 53:10*), which no mere creature could withstand.

But on Earth, Jesus *voluntarily* surrendered the *independent* exercise of divine powers like omniscience without His Father's authority. But Jesus never surrendered such *absolute* divine attributes as His perfect goodness, mercy, and (for our purposes), *truth*, so He would never teach something false. Furthermore, Jesus preached with the authority of God the Father (*John 5:30, 8:28*), who is always omniscient. Therefore, these theistic evolutionists really must charge God the Father with error as well.

WAVES OF GALAXIES—THE FALSEHOOD OF BIG BANG

Astronomers have recently claimed to detect a 'wave' pattern in the clustering of galaxies in the "Sloan Digital Sky Survey (SDSS)." They claim this pattern is a result of sound waves produced during the big bang. However, as with all things, it is important to distinguish between the **data** and the **interpretation.** The new discovery does not support the "big bang," and is in fact perfectly consistent with Biblical Creation.

Contextual

All the stars you observe in the night-time sky are part of the Milky Way Galaxy—a large spiral collection of over one-hundred billion stars. The universe contains many such galaxies: some smaller than ours, others much bigger.

Spider Web Galaxy Clusters

Galaxies are organized into clusters, which, in turn, are organized on an even larger scale forming a large non-uniform structure of filaments and voids. You can think of this like a gigantic, irregular spider-web; the galaxies exist primarily along the strands of the web, with fewer in between.

New Detection

Until recently, the galaxy clustering did not show any well-defined pattern or size scale; filaments of galaxies connect in seemingly random ways and come in many different sizes. But investigators have apparently discovered a weak 'pattern' in the arrangement of galaxies.

Galaxies have a very slight preference to be separated by 500 million light-years (five billion-million-million kilometers) than other distances, according to "Sloan Digital Sky Survey (SDSS)" astrophysicists. This pattern is extremely weak; you would not be able to see it by eye. The SDSS researchers have used some mathematical techniques to extract this ethereal pattern.

Data vs Interpretation

This subtle organization of galaxies is the **data**. The **interpretation** that many astronomers have offered is that sound waves from the big bang produced this pattern. Let's examine this interpretation.

In the "big bang" story of origins, the universe starts out very small and very dense. Some regions are slightly denser than others. This imbalance creates pressure waves (sound) which propagate through the early universe. Much like a rock thrown in a pond causes ripples to expand, imagine many rocks being thrown in at the same time. The interaction of all the waves would cause a complicated, irregular pattern of ripples. In the big bang model, the sound propagating in the early universe creates regions of greater density. Eventually, gravity causes these denser regions to collapse to form stars and galaxies as the universe expands. So, in essence, the sound waves act as 'seeds' for galaxies to form, Figure 10.19.

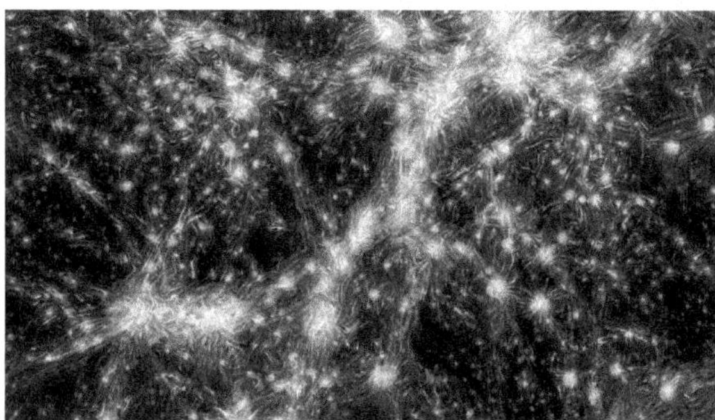

Fig.10.19: *Simulation of the Cosmic Web, Shock Waves along Filaments and around Clusters Emit Radio light (pink) as They Ripple through Magnetic Fields (cyan).*

Secular astronomers believe that the weak pattern detected in galaxy locations (the data) is a result of the sound waves from the big bang (the interpretation). Notice that this interpretation simply ***assumes*** that the big bang is true. The biases of the researchers have affected their interpretation of the data. The evidence has been interpreted to match their beliefs.

The big bang, however, has been **refuted** on the basis of both **Scripture** and good **science**. For example, the big bang is not compatible with the order, timescale and cause of the events of creation as recorded in Genesis. Really, the big bang is a secular opponent of the biblical framework.

So, this weak cluster-pattern of galaxies does *not* support the big bang with its billions of years. On the contrary, the **big bang** is simply **assumed** in order to explain this clustering within a naturalistic framework.

Furthermore, the big bang is not the only unwarranted assumption involved in the 'sound waves' interpretation. The secular explanation also assumes that stars and galaxies can form from regions of high density. But this has never been observed. **In fact, no galaxy has ever been observed to form at all!** And there are substantial scientific difficulties in getting stars to form from collapsing gas clouds. No wonder that even many secular scientists blast the big bang.

Consistent with creation

From a biblical creation view, there is no reason to think that the clusters of galaxies were formed by sound waves at all. We know from Scripture that **God made the stars** (and thus the galaxies, which are composed of stars) on **Day 4** of the Creation Week (*Genesis 1:16*). It may be that the galaxies were organized in a non-random way by the Creator's hand for His pleasure. The subtle pattern of galaxy locations (if confirmed) would be perfectly consistent with the order and creativity we have come to expect from the God of Scripture.

THE UNIVERSE HAD A BEGINNING

As an idea, the 'big bang' just doesn't cut it—that is, it is no longer sufficient for its intended task. In fact, it never was.

What *was* its intended task? To provide a godless means of explaining the origin of the cosmos. However, as we have earlier reported, even diehard atheist physicists are abandoning the big bang, given its increasingly evident failures to fit the known facts of the universe.

Nonetheless, there's another reason why "many physicists have been fighting a rearguard action against it for decades", as a recent editorial in *New Scientist* explained.

That is, its alleged "theological overtones". The *New Scientist* editorial put it this way:

"If you have an instant of creation, don't you need a creator?"

The editorial then outlines how cosmologists had, over a number of years, come up with "several different models of the universe that dodge the need for a beginning while still requiring a big bang. But recent research has shot them full of holes. It now seems certain that the universe did have a beginning."

"Lisa Grossman" elaborated in an article entitled/subtitled: "Death of the eternal cosmos—From the cosmic egg to the infinite multiverse, every model of the universe has a beginning."

Her writing relates how physicists such as Stephen Hawking "tend to shy away from cosmic genesis" in order to avoid "the thorny question", i.e., the need for, a supernatural creator. They had relied upon models designed to dodge the origins problem, such as an eternally inflating or cyclic universe, which give the appearance of time continuing indefinitely in the past as well as the future.

Counter-intuitively these models had been constructed to be compatible with the big bang. I.e., the big bang was not the beginning. However, as cosmologist "Alexander Vilenkin" of Tufts University in Boston explained earlier this year, hope in these ideas "may now be dead". Vilenkin showed that all these models still demand a beginning. They do not stretch back in true infinity; rather, there is still a "start of everything."

Of the eternal inflation model, Vilenkin says that, "You can't construct a space-time with this property"— the equations simply don't work. "It can't possibly be eternal in the past. There must be some kind of boundary."

Priest Lemaître - Hopelessly Wrong Model

And Vilenkin said that while cyclic universes have an "irresistible poetic charm and bring to mind the Phoenix" (quoting the late Belgian astronomer and priest Georges Lemaître), the *model was hopelessly wrong* in its predictions of the universe's level of order today. If there had indeed already been an infinite number of cycles, the universe today should be in a state of *maximum disorder*. As the *New Scientist* article pointed out:

"*Such a universe would be uniformly lukewarm and featureless, and definitely lacking such complicated beings as stars, planets and physicists—nothing like the one we see around us.*"

The attempted rescue suggestion, viz. that the universe just gets bigger with every cycle, therefore isn't yet at maximum disorder, also fails on the same point as the eternal inflation model. I.e., "*if your universe keeps getting bigger, it must have started somewhere.*"

Cracked Cosmic Egg – Big Bang

And as for a third idea that the cosmos previously "existed eternally in a static state called the cosmic egg", which later 'cracked' to create the big bang, Vilenkin and research associates have 'cracked' that notion, too. Their work showed that quantum instabilities would force such an 'egg' to collapse after a finite amount of time. And even if it cracked beforehand, leading to the supposed big bang, then this must have happened before it collapsed, and therefore also within a finite amount of time. "This is also not a good candidate for a beginningless universe", says Vilenkin. "All the evidence we have says that the universe had a beginning."

And so, in that light, as the editorial of that issue of *New Scientist* muses, physicists must answer the problem: "How do you get a universe, complete with the laws of physics, out of nothing?"

From Everlasting to Everlasting

The One who really is eternal, who exists "from everlasting to everlasting", and whose creation of the space-time construct presents such a conundrum to cosmologists who would deny Him, has already spoken much about the beginning—for those willing to listen. Sadly, even some Christian apologists (such as Hugh Ross and William Lane Craig) have naively been duped into advocating 'God used the big bang and other godless origins ideas'—untenable on all counts. The only way to avoid such error is by believing God's Word—from the very first verse.

SCIENTISTS REFUTE THE BIG BANG

Naïve Apologetics

It's amazing to see how many Christian leaders have not merely tolerated the 'big bang' idea, but embraced it wholeheartedly. To hear their pronouncements, believers should welcome it as a major plank in our defense of the faith. 'At last, we can use science to prove there's a creator of the universe.'

However, the price of succumbing to the lure of secular acceptability, at least in physics and astronomy, has been heavy. We have long warned that adopting the big bang into Christian thought is like bringing the wooden horse within the walls of Troy. This is because:

- The big bang forces acceptance of a sequence of events totally incompatible with the Bible (e.g. earth after sun instead of earth before sun—see Two worldviews in conflict and How could the days of *Genesis 1* be literal if the Sun wasn't created until the fourth day?)

- The big bang's billions of years of astronomical evolution are not only based on naturalistic assumptions, they are contrary to the words of Jesus Himself, who said people were there from the beginning, not towards the end of an interminably long 'creation' process (*Mark 10:6*)

- The slow evolution of the stars, then solar system and planets (including earth) in big bang thinking means that 'big bang Christians' are invariably dragged into accepting 'geological evolution' (millions of years for the earth's fossil-bearing rocks to be laid down). So, they end up denying the global Flood, and accepting death, bloodshed and disease (as seen in the fossils) before Adam. This removes the Fall and the Curse on creation from any effect on the real world, as well as removing *the* biblical answer Christians have always had to the problem of suffering and evil (God made a perfect world, ruined by sin).

- Marrying one's theology to today's science means that one is likely to be widowed tomorrow.

In fact, the signs are strong that exactly that is happening, and that those who have 'bought' the big bang for its allegedly irrefutable science have been 'sold a pup'. A bombshell 'Open Letter to the Scientific Community' by 33 leading scientists has been published on the internet (Cosmology statement) and in *New Scientist* (Lerner, E., Bucking the big bang, *New Scientist* **182** (2448)20, 22 May 2004). An article on www.rense.com titled 'Big bang theory busted by 33 top scientists' (27 May 2004) says, 'Our ideas about the history of the universe are dominated by big bang theory. But its dominance rests more on funding decisions than on the scientific method, according to "Eric Lerner," mathematician "Michael Ibison" of Earthtech.org, and dozens of other scientists from around the world.'

The open letter includes statements such as:

- The big bang today relies on a growing number of hypothetical entities, things that we have never observed—inflation, dark matter and dark energy are the most prominent examples. Without them, there would be a fatal contradiction between the observations made by astronomers and the predictions of the big bang theory.'

- 'But the big bang theory can't survive without these fudge factors. Without the hypothetical inflation field, the big bang does not predict the smooth, isotropic cosmic background radiation that is observed, because there would be no way for parts of the universe that are now more than a few degrees away in the sky to come to the same temperature and thus emit the same amount of microwave radiation. … Inflation requires a density 20 times larger than that implied by big bang nucleosynthesis, the theory's explanation of the origin of the light elements.' [This refers to the *horizon problem*, and supports what we say in Light-travel time: a problem for the big bang.]

- 'In no other field of physics would this continual recourse to new hypothetical objects be accepted as a way of bridging the gap between theory and observation. It would, at the least, *raise serious questions about the validity of the underlying theory* [emphasis in original].'

- 'What is more, the big bang theory can boast of no quantitative predictions that have subsequently been validated by observation. The successes claimed by the theory's supporters consist of its ability to retrospectively fit observations with a steadily increasing array of adjustable parameters, just as the old Earth-centered cosmology of Ptolemy needed layer upon layer of epicycles.'

The dissidents say that there are other explanations of cosmology that do make some successful predictions. These other models don't have all the answers to objections, but, they say, 'That is scarcely surprising, as their development has been severely hampered by a complete lack of funding. Indeed, such questions and alternatives cannot even now be freely discussed and examined.'

Those who urge Christians to accept the big bang as a 'science fact' point to its near-universal acceptance by the scientific community. However, the 33 dissidents describe a situation familiar to many creationist scientists: 'An open exchange of ideas is lacking in most mainstream conferences … doubt and dissent are not tolerated, and young scientists learn to remain silent if they have something negative to say about the standard big bang model. Those who doubt the big bang fear that saying so will cost them their funding.'

Evolutionist and historian of science, "Evelleen Richards," has noticed that it's hard even for rival *evolutionary* theories to get a hearing when challenging the ruling paradigm—see Science … a reality check. This should give some idea of the difficulties biblical creationists face.

But don't we read, even in the daily newspapers, about many 'observations' that only ever seem to support the big bang? In fact, these prominent secular scientists say:

'*Even observations are now interpreted through this biased filter, judged right or wrong depending on whether or not they support the big bang. So discordant data on red shifts, lithium and helium abundances, and galaxy distribution, among other topics, are ignored or ridiculed.*'

Science is a wonderful human tool, but it needs to be understood, not worshipped. It is fallible, changing, and is severely limited as to what it can and cannot determine. As Christian believers have often pointed out, instead of a scientific concept, the big-bang idea is more a dogmatic religious one—based on the religion of humanism. As these big-bang opposers point out:

'Giving support only to projects within the big bang framework undermines a fundamental element of the scientific method—the constant testing of theory against observation. Such a restriction makes unbiased discussion and research impossible.'

Furthermore, contrary to the naïve pronouncements of many who should know better, it is not in any sense a matter of 'looking into a telescope and "seeing"? the big bang billions of years ago.' As always, observations are interpreted and filtered through worldview lenses. Those who developed the big bang were guided by secular worldview filters just as much as those who are now crying that the emperor has no clothes. They wanted a universe that created itself; their opponents want an eternal, uncreated universe. From a Christian perspective, both are in open defiance of their Creator's account of what really happened.

With Darwinism on the run, the Enemy of souls is seeking to seduce believers into embracing a more subtle, yet far deadlier way of evading the authority of the Bible. With progressive creationism/big-banger rampaging through the evangelical community, he must think he is on a winner.

For a powerful, profound exposition of all of the issues involved in this, today's most important evangelical compromise position, my colleague has released books emphasizing on *Refuting Compromise* are not just for casual readings. They provide illumination to the fallacious big bang logic by ignoring many scientific problems— The writings in some books also show how one can use a 'first cause' argument without needing the big bang. The books are in fact destined to become Christian classics, a culture-changing colossus of 'cut-through-the-smokescreen' clarity and logic.

AGE OF THE UNIVERSE/EARTH

Five-Scores of Evidences for a Young Universe/Earth

Would Science Prove the Age of the Universe/Earth?

The widely accepted age of the universe is currently 13.77 billion years and for the solar system (including Earth) it is 4.543 billion years. However, no scientific method can **prove** the age of the earth and the universe, and that includes the ones we have listed here that strongly suggest that these accepted ages are in serious error. Although age indicators are called '**clocks**' they aren't, because all ages result from calculations that necessarily involve making assumptions about the past. The starting time of the 'clock' has always to be assumed as well as the way in which the speed of the clock has varied over time. Further, it has to be assumed that the clock was never disturbed, Figure 10.20.

There is no independent **natural clock** against which those assumptions can be tested. For example, the amount of cratering on the moon, based on currently observed cratering rates, would suggest that the moon is quite old. However, to draw this conclusion we have to assume that the rate of cratering has been the

Fig.10.20: There are many categories of evidence for the age of the cosmos and the earth that indicate they are much younger than is generally asserted today.

same in the past as it is now. And there are now good reasons for thinking that it might have been quite intense in the past, in which case the craters do not indicate an old age at all.

Ages of millions of years are all calculated by assuming the rates of change of processes in the past were the same as we observe today—called the principle of uniformitarianism. If the age calculated from such assumptions disagrees with what they think the age should be, they conclude that their assumptions did not apply in this case, and adjust them accordingly. If the calculated result gives an acceptable age, the investigators publish it.

Examples of *young* ages listed here are also obtained by applying the same principle of uniformitarianism. Long-age proponents will dismiss this sort of evidence for a young age of the earth by arguing that the assumptions about the past do not apply in these cases. In other words, age is not really a matter of scientific observation but an argument about our assumptions about the unobserved past.

The assumptions behind the evidences presented here cannot be proved, but the fact that such a wide range of different phenomena all *suggest* much younger ages than are currently generally accepted, provides a strong case for questioning the accepted ages.

Also, a number of the evidences, rather than giving any estimate of age, challenge the assumption of slow-and-gradual uniformitarianism, upon which all deep-time dating methods depend.

Many of these indicators for younger ages were discovered when faithful Christian astrophysicists started researching things that were supposed to 'prove' long ages. The lesson here is clear: when the evolutionists throw up some new challenge to the Bible's timeline, don't fret over it. Sooner or later that supposed evidence will be turned on its head and will even be added to this list of evidences for a younger age of the earth. On the other hand, some of the evidences listed here might turn out to be ill-founded with further research and will need to be modified. Such is the nature of science, especially historical science, because we cannot do experiments on past events.

Science is based on observation, and the only reliable means of telling the age of anything is by the testimony of a reliable witness who observed the events. The Bible claims to be the communication of the only One who witnessed the events of Creation: The Creator himself. As such, the Bible is the only reliable means of knowing the age of the earth and the cosmos. In the end we believe that the Bible will stand vindicated and those who deny its testimony will be confounded.

1. Biological Evidence for a Young Age Universe/Earth

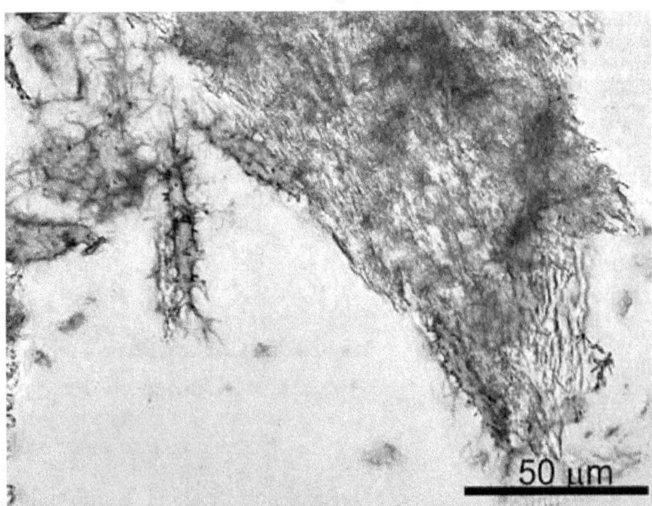

Fig.10.21: The finding of pliable blood vessels, blood cells, animal proteins, and even DNA in dinosaur bone is consistent with an age of thousands of years for the fossils, not the 65+ million years claimed by the paleontologists – Curtsy Dr Mary Schweitzer

1. **Evidence 1**: DNA in 'ancient' fossils. DNA extracted from bacteria that are supposed to be 425 million years old brings into question that age, because DNA could not last more than thousands of years, Figure 10.21.

2. **Evidence 2**: Lazarus bacteria—bacteria revived from salt inclusions supposedly 250 million years old, suggest the salt is not millions of years old.

3. **Evidence 3**: The decay in the human genome due to multiple slightly harmful mutations each generation is consistent with an origin several thousand years ago. "Sanford, J.," *Genetic entropy and the mystery of the genome*, "Ivan Press," 2005; see review of the book and the interview with the author in *Creation* **30** (4):45–47, September 2008. This has been confirmed by realistic modelling of population genetics, which shows that genomes are young, in the order of thousands of years. See Sanford, J., Baumgardner, J., Brewer, W., Gibson, P. and Remine, W., Mendel's Accountant: A biologically realistic forward-time population genetics program, *SCPE* **8** (2):147–165, 2007.

4. **Evidence 4**: The data for 'mitochondrial Eve' are consistent with a common origin of all humans several thousand years ago.

5. **Evidence 5**: Very limited variation in the DNA sequence on the human Y-chromosome around the world is consistent with a recent origin of mankind, thousands not millions of years.

6. **Evidence 6**: Many fossil bones 'dated' at many millions of years old are hardly mineralized, if at all. This contradicts the widely believed old age of the earth. See, for example, Dinosaur bones just how old are they really? Tubes of marine worms, 'dated' at 550 million years old, that are soft and flexible and apparently composed of the original organic compounds hold the record.

7. **Evidence 7**: Dinosaur blood cells, blood vessels, proteins (hemoglobin, osteocalcin, collagen, histones) and DNA are not consistent with their supposed more than -65million-year age, but make more sense if the remains are thousands of years old, at most, Figure 10.22.

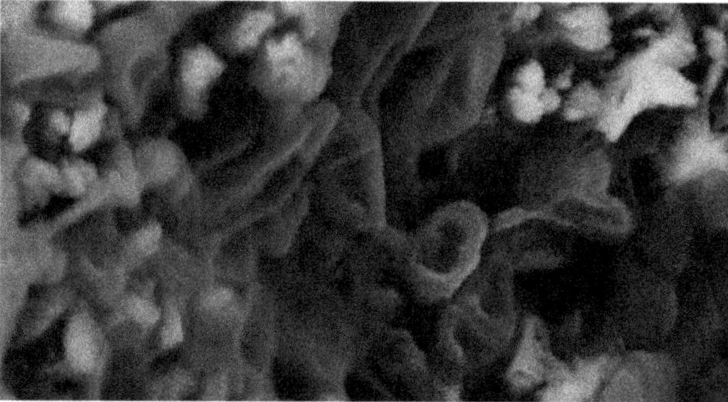

Fig.10.22: *Dinosaur blood cells, blood vessels, proteins (hemoglobin, osteocalcin, collagen, histones) and DNA are not consistent with their supposed more than -65million-year age, but make more sense if the remains are thousands of years old, at most.*

8. **Evidence 8**: Lack of 50:50 racemization of amino acids in fossils 'dated' at millions of years old, whereas complete racemization would occur in thousands of years.

9. **Evidence 9**: Living fossils—jellyfish, graptolites, coelacanth, stromatolites, Wollemi pine and hundreds more. That many hundreds of species could remain so unchanged, for even up to billions of years in the case of stromatolites, speaks against the millions and billions of years being real, Figure 10.23.

Fig.10.23: Wollemi pine – Young Universe/Earth

10. **Evidence 10:** Discontinuous fossil sequences. E.g., Coelacanth, Wollemi pine and various 'index' fossils, which are present in supposedly ancient strata, missing in strata representing many millions of years since, but still living today. Such discontinuities speak against the interpretation of the rock formations as vast geological ages—how could Coelacanths have avoided being fossilized for 65 million years, for example? See The 'Lazarus effect': rodent 'resurrection,' Figure 10.24!

11. **Evident 11:** The ages of the world's oldest living organisms, trees, are consistent with an age of the earth of thousands of years.

Fig.10.24: How could Coelacanths have Avoided being Fossilized for 65 million years? Impossible!

II. Geological Evidence for a Young Age of the Universe/Earth

Fig.10.25: *Radical folding at Eastern Beach, near Auckland in New Zealand, indicates that the sediments were soft and pliable when folded, inconsistent with a long time for their formation. Such folding can be seen world-wide and is consistent with a young age of the earth. Curtsy Photo by Don Batten*

12. **Evident 12:** Scarcity of plant fossils in many formations containing abundant animal / herbivore fossils. E.g., the Morrison Formation (Jurassic) in Montana. See *Origins* 21(1):51–56, 1994. Also, the Coconino sandstone in the Grand Canyon has many track-ways (animals), but is almost devoid of plants, Figure 10.25. **Implication:** these rocks are *not* ecosystems of an 'era' buried *in situ* over eons of time as evolutionists claim. The evidence is more consistent with catastrophic transport then burial during the massive global Flood of Noah's day. This eliminates supposed evidence for millions of years, Figure 10.26.

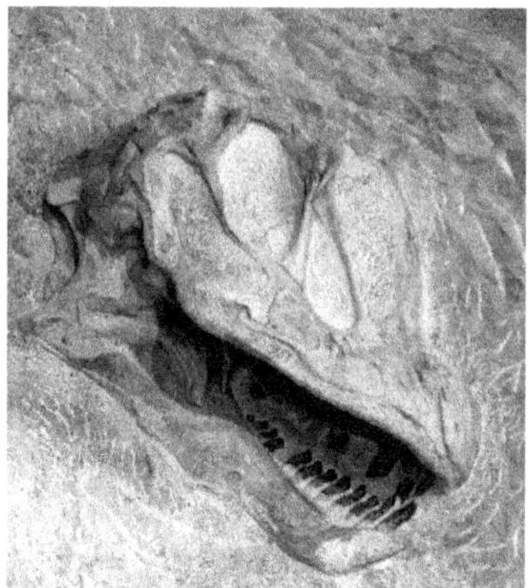

Fig.10.26: *Scarcity of plant fossils in many formations containing abundant animal / herbivore fossils. E.g., the Morrison Formation (Jurassic) in Montana*

13. **Evident 13:** Thick, tightly bent strata without sign of melting or fracturing. E.g., the Kaibab Upwarp in Grand Canyon indicates rapid folding before the sediments had time to solidify (the sand grains were not elongated under stress as would be expected if the rock had hardened). This wipes out hundreds of millions of years of time and is consistent with extremely rapid formation during the biblical Flood. See Warped earth (written by a geophysicist).

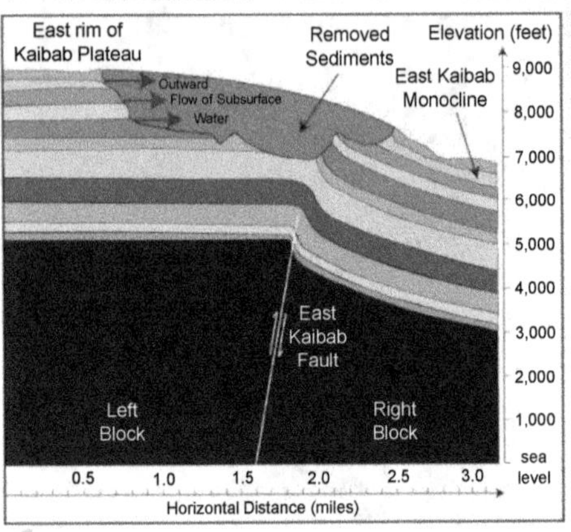

Fig.10.27: The Kaibab Upwarp in Grand Canyon Indicates Rapid Folding before the Sediments had time to Solidify

14. **Evident 14** - Polystrate fossils—tree trunks in coal (*Araucaria* spp. king Billy pines, celery top pines, in southern hemisphere coal). There are also polystrate tree trunks in the Yellowstone fossilized forests and Joggings, Nova Scotia and in many other places. Polystrate fossilized lycopod trunks occur in northern hemisphere coal, again indicating rapid burial / formation of the organic material that became coal, Figure 10.28.

Fig.10.28: Polystrate Fossils—Tree Trunks in Coal (Araucaria spp. King Billy Pines

15. **Evident 15**: Experiments show that with conditions mimicking natural forces, coal forms quickly; in weeks for brown coal to months for black coal. It does not need millions of years. Furthermore, long time periods could be an impediment to coal formation because of the increased likelihood of the permineralization of the wood, which would hinder coalification.

16. **Evident 16**: Experiments show that with conditions mimicking natural forces, oil forms quickly; it does not need millions of years, consistent with an age of thousands of years.

17. **Evident 17**: Experiments show that with conditions mimicking natural forces, opals form quickly, in a matter of weeks, not millions of years, as had been claimed.

18. **Evident 18**: Evidence for rapid, catastrophic formation of coal beds speaks against the hundreds of millions of years normally claimed for this, including Z-shaped seams that point to a single depositional event producing these layers.

19. **Evident 19**: Evidence for rapid petrifaction of wood speaks against the need for long periods of time and is consistent with an age of thousands of years.

20. **Evident 20**: Clastic dykes and pipes (intrusion of sediment through overlying sedimentary rock) show that the overlying rock strata were still soft when they formed. This drastically compresses the time scale for the deposition of the penetrated rock strata. See, Walker, T., Fluidisation pipes: Evidence of large-scale watery catastrophe, *J. Creation (TJ)* **14**(3):8–9, 2000.

21. **Evident 21**: Para (pseudo)conformities—where one rock stratum sits on top of another rock stratum but with supposedly millions of years of geological time missing, yet the contact plane lacks any significant erosion; that is, it is a 'flat gap'. E.g., Coconino sandstone / Hermit shale in the Grand Canyon (supposedly a 10-million-year gap in time), Figure 10.29. The thick Schnebly Hill Formation (sandstone) lies *between* the Coconino and Hermit in central Arizona. See Austin, S.A., *Grand Canyon, monument to catastrophe*, ICR, Santee, CA, USA, 1994 and Snelling, A., The case of the 'missing' geologic time, *Creation* **14**(3):31–35, 1992.

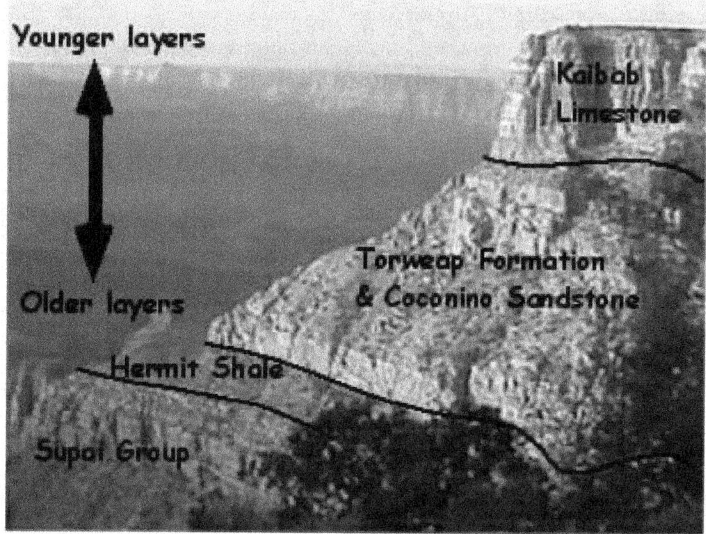

Fig.10.29: *Coconino Sandstone / Hermit Shale in the Grand Canyon (supposedly a 10-million-year gap in time).*

22. **Evident 22**: The presence of ephemeral markings (raindrop marks, ripple marks, animal tracks) at the boundaries of paraconformities show that the upper rock layer has been deposited immediately after the lower one, eliminating many millions of years of 'gap' time. See references in Para(pseudo) conformities, Figure 10.30.

Fig.10.30: *The Presence of Ephemeral Markings (raindrop marks, ripple marks, animal racks) at the Boundaries of Paraconformities*

23. **Evident 23:** Inter-tonguing of adjacent strata that are supposedly separated by millions of years also eliminates many millions of years of supposed geologic time. The case of the 'missing' geologic time; Mississippian and Cambrian strata interbedding: 200 million years hiatus in question, *CRSQ* **23**(4):160–167.

24. **Evident 24:** The lack of bioturbation (worm holes, root growth) at paraconformities (flat gaps) reinforces the lack of time involved where evolutionary geologists insert many millions of years to force the rocks to conform with the 'given' timescale of billions of years.

25. **Evident 25:** The almost complete lack of clearly recognizable soil layers anywhere in the geologic column. Geologists do claim to have found lots of 'fossil' soils (paleosols), but these are quite different to soils today, lacking the features that characterize soil horizons; features that are used in classifying different soils. Every one that has been investigated thoroughly proves to lack the characteristics of proper soil. If 'deep time' were correct, with hundreds of millions of years of abundant life on the earth, there should have been ample opportunities many times over for soil formation. See Klevberg, P. and Bandy, R., *CRSQ* **39**:252–68; *CRSQ* **40**:99–116, 2003; Walker, T., Paleosols: digging deeper buries 'challenge' to Flood geology, *J. Creation* **17**(3):28–34, 2003.

26. **Evident 26:** Limited extent of unconformities (unconformity: a surface of erosion that separates younger strata from older rocks). Surfaces erode quickly (e.g., Badlands, South Dakota), but there are very limited unconformities. There is the 'great unconformity' at the base of the Grand Canyon, but otherwise there are supposedly ~300 million years of strata deposited on top without any significant unconformity. This is again consistent with a much shorter time of deposition of these strata. See Para(pseudo)conformities.

27. **Evident 27:** The Arches National Park (USA) has over 2,000 rock arches. If 43 have collapsed since 1970 and the linked article was written in 2015, that's 45 years, giving a rate of collapse of ~1 per year, which means that all would be gone in ~2,000 years, Figure 10.30. This is thoroughly consistent with the biblical timeframe but not the evolutionary one of millions of years (5 million?). Historical records of the '12

Apostles' in southern Australia should allow a similar 'clock' to be calculated, albeit coarser than this USA park one, Figure 10.31.

Fig.10.31: The Arches National Park (USA) has Over 2,000 Rock Arches

28. **Evident 28:** The discovery that underwater landslides ('turbidity currents') travelling at some 50 km/h can create huge areas of sediment in a matter of hours (Press, F., and Siever, R., *Earth*, 4th ed., Freeman & Co., NY, USA, 1986). Sediments thought to have formed slowly over eons of time are now becoming recognized as having formed extremely rapidly. See for example, A classic tillite reclassified as a submarine debris flow (Technical).

29. **Evident 29:** Flume tank research with sediment of different particle sizes show that layered rock strata that were thought to have formed over huge periods of time in lake beds actually formed very quickly. Even the precise layer thicknesses of rocks were duplicated after they were ground into their sedimentary particles and run through the flume. See Experiments in stratification of heterogeneous sand mixtures, Sedimentation Experiments: Nature finally catches up! and Sandy Stripes Do many layers mean many years? Also, very fine particles have been shown to settle far more quickly than previously thought, enabling the rapid formation of mudstone deposits.

30. **Evident 30:** Observed examples of rapid canyon formation; for example, Providence Canyon in southwest Georgia, Burlingame Canyon near Walla Walla, Washington, and Lower Loowit Canyon near Mount St Helens. The rapidity of the formation of these canyons, which look similar to other canyons that supposedly took many millions of years to form, brings into question the supposed age of the canyons that no one saw form.

31. **Evident 31:** Observed examples of rapid island formation and maturation, such as Surtsey, which confound the notion that such islands take long periods of time to form. See also, Tuluman—A Test of Time.

32. **Evident 32:** Rate of erosion of coastlines, horizontally. E.g., Beachy Head, UK, loses a meter of coast to the sea every six years.

33. **Evident 33:** Rate of erosion of continents vertically is not consistent with the assumed old age of the earth.

34. **Evident 34:** Existence of significant flat plateaux that are 'dated' at many millions of years old ('elevated paleoplains'). An example is Kangaroo Island (Australia). C.R. Twidale, a famous Australian physical geographer wrote: "the survival of these paleoforms is in some degree an embarrassment

to all the commonly accepted models of landscape development." Twidale, C.R. On the survival of paleoforms, *American J. Science* **5**(276):77–95, 1976 (quote on p. 81). See Austin, S.A., Did landscapes evolve? *Impact* **118**, April 1983.

35. **Evident 35:** The recent and almost simultaneous origin of all the high mountain ranges around the world—including the Himalayas, the Alps, the Andes, and the Rockies—which have undergone most of the uplift to their present elevations beginning 'five million' years ago, whereas mountain building processes have supposedly been around for up to billions of years. See Baumgardner, J., Recent uplift of today's mountains. *Impact* **381**, March 2005.

36. **Evident 36:** Water gaps. These are gorges cut through mountain ranges where rivers run, Figure 10.32. They occur worldwide and are part of what evolutionary geologists call 'discordant drainage systems.' They are 'discordant' because they don't fit the deep time belief system. The evidence fits them forming rapidly in a much younger age framework where the gorges were cut in the recessive stage / dispersive phase of the global Flood of Noah's day. See Oard, M., Do rivers erode through mountains? Water gaps are strong evidence for the Genesis Flood.

Fig.10.32: Water gaps. These are gorges cut through mountain ranges where rivers run

Fig.10.33: Erosion rates at places like Niagara Falls are consistent with a time frame of several thousand years since Noah's Flood.

37. **Evident 37:** Erosion at Niagara Falls and other such places is consistent with just a few thousand years since the biblical Flood. However, much of the Niagara Gorge likely formed very rapidly with the catastrophic drainage of glacial Lake Agassiz; see: Climate change, Niagara and catastrophe, Figure 10.33.

38. **Evident 38:** River delta growth rate is consistent with thousands of years since the biblical Flood, not vast periods of time. E.g. 1. Mississippi—*Creation Research Quarterly (CRSQ)* **9**(2):96–114, 1972; *CRSQ* **14**(2):87, 1977; *CRSQ* **25**(3):121–123, 1988. E.g., 2 Tigris–Euphrates: *CRSQ* **14**(2):87, 1977.

39. **Evident 39:** Underfit streams. River valleys are too large for the streams they contain. Dury speaks of the "continent-wide distribution of underfit streams". Using channel meander characteristics, Dury concluded that past streams frequently had 20–60 times their current discharge. This means that the river valleys would have been carved very quickly, not slowly over eons of time. See Austin, S.A., Did landscapes evolve? *Impact* 118, 1983.

40. **Evident 40:** Amount of salt in the sea. Even ignoring the effect of the biblical Flood and assuming zero starting salinity and all rates of input and removal so as to maximize the time taken to accumulate all the salt, the *maximum* age of the oceans, 62 million years, is less than 1/50 of the age evolutionists claim for the oceans. This suggests that the age of the earth is radically less also.

41. **Evident 41:** The amount of sediment on the sea floors at current rates of land erosion would accumulate in just 12 million years; a blink of the eye compared to the supposed age of much of the ocean floor of up to 3 billion years. Furthermore, long-age geologists reckon that *higher* erosion rates applied in the past, which shortens the time frame. From a biblical point of view, at the end of Noah's Flood lots of sediment would have been added to the sea with the water coming off the unconsolidated land, making the amount of sediment perfectly consistent with a history of thousands of years.

42. **Evident 42:** Iron-manganese nodules (IMN) on the sea floors. The measured rates of growth of these nodules indicates an age of only thousands of years. Lalomov, A.V., 2006. Mineral deposits as an example of geological rates. *CRSQ* **44**(1):64–66. Related to this is the concentration of nickel in the oceans.

43. **Evident 43:** The age of placer deposits (concentrations of heavy metals such as tin in modern sediments and consolidated sedimentary rocks). The measured rates of deposition indicate an age of thousands of years, not the assumed millions. See Lalomov, A.V., and Tabolitch, S.E., 2000. Age determination of coastal submarine placer, Val'cumey, northern Siberia. *J, Figure 10.34.*

Fig.10.34: *The age of placer deposits (concentrations of heavy metals such as tin in modern sediments and consolidated sedimentary rocks).*

44. **Evident 44:** Pressure in oil / gas wells indicate the recent origin of the oil and gas. If they were many millions of years old, we would expect the pressures to equilibrate, even in low permeability rocks. "Experts in petroleum prospecting note the impossibility of creating an effective model given long and slow oil generation over millions of years (Petukhov, 2004). In their opinion, if models demand the standard multimillion-years geochronological scale, the best exploration strategy is to drill wells on a random grid." —Lalomov, A.V., 2007. Mineral deposits as an example of geological rates. *CRSQ* **44**(1):64–66.

45. **Evident 45:** Direct evidence that oil is forming today in the Guaymas Basin and in Bass Strait is consistent with a young earth (although not *necessary* for a young earth).

46. **Evident 46:** Rapid reversals in paleomagnetism undermine use of paleomagnetism in long ages dating of rocks and speak of rapid processes, compressing the long-age time scale enormously.

47. **Evident 47:** The pattern of magnetization in the magnetic stripes where magma is welling up at the mid-ocean trenches argues against the belief that reversals take many thousands of years and rather indicates rapid sea-floor spreading as well as rapid magnetic reversals, consistent with a young earth (Humphreys,

D.R., Has the Earth's magnetic field ever flipped, Figure 10.35? *Creation Research Quarterly* **25**(3): 130–137, 1988).

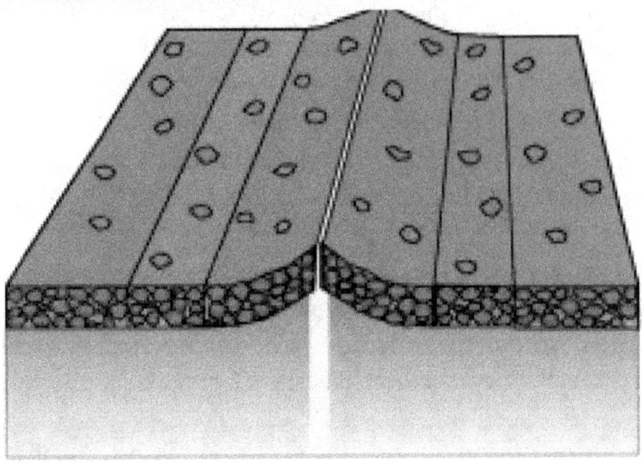

Fig.10.35: *Along the mid-ocean ridges, the detailed pattern of magnetic polarisation, with islands of differing polarity, speaks of rapid changes in direction of Earth's magnetic field because of the rate of cooling of the lava. This is consistent with a young Earth.*

48. **Evident 48:** Measured rates of stalactite and stalagmite growth in limestone caves are consistent with a young age of several thousand years. See also articles on limestone cave formation.

49. **Evident 49:** The decay of the earth's magnetic field. Exponential decay, with fluctuations especially during and after the Flood, is evident from historical measurements and is consistent with the hypothesis of free decay since creation, suggesting an age of the earth of only thousands of years. For further evidence that it follows exponential decay with a time constant of 1611 years (±10) see: Humphreys, R., Earth's magnetic field is decaying steadily—with a little rhythm, Figure 10.36 - *CRSQ* **47**(3):193–201; 2011.

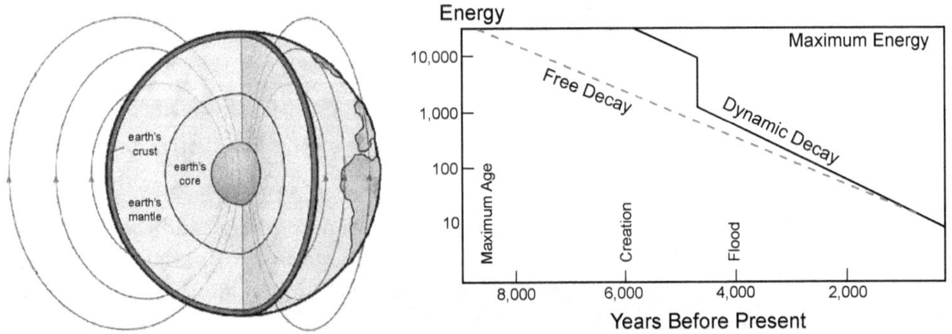

Fig.10.36: *Rapidly Decaying Magnetic Field, Answer to Genesis*

1. **Evident 50:** Excess heat flow from the earth is consistent with a young age rather than billions of years, even taking into account heat from radioactive decay, Figure 10.37. See Woodmorappe, J., 1999. Lord Kelvin revisited on the young age of the earth, *J. Creation (TJ)* **13**(1):14, 1999.

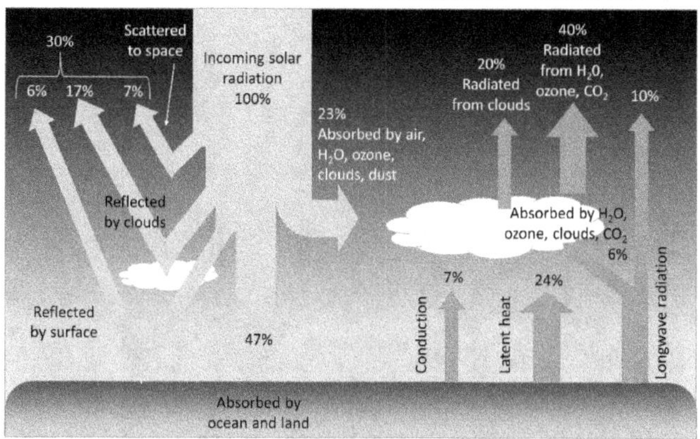

Fig.10.37: Excess heat flow from the earth is consistent with a young age rather than billions of years, even taking into account heat from radioactive decay

III. Radiometric Dating and the Age of the Earth

51. **Evident 51:** Carbon-14 in coal suggests ages of thousands of years and clearly contradict ages of millions of years, Figure 10.38.

article highlights

- Scientists use various dating methods to estimate Earth's age, but most provide results that are too young for the evolutionary story.
- Continental erosion, ocean salt accumulation, Earth's magnetic field decay, radiocarbon in "old" specimens, and belium in zircon crystals yield age estimates that contradict evolution but are consistent with biblical creation.
- These five evidences are strong arguments for recent creation.

Fig.10.38: Five Global Evidences for a Young Earth

52. **Evident 52:** Carbon-14 in oil again suggests ages of thousands, not millions, of years.
53. **Evident 53:** Carbon-14 in fossil wood also indicates ages of thousands, not millions, of years.
54. **Evident 54:** Carbon-14 in diamonds suggests ages of thousands, not billions, of years. Note that attempts to explain away carbon-14 in diamonds, coal, etc., such as by neutrons from uranium decay converting nitrogen to C-14 do not work, Figure 10.39.

Fig.10.39: Carbon-14 in Diamonds Suggests ages of Thousands, not Billions, of Years

55. **Evident 55:** Incongruent radioisotope dates using the same technique argue against trusting the dating methods that give millions of years.

56. **Evident 56:** Incongruent radioisotope dates using different techniques argue against trusting the dating methods that give millions of years (or billions of years for the age of the earth).

57. **Evident 57:** Demonstrably non-radiogenic 'isochrons' of radioactive and non-radioactive elements undermine the assumptions behind isochron 'dating' that gives billions of years. 'False' isochrons are common.

58. **Evident 58:** Different faces of the same zircon crystal and different zircons from the same rock giving different 'ages' undermine all 'dates' obtained from zircons

59. **Evident 59:** Evidence of a period of rapid radioactive decay in the recent past (lead and helium concentrations and diffusion rates in zircons) point to a young earth explanation, Figure 10.40.

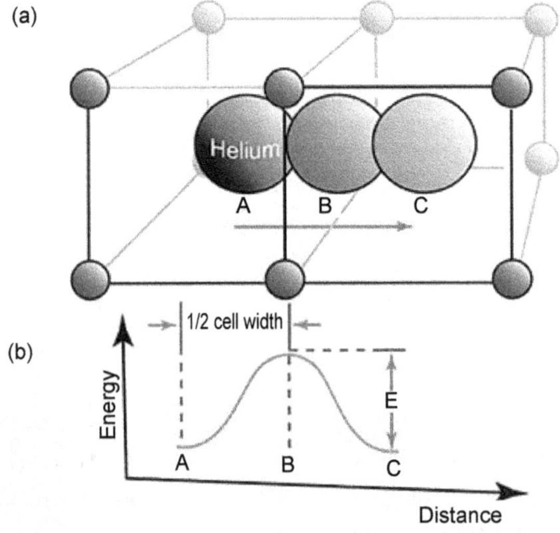

Fig.10.40: Helium Retention in Zircons Demonstrates a Young Earth

60. **Evident 60:** The amount of helium, a product of alpha-decay of radioactive elements, retained in zircons in granite is consistent with an age of 6,000±2000 years, not the supposed billions of years. See:

Humphreys, D.R., Young helium diffusion age of zircons supports accelerated nuclear decay, Chapter 2 (pages 25–100) in: Vardiman, Snelling, and Chaffin (eds.), *Radioisotopes and the Age of the Earth: Results of a Young Earth Creationist Research Initiative, Volume II*, Institute for Creation Research and Creation Research Society, 2005.

61. **Evident 61:** Lead in zircons from deep drill cores vs. shallow ones. They are similar, but there should be less in the deep ones due to the higher heat causing higher diffusion rates over the usual long ages supposed. If the ages are thousands of years, there would not be expected to be much difference, which is the case (Gentry, R., *et al.*, Differential lead retention in zircons: Implications for nuclear waste containment, *Science* **216**(4543):296–298, 1982; DOI: 10.1126/science.216.4543.296).

62. **Evident 62:** Pleochroic halos produced in granite by concentrated specks of short half-life elements such as polonium suggest a period of rapid nuclear decay of the long half-life parent isotopes during the formation of the rocks and rapid formation of the rocks, both of which speak against the usual ideas of geological deep time and a vast age of the earth. See, Radiohalo's: Startling evidence of catastrophic geologic processes, *Creation* **28**(2):46–50, 2006.

63. **Evident 63:** Squashed pleochroic halos (radiohalos) formed from decay of polonium, a very short half-life element, in coalified wood from several geological eras suggest rapid formation of all the layers about the same time, in the same process, consistent with the biblical 'young' earth model rather than the millions of *years claimed* for these events.

64. **Evident 64:** Australia's 'Burning Mountain' speaks against radiometric dating and the millions of years belief system (according to radiometric dating of the lava intrusion that set the coal alight, the coal in the burning mountain has been burning for ~40 million years, but clearly this is not feasible).

IV. Astronomical evidence for a young(er) age of the earth and the universe

Fig.10.41: Saturn's Rings are Increasingly Recognized as being Relatively Short-Lived rather than Essentially Changeless over Millions of Years. Curtsy NASA

Saturn's Rings are Increasingly Recognized as being Relatively Short-Lived rather than Essentially Changeless over Millions of Years. Figure 10.41.

65. **Evident 65:** Evidence of recent volcanic activity on Earth's moon is inconsistent with its supposed vast age because it should have long since cooled if it were billions of years old. See: Transient lunar

phenomena: a permanent problem for evolutionary models of Moon formation and Walker, T., and Catchpoole, D., Lunar volcanoes rock long-age timeframe, Figure 10.42, *Creation* **31**(3):18, 2009. See further corroboration: "At Long Last, Moon's Core 'Seen'"; www.sciencemag.org/news/2011/01/long-last-moons-core-seen.

Fig.10.42: *Evidence of recent volcanic activity on Earth's moon is inconsistent with its supposed vast age because it should have long since cooled if it were billions of years old*

66. **Evident 66:** Recession of the moon from the earth. Tidal friction causes the moon to recede from the earth at 4 cm per year. It would have been greater in the past when the moon and earth were closer together. The moon and earth would have been in catastrophic proximity (Roche limit) at less than a quarter of their supposed age.

67. **Evident 67:** The moon's former magnetic field. Rocks sampled from the moon's crust have residual magnetism that indicates that the moon once had a magnetic field much stronger than earth's magnetic field today. No plausible 'dynamo' hypothesis could account for even a weak magnetic field, let alone a strong one that could leave such residual magnetism in a billions-of-years' time-frame. The evidence is much more consistent with a recent creation of the moon and its magnetic field and free decay of the magnetic field in the 6,000 years since then. Humphreys, D.R., The moon's former magnetic field—still a huge problem for evolutionists, *J. Creation* **26**(1):5–6, 2012.

68. **Evident 68:** Ghost craters on the moon's *maria* (singular *mare*: dark 'seas' formed from massive lava flows) are a problem for the assumed long ages. Enormous impacts evidently caused the large craters and lava flows within those craters, and this lava partly buried other, smaller impact craters within the larger craters, leaving 'ghosts'. But this means that the smaller impacts can't have been too long after the huge ones, otherwise the lava would have flowed into the larger craters before the smaller impacts. This suggests a very narrow time frame for all this cratering, and by implication the other cratered bodies of our solar system. They suggest that the cratering occurred quite quickly. See Fryman, H., Ghost craters in the sky, *Creation Matters* **4**(1):6, 1999; A biblically based cratering theory (Faulkner); Lunar volcanoes rock long-age timeframe.

69. **Evident 69:** The presence of a significant magnetic field around Mercury is not consistent with its supposed age of billions of years. A planet so small should have cooled down enough so any liquid core would solidify, preventing the evolutionists' 'dynamo' mechanism. See also, Humphreys, D.R., Mercury's magnetic field is young! *J. Creation* **22**(3):8–9, 2008.

60. **Evident 60:** The outer planets Uranus and Neptune have magnetic fields, but they should be long 'dead' if they are as old as claimed according to evolutionary long-age beliefs. Assuming a solar system age of thousands of years, physicist Russell Humphreys successfully predicted the strengths of the magnetic fields of Uranus and Neptune.

71. **Evident 71:** Jupiter's larger moons, Ganymede, Io, and Europa, have magnetic fields, which they should not have if they were billions of years old, because they have solid cores and so no dynamo could generate the magnetic fields. This is consistent with creationist Humphreys' predictions. See also, Spencer, W., Ganymede: the surprisingly magnetic moon, *J. Creation* **23**(1):8–9, 2009.

72. **Evident 72:** Volcanically active moons of Jupiter (Io) are consistent with youthfulness (Galileo mission recorded 80 active volcanoes). If Io had been erupting over 4.5 billion years at even 10% of its current rate, it would have erupted its entire mass 40 times. Io looks like a young moon and does not fit with the supposed billions of year's age for the solar system. Gravitational tugging from Jupiter and other moons accounts for only some of the excess heat produced.

73. **Evident 73:** The surface of Jupiter's moon Europa. Studies of the few craters indicated that up to 95% of small craters, and many medium-sized ones, are formed from debris thrown up by larger impacts. This means that there have been far fewer impacts than had been thought in the solar system and the age of other objects in the solar system, derived from cratering levels, have to be reduced drastically (see Psarris, Spike, *What you aren't being told about astronomy, volume 1: Our created solar system* DVD, available from CMI).

74. **Evident 74:** Methane on Titan (Saturn's largest moon)—the methane should all be gone because of UV-induced breakdown. The products of photolysis should also have produced a huge sea of heavier hydrocarbons such as ethane. An *Astrobiology* item titled "The missing methane" cited one of the Cassini researchers, Jonathan Lunine, as saying, "If the chemistry on Titan has gone on in steady-state over the age of the solar system, then we would predict that a layer of ethane 300 to 600 meters thick should be deposited on the surface." No such sea is seen, which is consistent with Titan being a tiny fraction of the claimed age of the solar system (needless to say, Lunine does not accept the obvious young age implications of these observations, so he speculates, for example, that there must be some unknown source of methane).

75. **Evident 75:** The rate of change / disappearance of Saturn's rings is inconsistent with their supposed vast age; they speak of youthfulness.

76. **Evident 76:** Enceladus, a moon of Saturn, looks young. Astronomers working in the 'billions of years' mindset thought that this moon would be cold and dead, but it is a very active moon, spewing massive jets of water vapor and icy particles into space at supersonic speeds, consistent with a much younger age. Calculations show that the interior would have frozen solid after 30 million years (less than 1% of its supposed age); tidal friction from Saturn does not explain its youthful activity (Psarris, Spike, *What you aren't being told about astronomy, volume 1: Our created solar system* DVD; Walker, T., Enceladus: Saturn's sprightly moon looks young, *Creation* **31**(3):54–55, 2009), Figure 10.42.

Fig.10.42: Saturn's Enceladus Looks Younger than Ever

77. **Evident 77:** Miranda, a small moon of Uranus, should have been long since dead, if billions of years old, but its extreme surface features suggest otherwise. See Revelations in the solar system.

78. **Evident 78:** Neptune should be long since 'cold', lacking strong wind movement if it were billions of years old, yet Voyager II in 1989 found it to be otherwise—it has the fastest winds in the entire solar system. This observation is consistent with a young age, not billions of years. See Neptune: monument to creation.

79. **Evident 79:** Neptune's rings have thick regions and thin regions. This unevenness means they cannot be billions of years old, since collisions of the ring objects would eventually make the ring very uniform. Revelations in the solar system.

80. **Evident 80:** Young surface age of Neptune's moon, Triton—less than 10 million years, even with evolutionary assumptions on rates of impacts (see Schenk, P.M., and Zahnle, K. On the negligible surface age of Triton, *Icarus* **192**(1):135–149, 2007. <doi:10.1016/j.icarus.2007.07.004>.

81. **Evident 81:** Uranus and Neptune both have magnetic fields significantly off-axis, which is an unstable situation. When this was discovered with Uranus, it was assumed by evolutionary astronomers that Uranus must have just happened to be going through a magnetic field reversal. However, when a similar thing was found with Neptune, this *AD hoc* explanation was upset. These observations are consistent with ages of thousands of years rather than billions, Figure 10.43.

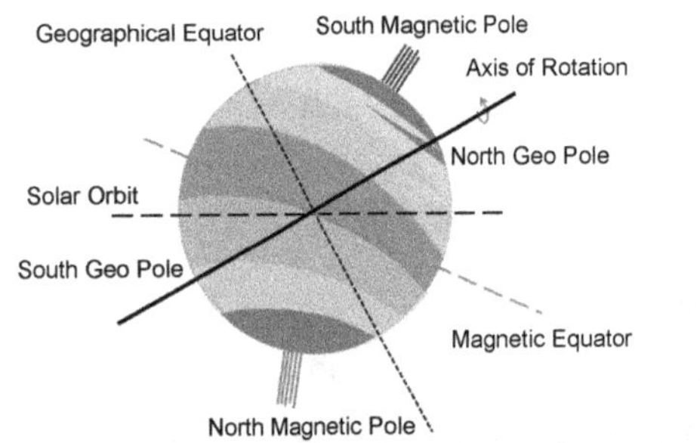

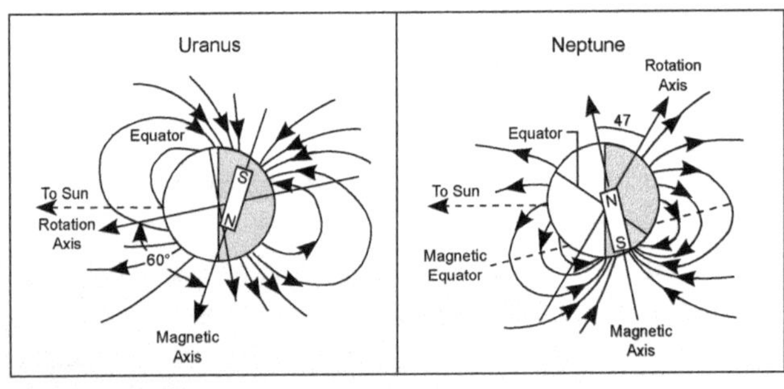

Fig.10.43: Uranus and Neptune both have Magnetic Fields Significantly Off-Axis, which is an Unstable Situation

82. **Evident 82:** The orbit of Pluto is chaotic on a 20 million year time scale and affects the rest of the solar system, which would also become unstable on that time scale, suggesting that it must be much younger. (See: Rothman, T., God takes a nap, *Scientific American* **259**(4):20, 1988).

83. **Evident 83:** The existence of short-period comets (orbital period less than 200 years), e.g., Halley, which have a life of less than 20,000 years, is consistent with an age of the solar system of less than 10,000 years. *ad hoc* hypotheses have to be invented to circumvent this evidence (see **Kuiper Belt**, Figure 10.44.

Fig.10.44: Halley Comet, has a Life of Less than 20,000 Years

84. **Evident 84:** "Near-infrared spectra of the Kuiper Belt Object, Quaoar and the suspected Kuiper Belt Object, Charon, indicate both contain crystalline water ice and ammonia hydrate. This watery material cannot be much older than 10 million years, which is consistent with a young solar system, not one that is 5 billion years old."

85. **Evident 85:** Lifetime of long-period comets (orbital period greater than 200 years) that are sun-grazing comets or others like Hyakutake or Hale–Bopp means they could not have originated with the solar system 4.5 billion years ago. However, their existence is consistent with a young age for the solar system. Again an *ad hoc* Oort Cloud was invented to try to account for these comets still being present after billions of years. See, Comets and the age of the solar system.

86. **Evident 86:** The maximum expected lifetime of near-earth asteroids is of the order of one million years, after which they collide with the sun. And the Yarkovsky effect moves main belt asteroids into near-earth orbits faster than had been thought. This brings into question the origin of asteroids with the formation of the solar system (the usual scenario), or the solar system is much younger than the 4.5 billion years claimed. Henry, J., The asteroid belt: indications of its youth, *Creation Matters* **11**(2):2, 2006.

87. **Evident 87:** The lifetime of binary asteroids—where a tiny asteroid 'moon' orbits a larger asteroid— in the main belt (they represent about 15–17% of the total): tidal effects limit the life of such binary systems to about 100,000 years. The difficulties in conceiving of any scenario for getting binaries to form in such numbers to keep up the population, led some astronomers to doubt their existence, but space probes confirmed it (Henry, J., The asteroid belt: indications of its youth, *Creation Matters* **11**(2):2, 2006).

88. **Evident 88:** The observed rapid rate of change in stars contradicts the vast ages assigned to stellar evolution. For example, Sakurai's Object in Sagittarius: in 1994, this star was most likely a white dwarf in the center of a planetary nebula; by 1997 it had grown to a bright yellow giant, about 80 times wider than the sun (*Astronomy & Astrophysics* **321**:L17, 1997). In 1998, it had expanded even further, to a red supergiant 150 times wider than the sun. But then it shrank just as quickly; by 2002 the star itself was invisible even to the most powerful optical telescopes, although it is detectable in the infrared, which shines through the dust (Muir, H., 2003, Back from the dead, *New Scientist* **177**(2384):28–31).

89. **Evident 89:** The faint young sun paradox. According to stellar evolution theory, as the sun's core transforms from hydrogen to helium by means of nuclear fusion, the mean molecular weight increases, which would compress the sun's core increasing fusion rate. The upshot is that over several billion years, the sun ought to have brightened 40% since its formation and 25% since the appearance of life on earth. For the latter, this translates into a 16–18 °C temperature increase on the earth. The current average temperature is 15 °C, so the earth ought to have had a -2 °C or so temperature when life appeared. See: Faulkner, D., The young faint Sun paradox and the age of the solar system, *J. Creation (TJ)* **15**(2):3–4, 2001. As of 2010, the faint young sun remains a problem: Kasting, J.F., Early Earth: Faint young Sun redux, *Nature* **464**:687–689, 1 April 2010; doi:10.1038/464687a; www.nature.com/nature/journal/v464/n7289/full/464687a.html

90. **Evident 90:** Evidence of (very) recent geological activity (tectonic movements) on the moon is inconsistent with its supposed age of billions of years and its hot origin. Watters, T.R., *et al.*, Evidence of Recent Thrust Faulting on the Moon Revealed by the Lunar Reconnaissance Orbiter Camera, *Science* **329**(5994):936–940, 20 August 2010; DOI: 10.1126/science.1189590 ("This detection, coupled with the very young apparent age of the faults, suggests global late-stage contraction of the Moon.") NASA pictures support biblical origin for Moon.

91. **Evident 91:** The giant gas planets Jupiter and Saturn radiate more energy than they receive from the sun, suggesting a recent origin. Jupiter radiates almost twice as much energy as it receives from the sun, indicating that it may be less than 1 % of the presumed 4.5 billion years old solar system. Saturn radiates nearly twice as much energy per unit mass as Jupiter. See The age of the Jovian planets.

92. **Evident 92:** Speedy stars are consistent with a young age for the universe. For example, many stars in the dwarf galaxies in the Local Group are moving away from each other at speeds estimated at to 10–12 km/s. At these speeds, the stars should have dispersed in 100 Ma, which, compared with the supposed 14,000 Ma age of the universe, is a short time. See Fast stars challenge big bang origin for dwarf galaxies.

93. **Evident 93:** The ageing of spiral galaxies (much less than 200 million years) is not consistent with their supposed age of many billions of years. The discovery of extremely 'young' spiral galaxies highlights the problem of this evidence for the evolutionary ages assumed, Figure 10.45.

Fig.10.45: Deep-Space Objects Are Young - Secular astronomers claim our universe is unimaginably ancient—almost 14 billion years old. Yet the Bible clearly teaches that God created the universe in the relatively recent past, about 6,000 years ago.

94. **Evident 94:** The number of types I supernova remnants (SNRs) observable in our galaxy is consistent with an age of thousands of years, not millions or billions. See Davies, K., *Proc. 3prd ICC*, pp. 175–184, 1994.

95. **Evident 95:** There is a great paucity of highly expanded SNRs compared to what is expected under evolutionary cosmogony. See supernova remnants.

V. Human History is Consistent with a Young Age of the Earth

96. **Evident 96:** Human population growth. Less than 0.5% p.a. growth from six people 4,500 years ago would produce today's population, Figure 10.46. Where are all the people? if we have been here much longer?

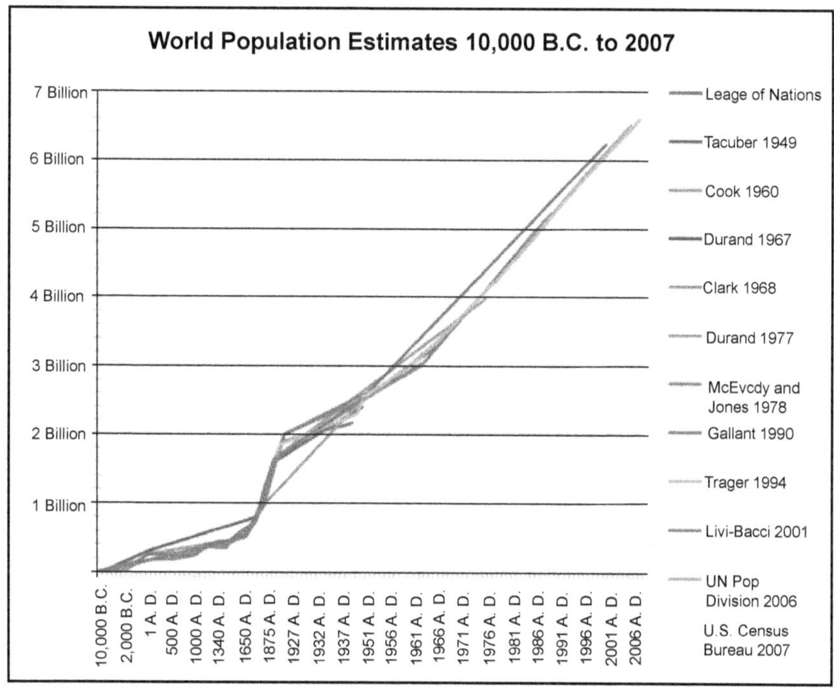

Fig.10.46: Population Distribution Since Noah

97. **Evident 97:** 'Stone age' human skeletons and artefacts. There are not enough for 100,000 years of a human population of just one million, let alone more people (10 million?). See Where are all the people?

98. **Evident 98:** Length of recorded history. Origin of various civilizations, writing, etc., all about the same time several thousand years ago. See Evidence for a young world.

99. **Evident 99:** Languages. Similarities in languages claimed to be separated by many tens of thousands of years speaks against the supposed ages (e.g. compare some aboriginal languages in Australia with languages in south-eastern India and Sri Lanka). See The Tower of Babel account affirmed by linguistics.

100. **Evident 100:** Common cultural 'myths' speak of recent separation of peoples around the world. An example of this is the frequency of stories of an earth-destroying flood.

101. **Evident 101:** Origin of agriculture. Secular dating puts it at about 10,000 years and yet that same chronology says that modern man has supposedly been around for at least 200,000 years. Surely someone would have worked out much sooner how to sow seeds of plants to produce food, Figure 10.47.

Fig.10.47: *Evidence of Advanced Agricultural Knowledge Dating Back 8,000 Years*

APPENDIX

a.   Many Christians who compromise with billions of years assert that the sun and other heavenly bodies were not really 'made' on the fourth 'day' (millions of years long). Rather, they 'appeared' to a hypothetical observer on earth when a dense cloud layer dissipated after millions of years. But this (mis)interpretation is not allowed by the Hebrew words used. The word *'asah* means 'make' throughout *Genesis 1*, and is sometimes used interchangeably with 'create' (*bara'*), e.g. in *Gen. 1:26–27*. It is pure desperation to apply a different meaning to the same word in the same grammatical construction in the same passage, just to fit in with atheistic evolutionary ideas like the big bang.

   i.   If God had *meant* 'appeared', then He presumably would have *used* the Hebrew word for appear (*ra'ah*), as when the dry land 'appeared' as the waters gathered in one place on Day 3 (Gen. 1:9). This is supported by Hebrew scholars who have translated the Bible into English. Over 20 major translations were checked, and all clearly teach that the sun, moon and stars were *made* on the fourth day. Return to text.

b.   Although the 'parent' galaxy has not actually been observed, many astronomers believe that supermassive black holes are directly associated with galactic nuclei. That is, they are found at the centers of large galaxies. Our own galaxy is believed to have a supermassive black hole, several million times the mass of the sun, at its center.

c.   From the Greek, ἐκένωσεν *ekenōsen*

d.   Sound cannot travel through empty space because sound waves are compressions of a material medium. However, the early universe (according to the big bang cosmology) would have been very dense. It would not have been 'empty' and this would have allowed sound to travel.

e.   Magnetic fields often split spectral lines—the Zeeman Effect—and this is detectable in sunspots.

f.   Astronomers measure the redshift of distant objects, called quasars, to calibrate the distances to objects such as the galaxies in this article.

REFERENCES

1. Levy, D., 'Blazar' illuminate's era when stars and galaxies formed, Stanford Report, 22 June 2004, online: <news-service.Stanford.edu/news/2004/july7/blazar-77.html>.

2. Romani, R.W. *et al.*, Q0906+6930: The highest-redshift blazar, <xxx.lanl.gov/abs/astro-ph/0406252>, 9 June 2004.

3. Hartnett, J.G., The heavens declare a different story! *Journal of Creation* 17(2):94–97, 2003.

4. Rigg, A., Young galaxies too old for the big bang, *Creation* 26(3):15, 2004.

5. Faintest spectra ever raise glaring question: Why do galaxies in the young universe appear so mature? *Gemini Observatory press release*, 5 January 2004, <www.gemini.edu/project/announcements/press/2004-1.html>.

6. De Nike, L., Glimpse at early universe reveals surprisingly mature galaxies, *Johns Hopkins Gazette*, 19 July 2004, <www.jhu.edu/~gazette/2004/19jul04/19early.html>.

7. Oard, M.J., The big bang problem of early maturity, *Journal of Creation* **18**(1):15–16, 2004.

8. For more information on redshift and problems with using it as a measure of distance, see reviews of Arp's books in *Journal of Creation* 14(3):39–45, 46–50, 2000.

9. Hartnett, J., Francis Filament: a large scale structure that is big, big, big bang trouble. Is it really so large? *Journal of Creation* 18(1):16–17, 2004.

10. 101 evidences for recent creation are provided in the article at creation.com/age.

11. Lisle, J., Blue stars confirm recent creation, *Acts & Facts* 41(9):16, 2012; icr.org.

12. Hubble observations cast further doubt on how globular clusters formed; astronomy.com; 20 November 2014.

13. Pfahl, E., Rappaport, S., and Podsiadlowski, P., A comprehensive study of neutron star retention in globular clusters, *Astrophysical Journal* 573:283–305, 2002; | doi:10.1086/340494.

14. This article is based on a paper by Keith Davies, Distribution of Supernova Remnants in the Galaxy, *Proceedings of the Third International Conference on Creationism*, Creation Science Fellowship, Pittsburgh, ed. E. Walsh, pp. 175–184, 1994.

15. See the article 'supernova', *Encyclopædia Britannica*, 15th Ed., 11:401, 1992.

16. A light year is a measure of distance, not time. It is the distance that light nowadays travels for one year in a vacuum—9.46 million million kilometres (5.87 million million miles).

17. Clark and Caswell, 1976. *Monthly Notices of the Royal Astronomical Society*, 174:267; cited in Ref. 1.

18. Mr Davies' original draft of Ref. 1 had 1.28 parsecs, which is 7 light years. Somehow in the final version, a typo occurred and 7 ly became 7 pcs, which would be 23 ly.

19. Keith Davies, Distribution of Supernova Remnants in the Galaxy, Ref. 1, has detailed observational limitation formulæ.

20. *Adiabatic* means 'not transferring heat to or from its surroundings'. During the second stage, the SNR loses very little thermal energy.

21. *Isothermal* means 'staying at the same temperature'. During the third stage, the SNR should stay at about the same temperature and radiate excess thermal (heat) energy.

22. Chown, M., What a star! *New Scientist* 162(2192):17, 1999.

23. Seife, C., Thank our lucky star, *New Scientist* 161(2168):15, 1999.

24. The researchers later theorised that such flares are triggered by the large magnetic field of a closely orbiting gas giant planet (Schaefer, B., reported in *Discover* 20(4):19, 1999). But they have not been seen, and the standard evolutionary accretion model forbids gas giants from forming that close to the star: they can grow large enough to attract gas only if they are cool enough to incorporate ice into the accreting body.

25. 'Bethe, Hans Albrecht', *The New Encyclopædia Britannica* 2:173, 15th Ed. 1992.

26. Four hydrogen atoms (mass = 1.008) convert to helium (mass 4.0039) losing 0.0281 atomic mass units (1 AMU = 1.66 x 10^{-27} kg), releasing 10×4.2^{-12} joules of energy.

27. Man-made hydrogen bombs use the heavy hydrogen isotopes deuterium and tritium, plus some lithium. The sun uses ordinary hydrogen—a reaction that requires higher temperatures. But Bethe calculated that carbon-12 nuclei in stars could catalyse the reaction, where nitrogen and oxygen also have a role, hence the *CNO cycle*. But the sun's core is not thought to be hot enough for the CNO cycle, and is thought to use the proton-proton (PP) chain instead.

28. The nett fusion reaction is $4^1H \rightarrow {}^4He + 2e^+ + 2\nu_e$, where e^+ is a positron or anti-electron, and ν_e is an electron-neutrino. If the sun were powered by nuclear fission (instead of fusion) or by radioactive decay of heavy elements, *antineutrinos* would be produced instead.

29. Snelling, A.A., *Solar neutrinos—the critical shortfall still elusive*, Journal of Creation 11(3):253–254, 1997.

30. See Dr Snelling's four-part study, *Creation* 11(1–4), 1989, including *That Matter of the Shrinking Sun*. Uniform shrinkage at this rate would mean that 100 million years ago the sun would have been too large for life on earth.

31. Eddy, J.A., quoted by Kazmann, R.G., It's about time: 4.5 billion years, *Geotimes* 23:18–20, 1978.

32. Reeves, H., The Origin of the Solar System, in: *The Origin of the Solar System*, Dermott, S.F., Ed., John Wiley & Sons, New York, p. 9, 1978.

33. Taylor, S.R., *Solar System Evolution: A New Perspective*, Cambridge University Press, p. 53, 1992.

34. See Spencer, W., *Revelations in the solar system*, Creation 19(3):26–29, 1997.

35. 'Galileo', *The New Encyclopædia Britannica* 19:638–640, 15th Ed. 1992.

36. Grigg, R., *The Galileo 'twist'*, Creation 19(4):30–32, 1997.

37. Sarfati, J., *Refuting Evolution* 5th ed., *ch. 7: Astronomy*, Creation Book Publishers, Atlanta, Georgia, USA, 2012.

38. Sarfati, J., *The moon: the light that rules the night*, Creation 20(4):36–39, 1998.

39. Weiss, P., The sun also writhes, *Science News* 153(13):200–202, 1999.

40. Irion, R. The great eclipse: Crown of fire, *New Scientist* 162(2188):30–33, 1999, discusses rapidly oscillating magnetic waves as a possible energy source.

41. 'Sun', *The New Encyclopædia Britannica* 11:387–388, 15th Ed. 1992.

42. 'Solar System, The', *The New Encyclopædia Britannica* 27:504–603, 15th Ed. 1992.

43. Synodic period is the time for the sun to return to the same orientation towards the earth.

44. Newton, R., *'Missing' neutrinos found! No longer an 'age' indicator*, Journal of Creation 16(3):123–125, 2002.

45. A Trace of the Corona, scientificamerican.com/article/a-trace-of-the-corona, 1 Dec 2000.

46. See Sarfati, J., The Incarnation: Why did God become Man?, December 2010; creation.com/incarnation.

47. See Sarfati, J., Why Bible history matters, *Creation* 33(4):18–21, 2011, as well as creation.com/nt and creation.com/gen-hist.

48. The extended passage cites Genesis 1:27 and 2:24 as real history, and about the same man and woman. In the parallel passage in Matthew 19:4–5, Jesus attributes Genesis 2:24 to the One who created them, i.e. to God himself.

49. See Sarfati, J., Biblical chronogenealogies, *J. Creation* 17(3):14–18, December 2003; creation.com/chronogenealogy.

50. In Australia, as in most British systems, 'professor' means head of a department. In the US, professor simply means someone who teaches at tertiary level, which could apply to someone, for example, who would be called a 'junior lecturer' in a British system.

51. ISCAST (Institute for the Study of Christianity in an Age of Science and Technology); see Sarfati, J., The Skeptics and their 'Churchian' Allies, November 1998; creation.com/iscast.

52. From Greek λοβός *lobos* = lobe (of the brain), and τομή tomē = slice/cut. A serious and irreversible operation that cuts certain connections to the cerebral cortex, the 'thinking' part of the brain.

53. Sparks, K., "After Inerrancy, Evangelicals and the Bible in the Postmodern Age, part 4" Biologos Forum, 26 June 2010. See also Cosner, L., Evolutionary syncretism: a critique of Biologos, 7 September 2010; creation.com/biologos.

54. Eisenstein, D.J. *et al.*, Detection of the baryon acoustic peak in the large-scale correlation function of SDSS luminous red galaxies, 10 January 2005.

55. The regions of higher density in the Cosmic Microwave Background are also supposedly produced in a similar fashion. However, the weakness of the ripples is highly problematic for big bang cosmology. See: Newton, R., Light Travel-time: a problem for the big bang, *Creation* 25(4):48–49, 2003.

56. Sarfati, J., *Refuting Compromise*, Master Books, Green Forest, AR, 2004, for an excellent refutation of the big bang and 'progressive creationism' (billions of years).

57. Chown, M., Let there be light, *New Scientist* 157(2120):26–30, 1998.See also: <creation.com/starform>.

58. Wieland, C., Secular scientists blast the big bang: what now for naïve apologetics? *Creation* 27(2):23–25, 2005.

59. Hartnett, J., New evidence: we really are at the centre of the universe, *Journal of Creation* 18(1):9, 2004. Humphreys, D.R., Our galaxy is the centre of the universe, 'quantized' red shifts show, *Journal of Creation* 16(2):95–104, 2002.

60. Lerner, E., Bucking the big bang, *New Scientist* 182(2448)20, 22 May 2004. See also the book by Alex Williams and Dr John Hartnett, *Dismantling the big bang*.

61. Editorial: In the beginning, *New Scientist* 213(2847):3, 14 January 2012.

62. Grossman, L., Death of the eternal cosmos, *New Scientist* 213(2847):6–7, 14 January 2012.

63. See Sarfati, J., *Refuting Compromise [updated and expanded] a biblical and scientific refutation of progressive creationism (billions of years) as popularized by astronomer Hugh Ross*, Creation Book Publishers, 2011.

THE UNIVERSE IS FINELY TUNED FOR LIFE

FINELY TUNED UNIVERSE

Strong evidence for a Designer comes from the fine-tuning of the universal constants and the solar system, Figure 11.1. The following examples are stated to substantiate our believe. They are consistent with Colossians 1: 16-17, for through him God created everything in the heavenly realms and on earth. He made the things we can see and the things we cannot see—such as thrones, kingdoms, rulers and authorities in the unseen world. Everything was created through him and for him. He existed before anything else, and he holds all creation together.

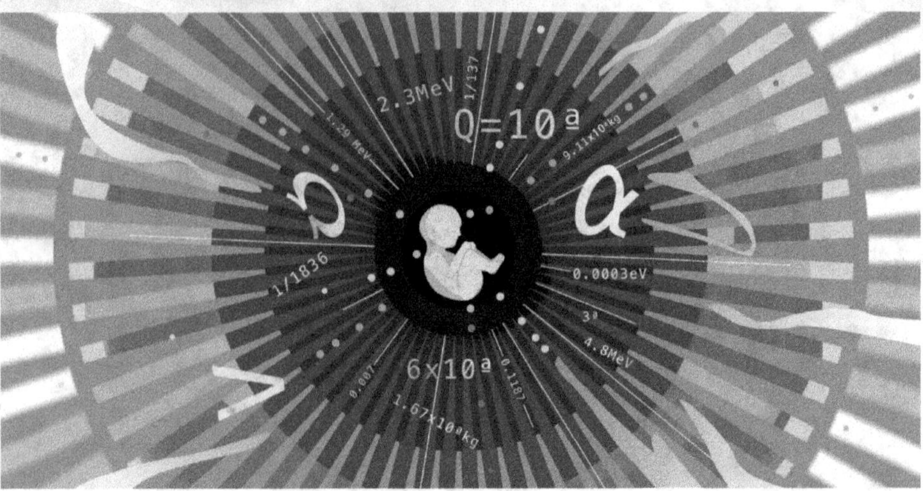

Fig.11.1: The Universe is Finely Tunes for Life – Curtsy John Templeton Foundation

1. The electromagnetic coupling constant binds electrons to protons in atoms. If it was smaller, fewer electrons could be held. If it was larger, electrons would be held too tightly to bond with other atoms.

2. Ratio of electron to proton mass (1:1836). Again, if this was larger or smaller, molecules could not form.

3. Carbon and oxygen nuclei have finely tuned energy levels, Figure 11.2.

4. Electromagnetic and gravitational forces are finely tuned, so the right kind of star can be stable.

5. Our sun is the right color. If it was redder or bluer, photosynthetic response would be weaker, Figure 11.3.

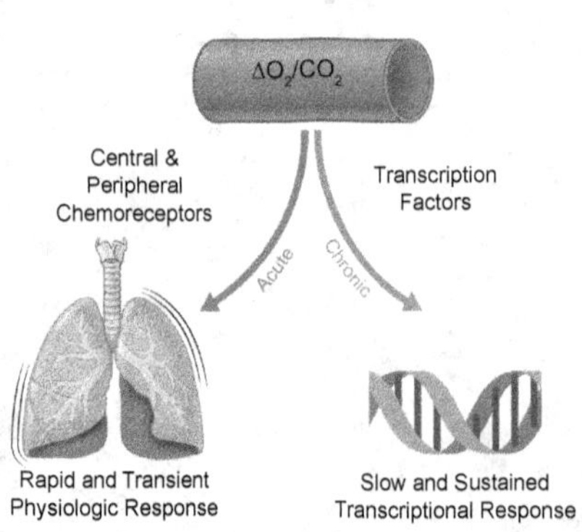

Fig.11.2: Mechanisms and Consequences of Oxygen and Carbon Dioxide Sensing in Mammals | Curtsy Physiological Reviews

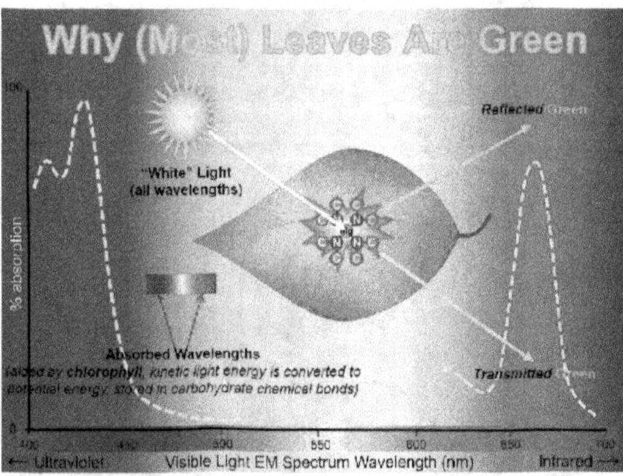

Fig.11.3: Our sun is the right color. If it was redder or bluer, photosynthetic response would be weaker – Curtsy Henry Norman

6. Our sun is also the right mass. If it was larger, its brightness would change too quickly and there would be too much high energy radiation. If it was smaller, the range of planetary distances able to support life would be too narrow; the right distance would be so close to the star that tidal forces would disrupt the planet's rotational period. UV radiation would also be inadequate for photosynthesis.

7. The earth's distance from the sun is crucial for a stable water cycle. Too far away, and most water would freeze; too close and most water would boil.

8. The earth's gravity, axial tilt, rotation period, magnetic field, crust thickness, oxygen/nitrogen ratio, carbon dioxide, water vapor and ozone levels are just right.

Former atheist Sir "Fred Hoyle" states, 'commonsense interpretation of the facts is that a super-intelligence has monkeyed with physics, as well as chemistry and biology, and that there are no blind forces in nature.'

GOD THE CAUSE

A Skeptic asks: "Why would a deity make a universe that's 13.8 billion light years in radius, in which practically none of it is useable? It's just a waste of stuff?"

There are, at least two reasons why God made the universe so huge:

1. According to *Psalm 19:1*, "The heavens declare the glory of God, and the sky above proclaims his handiwork." The fact that God created billions of stars and placed them in the trllions of galaxies that contain them demonstrates, for those willing to see it, the omnipotence and omniscience of Almighty God, Figure 11.4.

2. It also demonstrates that the worldview that says that all the mass and energy of the billions of stars in each of the 200 billion galaxies were once contained in a point of zero dimensions is **scientific nonsense**.

    Atheist: "Sam Harris" chips in with: "It is the most painfully wasteful system we could devise. Nowhere in that mayhem is there the suggestion that man is somehow central." This statement is wrong on two counts:

3. Dr Usama Hasan (Fellow of the Royal Astronomical Society) explains that, at least within our solar system, Jupiter is important for life on Earth because Jupiter's gravity sweeps up comets, meteorites and all kinds of debris which could possibly destroy Earth, Figure 11.5.

4. Earth is indeed the primary focus of God. According to the record in Genesis that God Himself has given us, God first prepared the earth to be a habitation for mankind on Days 1, 2, and 3, of Creation week, and this was *before* He made the stars and galaxies on Day 4. Furthermore, it was on Earth that the Son of God, the Lord Jesus Christ, was incarnated and died to pay the penalty for mankind's sin; not to pay for the sin of demons or imaginary aliens.

Fig.11.4: *The Heavens Declare The Glory of God*

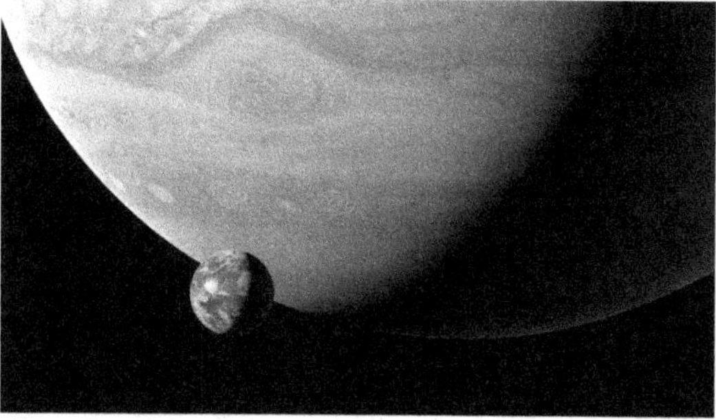

Fig.11.5: *Jupiter is Important for Life on Earth because Jupiter's Gravity Sweeps up Comets – Curtsy Jamie Carter*

Furthermore, it's tiresome to see atheists point to the immensity of the universe as if it were news. However, this truth has been well known for almost all the history of the Church. E.g., the Roman Christian philosopher Boëthius (AD c. 480–524/525), in prison awaiting trial and execution for an unjust charge of treason, wrote *The Consolation of Philosophy*, an imaginary dialogue between himself and 'Lady Philosophy.' She points out that just as the earth is just a point in space, how much more insignificant is any glory of any of its inhabitants:

As you may have heard from the demonstrations of the astronomers, in comparison to the vastness of the heavens, it is agreed that the whole extent of the earth has the value of a mere point; that is to say, were the earth to be compared to the vastness of the heavenly sphere, it would be judged to have no volume at all.

This was one of the most widely read and influential books in the West during most of the Middle Ages. Therefore, churchmen were well aware of how tiny the earth is, without considering it the slightest threat to faith. Furthermore, it was exactly this consideration that led some medieval scientists to propose that it was the tiny earth that rotated rather than the immense cosmos revolving around it.

Theist Professor "William Lane Craig" makes the point: "Since something cannot come out of nothing, since being does not come from non-being, there must exist some sort of transcendent Cause which brought the universe into existence, and this is the traditional concept of what theists have meant by God."

Atheist Professor "Steven Weinberg" responds: "We have to wait for an un-theistic explanation. Very often you have to wait a long time. But that's what science is. Science is looking for non-theistic explanations of what we see in nature." Notice the self-serving emphasis on the 'non-theistic' worldview. Today only materialistic notions are *allowed* to be entertained by scientists, and hence 'science' is hugely 'blinkered'.

This doesn't mean God does not exist—only that everything He has done is excluded by decree—not by evidence—as per this [in]famous quote by atheist evolutionist Professor "Richard Lewontin":

Our willingness to accept scientific claims that are against common sense is the key to an understanding of the real struggle between science and the supernatural. We take the side of science in spite of the patent absurdity of some of its constructs, in spite of its failure to fulfill many of its extravagant promises of health and life, in spite of the tolerance of the scientific community for unsubstantiated just-so stories, because we have a prior commitment, a commitment to materialism. It is not that the methods and institutions of science somehow compel us to accept a material explanation of the phenomenal world, but, on the contrary, that we are forced by our a priori adherence to material causes to create an apparatus of investigation and a set of concepts that produce material explanations, no matter how counter-intuitive, no matter how mystifying to the uninitiated. Moreover, that materialism is absolute, for we cannot allow a Divine Foot in the door.

Atheist Michael Shermer asks: "*If you posit a god that started it, I can just say, who created God? … If you say that God is that which does not need to be created, why can't the universe be that which does not need to be created?*"

The answer is that everything that has a beginning needs a creator. God did not have a beginning, so He *did not* need to be created. On the other hand, the universe did have a beginning, so it *did* need to be created.

WHO FINE-TUNED THE UNIVERSE?

Having metamorphosed the problem of how our one universe came into existence into the multi-faceted problem of how millions of hypothetical universes could have come into existence, without solving anything. The discussion is centered around why our universe obeys the laws of physics. These laws operate across the entire universe, impacting the movement of galaxies, and controlling events in the miniscule world of the atom, Figure 10.6.

Fig.11.6: Scientific Evidence for God: A Fine-Tuned Universe –
Curtsy Robert Clifton Robertson

Dr Mario Livio (Senior Astrophysicist, Space Telescope Science Institute, Baltimore) tells viewers that he doesn't think we will ever be able to answer the question why there are laws. But once again "Paul Davies" comes up trumps with the comment: "Science proceeds on the basis that the universe is ordered in a rational and intelligent way. It doesn't have to be, but that's the way it is. It sort of suggests that there is a lawmaker, an omnipotent, immutable being who imposed this order on the universe and presides over it."

"Paul Davies" adds: "To say that we need brilliant minds to figure out what is going on, but mind doesn't play a part in the structure and origin of that, seems to me to be rather peculiar."

Atheist Professor "Steven Weinberg" isolates "dark energy' as an extreme example of fine tuning. Concerning this, Professor Mario Livio says: "About 73% of the universe is in the form of this dark energy which, we see its effect, it's pushing the expansion of the universe to accelerate, but we really don't have a clue what it is, and we certainly don't see it directly."

Therefore, what is this incognito dark energy? It, along with 'dark matter', are terms used by astronomers to explain motions in the cosmos that they cannot explain by the laws of physics. Another term for these entities is 'fudge factors,' needed to make the big bang 'work'. The author will discuss the entities of Dark Matter and Dark Energy in a later chapter, based on Biblical understanding.

The formation of galaxies is a huge problem for big bangers, because what the big bang supposedly produced from an initial point was an *expanding mass of gas*. The problem is how to get it to stop expanding and start contracting in localized regions so as to allow stars and galaxies to form, but without everything disappearing back into the singularity. This, according to Prof Weinberg, requires fine tuning to the order of at least 56 decimal places, with the correct amount of dark matter to get the stars, and then the galaxies, to form.

Not being able to explain any of this, the scientists offer the atheists' panacea—multiverse theory: "With so many potential universes popping into existence, perhaps the conditions could be just right in one of them to have allowed stars and galaxies to form."

Professor Weinberg adds: "Maybe in every trillion, trillion, trillion, trillion [universes], Figure 11.7, there's one where the net dark energy, the sum of all the contributions, is small enough to allow the expansion to be slow enough for life to form."

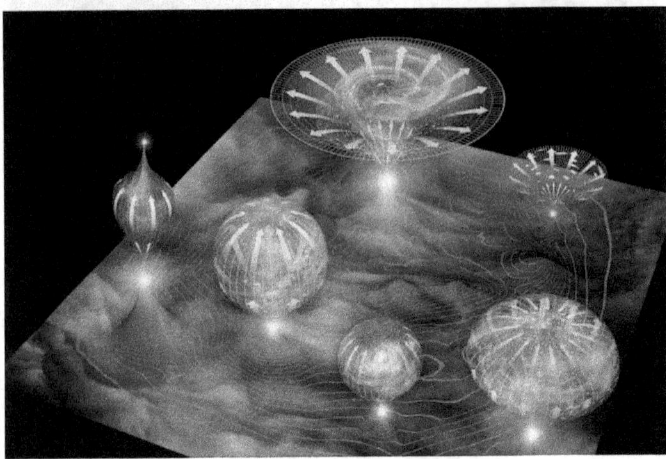

Fig.11.7: Big Bang Hypothesis Opened the Door to Multi-Universe Hypothesis – Curtsy Moonrunner Design

Indeed, it takes far more faith—at least in the unbiblical sense of credulity—to believe in the big bang than it takes to believe that Almighty God created everything by the power of His Word, in the way that He says He did in Genesis.

WHO FINELY TINED THE UNIVERSE?

The fact that the universe is fine-tuned to allow for life is no longer a matter of conjecture. What is at issue is whether this fine-tuning was caused by a Being or was the natural result of physics. Critics object to the idea that our universe was designed for life because most of the cosmos is completely inhospitable and hostile to life. When we understand what these fine-tuned physical constants really involve, it becomes clear that only under certain specific and rare conditions can physical life ever exist on a planet.

The scientific evidence we have today confirms that the clear intent of these physical constants was not to seed the entire universe with life, but to allow precise conditions only on specific planets like earth. By their very nature the fine-tuned physical constants of our universe would make most of the universe inhospitable to life, while providing a very narrow possibility of life on at least one planet.

Fig.11.8: The universe proclaims the existence of God. The skies display his craftsmanship. Day after day they continue to speak; night after night they make him known. They speak without a sound or word; their voice is never heard. Yet their message has gone throughout the earth, and their words to all the world

When persons claim that it was the God of the Bible who is responsible for the universe, it is because the text of this book describes God in this manner, as the source of all that exists. He is not described as a Creator who wanted to fill the universe with life, but chose just one planet to have the precise conditions necessary to support human beings.

The text of Psalms 19, written 3,000 years ago, states that the universe was created as a message to the people of earth: "*The universe proclaims the existence of God. The skies display his craftsmanship. Day after day they continue to speak; night after night they make him known. They speak without a sound or word; their voice is never heard. Yet their message has gone throughout the earth, and their words to all the world.*" Figure 11.8.

Other texts describe the universe as existing exactly as science describes it today, "stretched out." This is no arbitrary term, because it accurately describes the precise procedure for how the universe began and proceeded. During the initial explosion of pure energy, matter was distributed in one direction, while space was created simultaneously to accommodate the expanding matter. This expansion did not spread out circular as with most explosions, but moved out as one would expand a curtain—from one point to another. This is precisely how the Bible described the expansion of the universe, more than 3,000 years ago:

Fig.11.9: Who stretch out the heavens like a curtain

*O Lord my God, You are very great: You are clothed with honor and majesty, Who cover Yourself with light as with a garment, **Who stretch out the heavens like a curtain**.* ~Psalms 104:1-2. Figure 11.9.

The prophet Jeremiah wrote 2,700 years ago and described the expansion of the initial universe as taking place, "*at His discretion.*" In other words, the Creator knew beforehand that the initial expansion of pure energy had to be conducted in a specific manner or it would not produce results that would later allow for life on earth. He used His wisdom and discretion to determine how this process must proceed.

"*He has made the earth by His power, He has established the world by His wisdom, And has **stretched out the heavens at His discretion**.*" ~Jeremiah 10:12. Figure 11.10

The Hebrew word used to describe the stretching out of the universe when it began, is נָטָה nâṭâ, literally to expand from one point to another like one opens a curtain in the morning. This is incredible because during the time that Jeremiah lived on earth in 655 B.C., no one knew that the universe proceeded from its beginning in this manner. Only in the last 80 years has man acquired the knowledge that this is true.

This same description for the universe is repeated twice by the prophet Isaiah, also

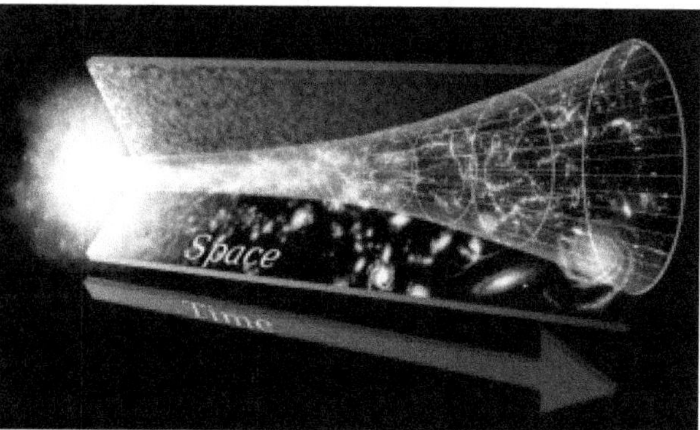

Fig.11.10: Expanding Universe - *He has made the earth by His power, He has established the world by His wisdom, And has Stretched out the Heavens at His Discretion*

writing 2,700 years ago: Isaiah 42:5 states: *"Thus says God the Lord, who created the heavens and **stretched them out**..."* and again in Isaiah 40:22: *"It is He who sits above the vault of the earth, and its inhabitants are like grasshoppers, who **stretches out the heavens like a curtain** and spreads them out."*

Theoretical Physicist, "Alan Guth," at the Massachusetts Institute of Technology, put forth the idea in the 80's, that if during the initial moments of the Big Bang, (Moment of Creation) the massive expansion of material was not uniform or controlled as it was stretched out, the resulting universe would have been quite disorderly. What we have discovered is that when the universe began, it was quite orderly with an extreme low state of entropy. Under these conditions in which the universe began, a low state of entropy would not have been possible by a natural process. Something or someone, acted upon these initial forces to cause a low state of entropy so that matter could properly form into clumps, making the production of galaxies and stars a reality much later.

Instead, as if by design, a microsecond later, the entire universe jumped in size by ten trillion-trillion (10^{25}). It was at this point that the entire expansion stopped and a normal rate of expansion began. This rapid and sudden expansion **stretched out** the irregularities of the initial disorderly explosion, exactly as the Bible described the universe over 2,700 years ago. Once the rapid expansion ceased and a normal rate of expansion started once again, the material of the universe could expand into an orderly and even universe.

1. Where did the matter for the universe come from since nothing existed prior to this moment?

2. What caused the initial sudden expansion?

3. What force made the expansion stop, only to resume in an orderly fashion?

4. How did the source of these essential procedures know exactly how and when they must take place?

According to the comments made by critics of a Being as the source for the universe, there are naturalistic explanations for these conditions and we don't need God.

"The orbits of the planets are not fine-tuned ellipses because of gods, but because of physics. Humans are not fine-tuned for their qualities because Yahweh crafted man from dust and woman from man's rib, but because of the extraordinary power of natural selection and other evolutionary pressures." ~Luke Muehlhauser

The problem with statements like these is that are made with such authority, but they are simply not true. Isaac Newton observed comets and planets moving in concentric orbits from many different positions, and concluded that this would be impossible apart from the design and engineering of an intelligence. In Newton's view, a transcendent Being of immense power created gravity to act on these planets and cause them to follow the same center point, ordered by the laws of physics.

ISAAC NEWTON AND GRAVITATIONAL FORCE

Sir Isaac Newton's work on the physics of time and space are considered the foundation for all scientific knowledge today. On July 5, 1687, Newton published three books that have been referred to as the *Principia,* a Latin term that describes his Mathematical Principles of Natural Philosophy, Figure 11.11.

Newton wrote in his Principia that the order and design which we observe in time and space are inseparable from the existence of God. Concerning the evidence presented by the universe through science, mathematics, and astronomy, Newton

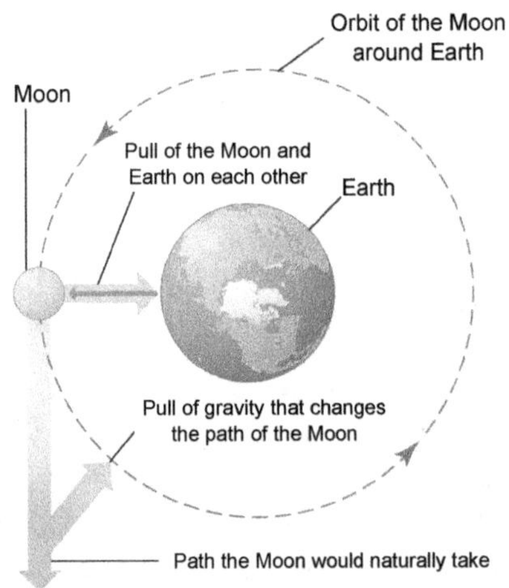

Fig.11.11: Gravity – Newton's Law of Gravity

said this, *"The most beautiful system of the sun, planets, and comets could only proceed from the counsel and dominion of an intelligent and powerful Being."*

"The true God is a living, intelligent, and powerful being. His duration reaches from eternity to eternity; His presence from infinity to infinity....He governs all things and knows all things that are or can be done. He is not eternity and infinity, but eternal and infinite; he is not duration or space, but he endures and is present. He endures forever, and is everywhere present; and, by existing always and everywhere, he constitutes duration and space. This most beautiful system of the sun, planets, and comets could only proceed from the counsel and dominion of an intelligent and powerful Being....He is omnipresent not virtually only, but also substantially; for virtue cannot subsist without substance. In him are all things contained and moved." [17] ~Isaac Newton

Isaac Newton viewed the evidence for God's existence as self-evident by the presence of the universe itself. In his observance of all the scientific and mathematical absolutes of the Cosmos, he said, *"the Creator cannot be denied in the presence of such a magnificent creation."* Newton found it preposterous that any intelligent person could imagine that the universe came into being without an unlimited intelligence as its cause. Newton came to these conclusions based on the evidence of science, mathematics, and astronomy.

Albert Einstein demonstrated that gravity occurs as a result of space-time bending. In the final conclusion of his theories, Einstein confirms Newton's judgment that orbits follow geodesic trajectories by design. In other words, both of these great men of science concluded that the universe has a design and a purpose.

Newton determined that without intervention from the Creator of the universe, the stars would collide with each other on a more frequent basis. By apparent design, God has limited the amount of motion that occurs in the universe from decay due to viscosity and friction. In many of Newton's writings, he implies that the force of gravity was influenced by something immaterial—an intelligence.

"It is inconceivable that inanimate brute matter should, without the mediation of something else which is not material, operate upon and effect matter without mutual contact."

It was the conclusion of Newton that God was the force that kept the planets in orbit—as they could not sustain their motion by themselves.

"This most beautiful system of the sun, planets, and comets, could only proceed from the counsel and dominion of an intelligent Being.

This writing is an examination of the scientific evidence that describes the 209 physical constants that cause the universe to be fine-tuned for life. There is no ambiguity today regarding the provable facts of science as to whether the universe is fine-tuned for life.

1. The universe as we know it, cannot be explained by natural forces.
2. The finely-tuned universe, as empirical proof for the existence of God, has no formal logical defects.
3. The only current explanation for the existence of the universe—which can be tested and verified is the finely-tuned and designed evidence of the universe.
4. Our universe contains the precise physical constants that have the exact values that are required to allow for complex structures, such as galaxies, stars, planets, and people to exist.
5. None of these values are possible under any circumstance, by any naturalistic process.

It is because the fine-tuned universe is not impeachable by provable science, that other alternatives for our universe have been postulated, but always in the theoretical. Among these ideas are the assertions that the fine-tuning of our universe was caused by other universes, or even an infinite number of universes. Although these ideas are put forth in an attempt to call into question the fine-tuning of our universe by a Creator, they actually cause even greater problems for those who present them as answers.

INFINITE NUMBER OF OTHER UNIVERSES

The first problem that occurs is that other worlds, or an infinite number of other universes, cannot be scientifically proven. They exist only as mathematical calculations that are postulated by theoretical physicists and cosmologists. We cannot see beyond our own universe or detect anything past our own universe that would prove there actually are other universes in existence.

All the scientific data that we have observed proves, is that ours is the only universe that exists.

The creation of these mathematical models to posit other universes, were created solely for the purpose of impeaching the idea that God is the Creator of our universe. Unfortunate for the critics of a fine-tuned universe by a Creator, is that science cannot validate the other world's theory—while it does provide evidence that an intelligent Being was necessary for these fine-tuned constants to exist. It is the facts of our fine-tuned universe that impeaches the idea there is no God.

FINE-TUNED CONSTANTS AND PROVABLE SCIENCE

In addition to the existence of 209 fine-tuned physical constants that exist in the universe, there are further problems in the naturalistic argument for the existence of the universe. The following are a few of these insurmountable problems that can only be overcome by stipulation that the universe exists because of intelligent actions of an unlimited mind.

First: Life is only possible on earth because of certain elements that exist and were able to bond together in order to form molecules. In the case of earth, there are more than forty of these elements. For molecular bonding to take place, there are two essentials that must also exist: the strength of force for electromagnetism and the ratio of the electron mass in relation to the proton mass. These two are so closely tuned that if their precise balance was changed even slightly, there never would have been life on earth.

1. If the electromagnetic force was larger, atoms would cling to electrons so closely that no allocation between electrons and other atoms would be possible.

2. If the electromagnetic force was weaker, atoms would not cling to electrons at all, and there would be no allocation between electrons and other atoms, preventing all molecules from forming, resulting in no life.

3. If we are to have multiple types of molecules that will exist, the electromagnetic force is even more delicately balanced. The size and stability of the electron orbits around the nuclei of atoms, greatly depends upon the proper ratio of the electron mass to the proton mass. Unless this ratio achieves a precise balance, the chemical bonding necessary for life chemistry would never have taken place.

Second: The Atoms that exist in all matter could not exist unless there were sufficient quantities of the elements necessary for life.

Atoms of many different sizes must be able to form and in order for this to be possible, there must be a delicate balance between the constants of physics the govern the strong and weak nuclear forces, as well as gravity, and the nuclear ground state energies, which are quantum energy levels that are important in the formation of elements from protons and neutrons, that come from several key elements.

In the strong nuclear force that governs the level at which protons and neutrons may bind together in atomic nuclei, we are able to observe this complex mechanism.

1. If the strong nuclear force was too weak, multi-proton nuclei would not remain bound together. If this had taken place in the early universe, hydrogen would have been the only element in the universe.

2. If the strong nuclear force was only slightly stronger than what we now see in the universe, protons and neutrons would have such great affinity towards each other that none would remain. These protons and

neutrons would have attached to other protons and neutrons and our universe would have produced no hydrogen, only heavy elements.

3. The chemistry necessary for life is not possible without hydrogen, but if the strong nuclear force at the commencement of the universe was only slightly stronger, hydrogen would have been the only element. This illustrates how fine-tuned the balance is for the strong nuclear force.

4. If at the inception of the universe the strong nuclear force was just four percent stronger that it was set at, the diproton, an atom that has two protons and no neutrons, would have formed. This event would have resulted in stars rapidly exhausting their nuclear fuel that no life later would have been possible. The control exercised in this early event allows us to see that these events were intentionally caused for the purpose of human life existing on earth much later.

5. If the strong nuclear force was decreased when the universe began, by only ten percent, carbon, oxygen, and nitrogen would have been unstable and life on earth today would never have been possible.

6. Imagine that you want to make a planet where human beings will be able to live, billions of years later. You cannot simply make a planet with all of the necessary elements to allow humans to thrive. You must also prepare a universe to allow a planet like this to form in the first place, and sustain this universe for billions of years before the planet is formed. There must be many fine-tuned constants pre-set before earth ever exists, or earth will not be able to produce and environment necessary for these beings to live. This requires tremendous foreknowledge of exactly how to start the universe and allow its expansion to continue so that all of these fine-tuned constants are set correctly.

7. These parameters are not only essential for life on earth, but any possible life of any kind requires life chemistry of the same nature any other place in the universe. What is truly amazing is that this fine balance is a condition that must exist universally on every planet in the universe if it will contain life.

8. We understand today that the strong nuclear force is the strongest attractive force in nature, as well as the strongest repulsive force in nature. The stunning reality that this is attractive on one length scale and repulsive on another length scale, defining this constant as particularly unusual and counterintuitive. In a natural process that was solely guided by chance, we would never expect to find such a fine-tuned parameter to make life possible. In spite of this, we find this exact precise setting that allows life to exist in our universe.

9. In order that life might be possible, it is essential that the strong nuclear force is only attractive over lengths no greater than 2.0 fermis and no less than 0.7 fermis. A single fermi is a quadrillionth of a meter. This force must manifest its greatest attractive force at near 0.9 fermis.

10. If these lengths were shorter than 0.7 fermis, the strong nuclear force must be strongly repulsive. This is due to the protons and neutrons existing as groups of more fundamental particles that are called quarks and gluons. Each of these proton packages are made up of two up quarks and one down quark, plus the relevant gluons. This is while each neutron also contains two down quarks and one up quark with their relevant gluons.

11. If the fine balance of strong nuclear force was not strongly repulsive on length scales below 0.7 fermis, the proton and neutron packages of quarks and gluons would merge together. This would produce no atoms, no molecules, resulting in no chemistry possible anywhere—or at any time—during the entire history of the universe.

12. This balance must be exquisitely fine-tuned, both in its length of operation throughout the history of the universe and the level of strength in its repulsion. Concerning the weak nuclear force which determines the rates of radioactive decay, if this force was stronger than currently observed, the matter of the universe would be swiftly converted into heavy elements. If this force was weaker, the universe would contain

only the lightest of elements. Essential elements necessary for life, like oxygen, carbon, phosphorus and nitrogen, would either be absent from the universe, or exist in quantities so small that life could not exist. What natural process in the production of the universe was capable of advanced determination to produce these precise balances?

13. It is insufficient to simply say that since we have these precise balances in the universe, this is proof that they were necessary, and no supernatural explanation is required. This conclusion is preposterous. Of course, science must examine how every process produced the necessary elements needed for life, and understand whether these processes happened by chance, or they were precisely controlled. In examination of all 209 physical constants of our universe, it becomes abundantly clear that the presence of these 209 constants, all at the same time, demands a causal explanation.

14. Science understands today that unless the weak nuclear force was precisely fine-tuned at one part in ten thousand, these essential elements required for life which are produced only in the core of a supergiant star, would have never been able to escape the boundaries of their cores, making supernovae impossible. If no supernova's occurred in our universe, no second-generation stars would have been created like earth, resulting in no human life to exist. There are far too many coincidences, if we are to conclude that nature is responsible for these results. The science of the universe proves intelligent actions took place to produce these results. The goal of these precise balances is clear: to allow human beings to live on earth.

Third: The gravitational force strength is determinative in the heat produced by the nuclear furnace of star cores. If the gravitational force was weaker during the expansion of the universe, stars would have been so hot that they would have burned up their nuclear fuel too quickly for life to exist much later. A planet that is capable of sustaining life must have a star that stable and long burning:

1. If the gravitational force was weaker during the inception of the universe, there would be no stars hot enough to ignite nuclear fusion. In a universe under these constraints, not elements heavier that hydrogen and helium would have been produced.

2. This fine-tuned parameter for the nuclear ground state energies of helium, carbon, beryllium, and oxygen were essential for life to exist. Hoyle discovered that these elements were balanced so precisely that any disruption to the early process of the universe would have resulted in these elements being far different from how they ended up. For Hoyle, this spoke volumes concerning the source of these balances, defining these results as impossible by a natural process, being caused by an outside source. Only a mind is capable of setting precise balances in physical processes so that essential elements can form as they have in our universe.

3. Hoyle reasoned that the ground state energies for these elements could not be higher or lower by more than four percent without producing a universe containing insufficient oxygen or carbon for life. Although Fred Hoyle has argued against the idea of theism , he concluded, from the evidence of the universe, that *"a super intellect has monkeyed with physics, as well as with chemistry and biology."* [3,4] In other words, Hoyle was not ready to concede to the existence of the Christian God, but he also could not dismiss the facts of the data that proved an intelligence must have acted upon these processes or they never would have achieved the results our universe displays.

4. Many other astrophysicists have demonstrated that the level of design present in electromagnetism and the strong nuclear force is much greater than what physicists previously had determined. This inherent design quality found in these two forces are mandatory for any physical life to exist in the universe. There must be an specific abundance of carbon and oxygen for life, and the only astrophysical sources that has significant quantities of carbon and oxygen are red giant stars. A red giant star that has used up all of its hydrogen fuel by nuclear fusion, can fuse helium into heavier elements. Mathematical models for red

giant stars that take on slightly different values in their strong nuclear force and electromagnetic force constants, produce insufficient carbon, oxygen. By adjusting these balances in tiny fractions, these two constants by just four percent larger of smaller, life become impossible.

5. When we examine the coupling constant for the strong nuclear force, we learn that had this constant been altered by just one percent smaller or larger, no life on earth would have ever been possible. Again, what process of nature or as a result of the physical laws, could produce this result? This precise limit on the strength of the strong nuclear force and electromagnetic force, demands even more stringent constraints on quark masses and the Higgs vacuum expectation value. These facts provide us with great evidence of specific design for the physics of the stars and planets in our universe and the tremendous mathematical design of their fundamental particles. It is the knowledge of these facts that have changed the minds of many Cosmologists in recent years from the idea that the universe is a product of natural events, to an awareness that our Cosmos was designed for life on at least one planet.

6. In achieving the correct nucleons for our universe, the physics must also be adjusted in order to achieve the correct elements for life. Adjustments were also necessary to cause these elements to bind together so they could form life molecules. Left to the randomness of a universe uncontrolled, the outcome would not obtain enough nucleons, or protons and neutrons, to form the necessary elements. This process had to be controlled or it would have produced and entirely different universe, incapable of supporting human life.

7. At the inception of the universe, there were about ten billion and one nucleons for every ten billion antinucleons. The ten billion antinucleons annihilated the ten billion nucleons, which generated a tremendous amount of energy. Every galaxy that exists today, every star that comprises the universe, was formed from these remaining nucleons.

8. If the number of excess nucleons in relation to antinucleons were less, there would not have been enough matter for galaxies, stars, and heavy elements to form. If the excess of nucleons to antinucleons was greater, galaxies would have formed, but with such inefficiency that they would have trapped radiation, preventing fragments from forming stars and planets.

Four: one important example of a fine-tuned universe that is understood by a critical requirement for life is the ratio between the electron and a proton of every atom in the universe.

ELECTRON AND PROTON

The ratio of mass between an electron and a proton is 1:1836, meaning that a proton is 1,836 times larger than an electron. Although there is a great degree of difference between the size of these two parts of an atom, both the electron and proton maintain the exact same electrical charge. With this massive difference in size, how is it possible that both still have the same electrical charge? It took a thinking mind to know the precise size difference that was necessary before the first atom existed. If the electrical charge of the electron is altered by just one part in 100 billion, the body of every human being on earth would explode.

Neutron and Proton

When we examine the neutron, we learn that it is 0.138% larger than a proton. This extra mass, causes the neutron to require more energy to produce than a proton. When the universe began to cool after the massive heat of the initial creation of the universe, there were nearly seven times more protons than neutrons. Had this number increased just 0.1%, the number of neutrons that remained after cooling would not have been sufficient to create the nuclei for the life-essential heavy elements. How is it that this precise limit was reached without exceeding the critical boundary that would have destroyed the universes' ability to produce life-essential heavy elements? It is impossible to conclude that this was a random act of the natural process.

The physical laws of the universe do not define the outcome of events, only they manner in which they must occur. It is the outcome of events defined by the 209 physical constants that require intelligence, or they could not have happened, could not exist today to sustain the universe. The physical laws merely define what has already happened when these constants to place and allowed every other event of our universe to proceed. The laws of physics could not produce precise balances of these constants, only described the manner in which they took place.

1. It is the larger mass of the neutron, relative to the size of the proton that determines the rate at which neutrons may decay into protons and protons into neutrons. One neutron decay into one proton + one electron + one neutrino. There is such a fine balance here that if the neutron was just 0.1% smaller, the result would produce such a great number of protons that all stars in the universe would rapidly collapse into neutron stars of black holes. For life to exist anywhere in the universe, the neutron mass must be fine-tuned to greater than 0.1%.

2. The decay rate for protons is so slow that as yet we have not been able to record a solitary decay event. This defines the decay process of protons to necessitate a fine-tuned parameter in order for life to exist anywhere. Theoretically, protons must decay into mesons at a rate that is close to the known experimental limits. If protons decay even slightly slower into mesons, our universe would not contain enough nucleons to make galaxies, stars, and eventually, planets. The same components that determines the decay of protons also determines the rate of decay for nucleons into antinucleons when the universe began. What natural process of law of physics is capable of determining this precise balance when at the beginning of the universe as -430th of a second, there were no laws or anything else to cause this balance. This precise rate of decay must have been known before the universe commenced and a Being had the ability to control the rate of decay precisely, or the rate of decay for nucleons would never have achieved the precise number required to later produce the correct type of stars.

3. If the decay rate at the beginning of the universe was slightly faster, the ratio of nucleons to antinucleons would be incorrect, producing far too much energy being released into the decay process, and this incorrect decay process would have destroyed life, not allow it to exist.

4. A precise number of Electrons must exist, equivalent to the number of protons with a tolerance of one part in 10^37. If this balance was off even slightly the electromagnetic force of the universe would have overpowered the gravitational force, resulting in non-production of galaxies, stars, and planets. The number: 10^37, is such a great impossibility that it is beyond the mathematical possibility of chance. No natural process can aim, it certainly cannot hit a target so small, at such a great distance. If we were to shoot at a target from earth that is at the other side of the universe and hit the bullseye the first time, this is equal to the balance of gravity and electromagnetism achieving this precise fine-tuned balance when the universe began.

The physics of the universe must be so meticulously fine-tuned, within such precise limits, that there is no possibility that any natural or accidental process could ever have achieved these results. We see the fingerprints of the Creator all over the physics that define our universe. His knowledge; His engineering skills: His power to achieve these results, are defining qualities that are easily observed once we understand the complicated and impossible process that was required to exact a universe like ours. There has never been any scientific explanation presented that proves the laws of the universe, or a purely natural process, was capable of producing these fine-tuned constants described here Their precise boundaries that allow these constants to operate, are exceedingly beyond the ability of unguided processes.

COSMIC EXPANSION

When we begin to understand the universe, the first junction we are brought to is how the universe has expanded since its inception. Fortunate for us today, scientists have been able to accurately measure this rate by examining the Cosmic Background Radiation that still exists from the beginning of the universe. When we compare this rate of expansion to the physics of galaxy and star formation, astrophysicists have discovered something truly amazing.

1. How fast or slow the universe has expanded, determines whether or not galaxies were able to form. If too fast, all matter would have dispersed quite efficiently and there would have not been the clumping or gathering of matter necessary to form galaxies. Without galaxies, there would never have been any stars. We need both first- and second-generation stars in order for a planet like earth to have the correct star to allow life on earth to exist. Had the universe expanded too rapidly when it began, the universe would never have produced galaxies and no one would have ever lived on any planet like earth.

2. If the universe had expanded too slowly, matter would have grouped together so well that the universe would have collapsed upon itself at the start into a super-dense lump of matter, never producing a single galaxy or star.

3. The creation event of the universe was not a mindless occurrence by random events that were from a purely natural process. Every important event that began at -430th of a second, had to happen under extreme control—such as the rate of expansion, and all the other events mentioned in this essay—or we would never have a universe like ours. The stunning revelation of all this is that if the universe occured by a purely natural process, the result would have been mass disorder, not mass order that is scientifically observed by the balance of these 209 fine-tuned physical constants.

Dark Energy

After the universe began, the cosmic mass density and dark energy density, began to modify the universe's expansion velocity in dissimilar processes. In order for the universe to produce stars and planets and later earth that would sustain physical life, the value of cosmic mass density had to be exquisitely fine-tuned. Again, trying to explain this by the laws of physics, always comes up short. There is no physical reason these fine-tuned constants could happen by a mere physical process. Guidance had to be instituted or the balance could not have been achieved.

1. In these early processes of the universe, dark energy played a critical role. It was the discovery that dark energy had to be exceptionally fine-tuned for the universe to exist that has brought many scientists to the conclusion that intelligence was required. The earliest source for dark energy had been at least 122 orders of magnitude larger than the amount astronomers have presently detected. This means that the original source or sources, had to cancel each other out so that they left just one part in 10^{122}.

2. **Think about this for a moment:** As the universe rapidly expands, so also does space to accommodate its expansion. Dark Energy is the agent that acts to allow the universe to stretch out into space. The more that space expands to accommodate the universe, the greater the power of dark energy to act as an anti-gravity factor. The larger that space grows in size, the greater the power of dark energy, acting as an "anti-gravity factor."

3. Dark Matter acts to make two very large bodies appear to repel each other, with their repelling force increasing the farther apart these two bodies get from each other. This is the opposite effect of gravity which acts to slow cosmic expansion. Science has now proven that the cosmic expansion rate has been accelerating during the last half of the universe, due to dark energy as a dominant component acting upon this expansion.

4. What this mean is that the presence of dark matter demands a recent cosmic beginning, so recent, a natural or accidental beginning of the universe or human life on earth is impossible. Critics of a Creation event by intelligence nearly always seek to refute dark matter because it proves a fine-tuned universe that resulted from intelligence, necessary by a thinking Being.

5. Theoretical physicists Lisa Dyson, Matthew Kleban, and Leonard Susskind stated that, "Arranging the cosmos as we think it is arranged . . . would have required a miracle." These scientists further stated that an "unknown agent intervened in the evolution of the Universe for reasons of its own."

6. Using mathematics and a knowledge of physics, scientists calculate that the force of dark energy should be incredibly large. The degree of its existence is so large that the universe should have expanded so quickly that galaxies and stars could not have formed and there would be no planets or life. The reason that dark energy does not overpower the universe and cause it to implode is there is an equally strong counterforce that restricts the impact of dark energy, so that stars and galaxies can form.

7. The precision required to balance dark energy is stunning. This setting has to be accurate from 56 decimal places (Septendecillion) to 122 decimal places (Vigintillion). This is a number so immense that there is no possibility that this physical constant for dark energy could exist unless a mind had known this precise balance was essential to allow a universe that could produce stars and planets. This fact becomes the strongest scientific argument for the existence of God amongst all of the other evidence that exists.

8. The universe came into existence from nothing, proceeded with order because of laws, and is understood by mathematics. Describe one naturally occurring phenomenon that can accomplish this? Can any non-thinking thing create everything that will exist and simultaneously create laws to cause it to proceed with specific order to make life possible, and then allow a human mind to understand how it all functions by mathematics, also created simultaneously with the universe One analogy that has been used, but still falls short of the reality of this massive number, one part in 10^{122}, is in trying to imagine a billion pencils all standing together at once on their sharpened points while on a smooth glass surface, without any visible support.

RELATIVITY, DIMENSIONALITY AND QUANTUM UNCERTAINTY:

The fine-tuning of the physical constants of the universe has already presented us with overwhelming evidence for a Creator, but we find additional evidence. When we examine the fundamental particles, energy, and the space-time dimensions that exist in the universe, these allow for the principles of quantum tunneling and special relativity to operate exactly as they do. In order for hemoglobin to have the capacity to transport the correct level of oxygen to living cells, quantum tunneling must operate efficiently.

Quantum tunneling must function within very narrow boundaries in transporting hemoglobin with the correct a quantity of oxygen to living cells in all vertebrates and a majority of invertebrate species. Extreme fine-tuning is required in the operation of this system of hemoglobin so that copper and vanadium complete their critical roles in the function of the nervous system, as well as the development of bones in all advanced species. For this process to work correctly, the Heisenberg uncertainty principle must be fine-tuned. In measuring the momentum of particles with the most precise precision, the position of the particle is now known to reside within tolerances of ± a third of a mile. When we increase the distance smaller than half a mile, hemoglobin will not function in advanced life forms. This precise process demonstrates that fine-tuning is so important in advanced life that it would be impossible for it to occur randomly. There is an exquisite design in all of life that is specifically created for the benefit of the beings themselves. This proves that the universe was created to benefit mankind specifically, removing the chance that these many fine-tuned constants are random occurrences.

Einstein demonstrated that relativity only works when certain proteins contain the correct balance of copper and vanadium and the velocity of light is fine-tuned to make this possible. When we examine the velocity of light at a value of 299,792.458 kilometers per second, we see by Einstein's equation,

$$E = mc^2$$

that even small changes in c (the velocity of light) lead to huge changes in E (the energy) or m (the mass).

When these minor changes in light's velocity take place, either too strong or too weak, life cannot exist, and stars cannot produce the correct elements for life. In order for stable orbits of planets to take place around their stars, as well as all electrons about the nuclei of atoms, are only possible in a universe that has three large and rapidly expanding dimensions of space. Also required, are six extremely small dimensions of space that are currently dormant— but actively expanded during the first 10^{43} seconds of the universe's history. These factors are essential for quantum mechanics and gravity to coexist. For this reason, physical life likely requires different fine-tuning in the number of effective dimensions—both in the present four: three of space plus one of time—and in the earliest moment when the universe began, specifically—ten dimensions, comprised of nine of space plus one of time.

If God, Why So Long?

The question of why God would wait so long, billions of years, to create human life? He is God, anything is possible, why wait so long instead of just creating a universe that is ready for earth and human life. The answer is understood by our comprehension of the physical laws that govern the universe and the physical constants that make life possible on earth. God chose to cause our universe within the constraints of laws and constants. In this model, it takes billions of years to fuse enough heavy elements in the nuclear furnaces of giant stars that make life chemistry possible.

Under the model God created for our universe, He seems interested in displaying great expanses of time, which for Him, are mere days. The Bible describes God as outside of time, space, and matter, not affected by its passing or constraints. These descriptions place God as only taking days to accomplish things that extend into billions of years for us in linear time. The Bible say that "in the beginning God created the heavens and the earth." It does not say when this beginning was—either for God, outside time, or for us, within time. For God it could have been days, for us in viewing the scientific evidence available, clearly billions of years were required.

In a universe that ages and this process allows the production of first-generation stars, which burn up their fuel and turn into Supernovas. This long process produces the necessary elements for life in a second-generation star like earth. If the universe was merely a few billion years old, no older second generations stars like our own would exit. This would make earth impossible and human life on earth none existent. A second problem for a young universe is that ours began with a very low state of entropy, which was essential to allow the early processes that took place. If the expected state of high entropy had been present, we would not have universe at all.

As the universe ages, entropy degrades from low to high, which is needed for life on earth. Without the low entropy that happened in the process of an aging universe, small stars were able to form, and carbon was created, from where all human life depends.

God wanted specific things to happen at the beginning of the universe, and so He ordered it as such so that an extremely low state of entropy existed, and then fine-tuned the universe to proceed and produce first- and second-generation stars. At an aged universe, God kept the physical constants in place, but increased entropy through the process of time, and all of us who live on this planet came into existence, also by creation from the same God who made the universe.

This brings us to the narrative of the Bible, where we learn that God created this world for us, and it was perfect and there was no evil, sin, suffering, or death on earth. As man violated the spiritual laws God created, the world fell into chaos as sin ruined the perfect creation God made for us. Not long after man's fall, God promises a Savior for the world, in Genesis 3:15, who would be God Himself, dying for the sins of the world, and making it possible for anyone who believes what God said, to return to a world where perfection and beauty never end.

Fine-Tuning the Universe – Further Evidence for a Designer – Sir Fred Hoyle

Sir Fred Hoyle the man who coined the term 'big bang,' died on Monday, August 20, 2001, from complications following a severe stroke.

Born in Yorkshire, England, in 1915, Hoyle was one of Britain's best-known mathematicians and astronomers in the last half of the 20th century. He spent decades searching for answers to questions of the origins of life and the origin and age of the universe. In the 1940s, he, along with "Hermann Bondi and Thomas Gold," mistakenly, proposed the 'steady state' theory, a belief that the universe had no beginning or end, but always existed and would continue to exist.

The Law of Conservation of Mass/Energy is Consistent with Genesis

All these men were strong humanists, so they rejected any theory that seemed to teach a beginning for the universe, because that would point to a Beginner—Their bias was so strong that they were even prepared to violate the fundamental Law of Conservation of Mass/Energy, which states that mass/energy in the universe can neither be created nor destroyed. Of course, this fundamental law is consistent with Genesis—God's creation of the space-time universe, and finely tuning it for life was *finished* after six days. But the Steady State Theory posits a continual spontaneous appearance of hydrogen atoms from nothing, Figure 11.12.

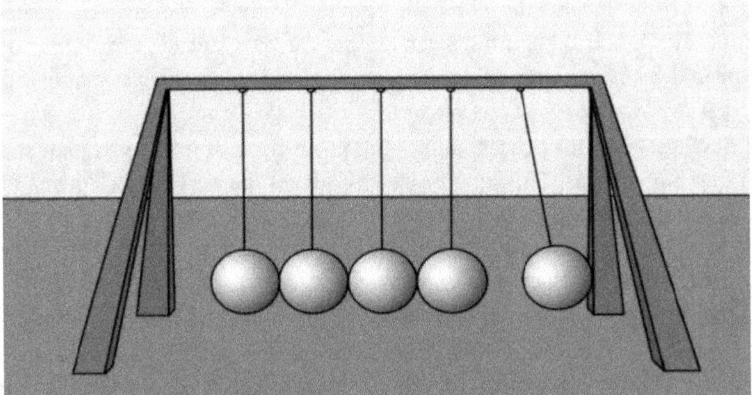

Fig.11.12: The Law of The Law of Conservation of Mass/Energy

However, because the evidence of the rapid expansion of the universe exceeded the predictions of Hoyle's theory, and because of the reluctance to believe that fundamental laws were violated, many astronomers began to postulate that an explosion of highly dense matter was the beginning of all space and time.

In his 1950 BBC radio series, *The Nature of the Universe*, Hoyle mockingly called this idea the 'big bang', considering it preposterous. Yet the theory—and the derisive term—have become mainstream, not only in astronomy but in society as well.

Hoyle readily saw through the fallacious assumptions behind the 'big bang' theory. In 1994 he wrote, 'Big-Bang cosmology refers to an epoch that cannot be reached by any form of astronomy, and, in more than two decades, it has not produced a single successful prediction.' Even though many people currently consider cosmic microwave background radiation a successful prediction of the 'big bang', this is very shaky, and would fit better with Dr Russ Humphreys' cosmological model that involves God having stretched out the cosmos (*Isaiah 42:5*).

This should be a lesson to 'big-bang' apologists, who are seduced by its apparent teaching of a beginning of the universe and simply ignore the contradictions with God's Word. What happens to their apologetic framework if the secular astronomical community goes along with Hoyle after all, and rejects the 'big bang'? Then the 'big-bang' apologists would need to reinterpret their reinterpretations of Genesis!

Also, commenting on the general state of mainstream cosmology, Hoyle and several colleagues wrote, 'Cosmology is unique in science in that it is a very large intellectual edifice based on very few facts. The strong tendency is to replace a need for more facts by conformity.'

Though Hoyle was not a Biblical creationist or even a Christian, he eventually recognized the impossibility of Darwinian evolution. Hoyle regularly took to task the Darwinian establishment for ignoring the complex sources of information and information processing programs (like DNA) needed for the creation and continuation of life. He realized that life couldn't have arisen by chance in a primordial soup on Earth. First, he tried to solve the problem by saying that if we had the whole universe to work with instead of Earth, then this might overcome the problem. Hoyle favored and popularized a view called *panspermia,* the notion that life originated somewhere else in the universe and was driven to earth by electromagnetic radiation pressure.

Simplest Living Cell is One in $10^{40,000}$ – Evidently Nonsense!

Nonetheless, eventually he realized that even this would be woefully inadequate as a materialistic explanation of life's origin. In his 1981 book *Evolution from Space* (co-authored with Chandra Wickramasinghe), he calculated that the chance of obtaining the required set of enzymes for even the simplest living cell was one in $10^{40,000}$ (one followed by 40,000 zeroes). Since the number of atoms in the known universe is infinitesimally tiny by comparison (10^{80}), even a whole universe full of primordial soup wouldn't have a chance, Figure 11.13.

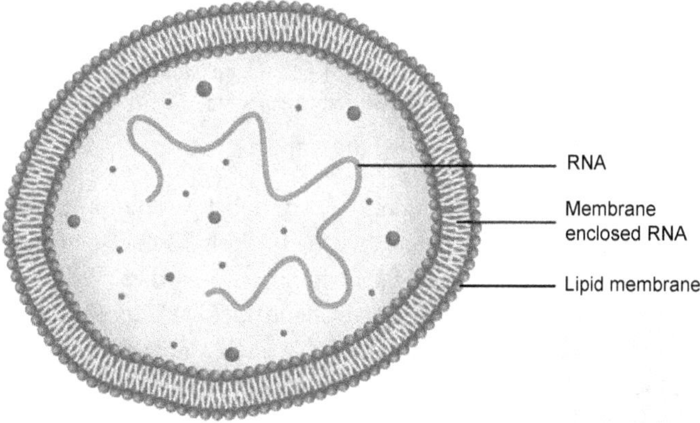

Fig.11.13: *Evolving Simplest Living Cell is One in $10^{40,000}$ – Evidently – Impossible*

Hoyle explained this in his typically lucid manner, and as with the 'big bang,' his turns of phrase have found their way into popular culture. For instance, he wrote, 'The notion that not only the biopolymer but the operating program of a living cell could be arrived at by chance in a primordial organic soup here on the Earth is evidently nonsense of a high order.' Hoyle originated the famous illustration comparing the random emergence of even the simplest cell to the likelihood that 'a tornado sweeping through a junk-yard might assemble a Boeing 747 from the materials therein.'

Hoyle also compared the chance of obtaining even a single functioning protein by chance combination of amino acids to a solar system full of blind men solving the Rubik's Cube simultaneously.

Hoyle eventually came to believe that the fine-tuning of the universe as a whole was further evidence for a designer:

'A common-sense interpretation of the facts suggests that a super intellect has monkeyed with physics … The numbers one calculates from the facts seem to me so overwhelming as to put this conclusion almost beyond question.'

FINE TUNING – AMAZING

The fine-tuning of fundamental constants is indeed amazing, but creationists must be cautious—some of the alleged 'fine-tuning' presupposes a big bang or other evolutionary cosmology.

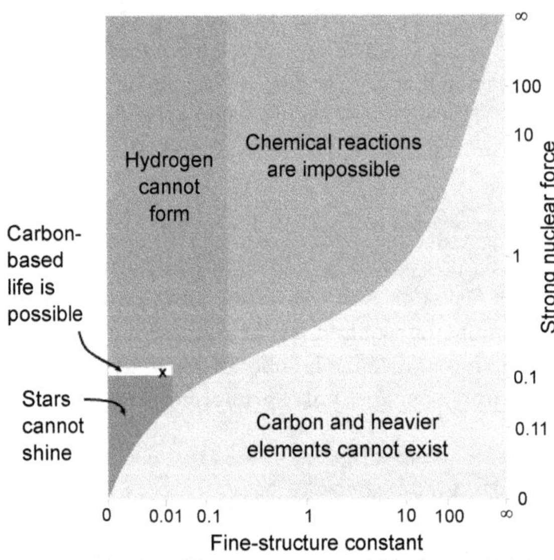

Fig.11.14: Fine Tuning the Universe – Amazing

Alas, Hoyle paid for his outright questioning of the materialist paradigm. In the 1950s, Hoyle had some ingenious ideas about stellar fusion, and predicted that the Carbon-12 nucleus would have a certain energy level (called a *resonance*) to enable helium to undergo fusion. His co-worker "William Fowler" eventually won the Nobel Prize for Physics in 1983 (with Subramanyan Chandrasekhar), but for some reason Hoyle's original contribution was overlooked, and many were surprised that such a notable astronomer missed out. Fowler himself in an autobiographical sketch affirmed Hoyle's immense contribution:

'*Fred Hoyle was the second great influence in my life. The grand concept of nucleosynthesis in stars was first definitely established by Hoyle in 1946.*'

But for all his ability to see through popular anti-God science, Hoyle's own views about God were equally un-Biblical. He still held onto panspermia, and in his last book, *A Different Approach to Cosmology*, Hoyle and his co-authors reaffirmed a quasi-steady-state theory for the universe, but this time one that requires ongoing episodic creation by some intelligent force within the universe (a complete denial of a six-day Creation *ex nihilo* by a transcendent, personal God), Figure 11.14.

HOYE – SCIENCE FICTION WRITER

Hoyle was also known as a science fiction writer. That he took to this sort of writing is not surprising, given his fascination with space and extraterrestrial life forces.

While Hoyle's comments on the 'big bang' theory and Darwinian evolution are helpful, it is sad to see that Hoyle died apparently having rejected the truth about Creation. God has revealed the truth for all to see in the Bible, the History Book of the Universe. All the answers about the origins of life and the finely tuned universe for life can be found right there in the first book, Genesis.

MULTIPLE UNIVERSES OUT OF NOTHING

Quantum Mechanics Creates

Some people ask: "If God didn't cause the universe to exist, what did?

Atheist physical chemist Professor Peter Atkins acknowledges the problem: "What science cannot yet do, and one day I think it will have to turn to doing, is thinking about how a universe can come into existence without intervention, from absolutely nothing, without a creator. And that's going to be really tough for science, Figure 11.15."

Well, it's going to be really tough for scientists, because science does not support such illogicality. Nevertheless, many scientists tell the public:

Fig.11.15: Galaxies, Exist; Multiverses, Non-Existence. Curtsy - Wikipedia.org

To understand how a universe might be created out of nothing, we must turn to the theory of quantum mechanics. It describes how in the tiny world of the atom small particles seem to appear and disappear for no obvious reason. Apparently absurd! Imagine if in our everyday world people appeared and disappeared *without obvious cause*. And yet the theory of quantum mechanics describes how this can happen at the atomic level, and that might offer an alternative to a creator. But why should a universe as fruitful as ours just pop out of nothing and somehow work? Imagine not just one tiny universe popping into existence but millions. Imagine that, among the millions, one of them by chance had the properties to support life. Maybe that universe was ours. This is the multiverse concept, which suggests that out of millions of possibilities we by chance are in the one universe that works.

We seem to have gotten off the subject. The question was 'how could our universe pop out of nothing?' And the answer is: 'imagine millions of universes popping out of nothing!' That's not an answer; that's millions more 'how' questions! And here are a few other specific ones: For quantum mechanics to work, the laws of quantum mechanics must already be in existence, so how did they form out of nothing? Another problem is: how does what seems to happen at the sub-atomic level in a laboratory experiment (called a quantum fluctuation) apply to the formation of all the stars in the 200 billion galaxies in our universe? If they all appeared *without obvious cause* because of quantum mechanics, why aren't they also disappearing *without obvious cause* due to quantum mechanics (as per the analogy given above)?

In laboratory experiments, virtual particles do appear within the vacuum of space. However, in the primordial singularity there was no space and so no vacuum. Another way of expressing this problem is: **What was it that quantum fluctuated before there was anything there to quantum fluctuate?**

It was correctly said: "But still the question remains: Could our universe really come from absolutely nothing?"

Professor "Alexander Vilenkin" (Director, Institute of Cosmology, Tufts University, Boston) sets the record straight: "Of course, if you ask for evidence that the universe began by this spontaneous nucleation out of nothing, I don't think we can produce any evidence for that." Good answer!

Atheist "Richard Dawkins" as usual muddies the water. He tells viewers:

The principle of the multiverse provides at least an interim satisfying explanation [how?] in a way that a creator couldn't possibly be a satisfying explanation [why not?]. Then having got ourselves into a universe which is capable of generating stars, capable of generating chemistry, and ultimately capable of generating the origin of life [says who?], then biological evolution takes over and now we have a clear run.

Professor Paul Davies exposes this nonsense with: "The idea of shunting the problem off from the universe to the multiverse, I think, is a fraud. You can just ask the same question, where did all that come from? You really don't explain things just by pushing it off a level. It's just a trick."

FOUR REASONS TO REJECT THE BIG BANG HYPOTHESIS

In addition to the many reasons contra the big bang we gave in our response to Part 1 of this DVD we submit the following:

1. **The theory lacks a credible mechanism.** The big bang universe begins with all matter, energy, space and time compressed into a point of infinite density. There is no known mechanism to start the universe expanding out of this singularity—equations in the theory only work *after* the expansion has begun.
2. **It depends on 'fudge factors'.** It has to violate physical laws and appeal to unknown forces (dark energy) and unknown substances (dark matter) to explain what we observe.
3. **The Multiverse concept is excluded.** Appeals to chance, e.g. via an infinite number of bubble universes, do not work because chance cannot accomplish what the laws of nature do not allow.

4. **Science cannot produce any final answers on the subject of origins**. Present-day observations, when applied to the past, are always based on the belief system of the person(s) making the claims, and hence express their worldview, not facts about how the past came to be that way.

I DON'T HAVE ANY HOPE

Professor "Peter Atkins" closed with this comment from atheist: "I would be ultimately rather pleased if there was a god, because then I would have a better lease of life. I would very much like to be wrong. I would very much like to be immortal, but I'm afraid I don't have any hope that that is the case."

We have news for you, Professor Atkins, and for all our readers. The big bang is certainly not a reliable foundation when considering one's eternal destiny. Nonetheless, anyone can come to terms with God in this life, to spend eternity with Him. The Bible stated that through faith is the infinite Eternal God/Jesus-Christ we can have eternal life, (*John 1:12*).

Many atheists have come to the full knowledge of truth!

Genesis—Fact Not Fancy

Philosopher "William Lane Craig" then takes it upon himself to attack the 'Young Earth' Creationist (Genesis-means-what-it-says, Figure 11.16) biblical position, when he says:

Faith and science ought not to conflict, if both are means of discovering truth about reality. I think that the impression that there is a conflict has largely arisen because of literalistic interpretations of the opening chapter of the book of Genesis in the Hebrew Bible. ... People have taken this to describe six consecutive 24-hour days, and that viewpoint has been exploded by modern science.

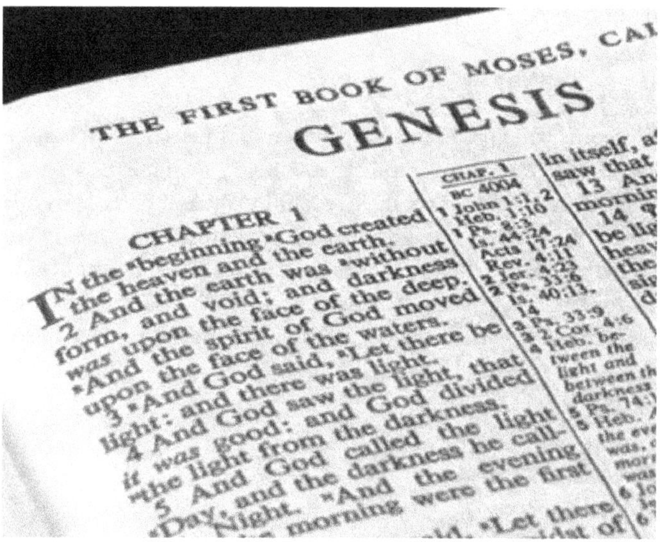

***Fig.11.16**: Genesis – The First Book in the Bible*

Conflict is not between faith and science, but between faith and scientists (both atheists and theists) who do not believe God's Word. Genesis is the true account of the world's history, which began with God creating the earth and everything in it in six consecutive days. The Bible clearly teaches:

• The days were 24 hours long.

- The whole universe was created during creation week, which was the same length as our working week.
- Mankind was created on Day 6, "from the beginning of creation" (*Mark 10:6*).
- The genealogies imply that Adam was created about 6,000 years ago.
- Death arose from Adam's sin, not before.
- There was a worldwide Flood that left so much evidence that scoffers are "willingly ignorant" (2 Peter 3:3–7),

This viewpoint is rejected by atheists, but it has certainly not been "exploded by modern science" as claimed by Craig. We believe all this because the text of the Bible says so, and Jesus and the New Testament writers took it to mean this. The proper way to describe this hermeneutic is 'historical grammatical', or perhaps 'classical literal', rather than 'literalistic'.

THE DAYS OF GENESIS 1

Meaning of *yôm*

When Moses, under the inspiration of God, compiled the account of creation in Genesis 1, he used the Hebrew word *yôm* for 'day'. He combined *yôm* with numbers ('first day', 'second day', 'third day', etc.) and with the words 'evening and morning', and the first time he employed it he carefully defined the meaning of *yôm* (used in this way) as being one night/day cycle (Genesis 1:5). Thereafter, throughout the Bible, *yôm* used in this way always refers to a normal 24–hour day. There is thus a *prima facie* case that, when God used the word *yôm* in this way, He intended to convey that the days of creation were 24 hours long.

Let us now consider what other words God could have used, if He had wanted to convey a much longer period of time than 24 hours.

IMPORTANT HEBREW TIME-WORDS

There are several Hebrew words which refer to a long period of time. These include *qedem* which is the main one–word term for 'ancient' and is sometimes translated 'of old'; *olam* means 'everlasting' or 'eternity' and is translated 'perpetual', 'of old' or 'forever'; *dor* means 'a revolution of time' or 'an age' and is sometimes translated 'generations'; *tamid* means 'continually' or 'forever'; *ad* means 'unlimited time' or 'forever'; *orek* when used with *yôm* is translated 'length of days'; *shanah* means 'a year' or 'a revolution of time' (from the change of seasons); *netsach* means 'forever'. Words for a shorter time span include *eth* (a general term for time); and *moed*, meaning 'seasons' or 'festivals.' Let us consider how some of these could have been used.

I. Event of Long Ago

If God had wanted to tell us that the creation events took place a long time in the past, there were several ways He could have said it:

yamim (plural of *yôm*) alone or with 'evening and morning', would have meant 'and it was days of evening and morning'. This would have been the simplest way, and could have signified many days and so the possibility of a vast age.

qedem by itself or with 'days' would have meant 'and it was from days of old'.

olam with 'days' would also have meant 'and it was from days of old'.

Accordingly, if God had intended to communicate an ancient creation to us, there were at least three constructions. He could have used to tell us this. **However, God chose *not* to use any of these.**

II. Continuing – Long Ago

If God had wanted to tell us that creation started in the past but continued into the future, meaning that creation took place by some sort of theistic evolution, there were several ways He could have said it:

dor used either alone or with 'days', 'days' and 'nights', or 'evening and morning,' could have signified 'and it was generations of days and 'nights.' This would have been the best word to indicate evolution's alleged eons, if this had been meant.

olam with the preposition *le*, plus 'days' or 'evening and morning' could have signified 'perpetual;' another construction *le olam va-ed* means 'to the age and onward' and is translated 'for ever and ever' in *Exodus 15:18*.

tamid with 'days', 'days' and 'nights', or 'evening' and 'morning', could have signified 'and it was the continuation of days.'

ad used either alone or with olam could have signified 'and it was forever.'

shanah (year) could have been used figuratively for 'a long time', especially in the plural.

yôm rab literally means 'a long day' (cf. 'long season' in *Joshua 24:7*, or 'long time' in the New American Standard Bible). This construction could well have been used by God if He had meant us to understand that the 'days' were long periods of time.

Thus, if God had wanted us to believe that he used a long–drawn–out creative process, there were several words He could have used to tell us this. **However, God chose not to use any of these.**

III. Ambiguous Time

If God had wanted to say that creation took place in the past, while giving no real indication of how long the process took, there were ways He could have done it:

yôm combined with 'light' and 'darkness', would have signified 'and it was a day of light and darkness'. This could be ambiguous because of the symbolic use of 'light' and 'darkness' elsewhere in the Old Testament. However, *yôm* with 'evening and morning', especially with a number preceding it, can never be ambiguous.

eth ('time') combined with 'day' and 'night' as in *Jeremiah 33:20* and *Zechariah 14:7* could have been ambiguous. Likewise, *eth* combined with 'light' and 'darkness' (a theoretical construction). If any of these forms had been used, the length of the 'days' of creation would have been widely open for debate. **However, God chose *not* to use any of these.**

GOD'S INTENTION

The following considerations show us what God intended us to understand:

1. The meaning of any part of the Bible must be decided in terms of the intention of the author. In the case of Genesis, the intention of its author clearly was to write a historical account. This is shown by the way in which the Lord Jesus Christ and the Apostle Paul regarded Genesis—that is, they quoted it as being absolute truth, not symbolic myth or parable. It was plainly not the author's intention to convey allegorical poetry, fantasy, or myth. Accordingly, what God, through Moses, said about creation in Genesis should not be interpreted in these terms.

    Moses did, in fact, use some of the above 'long-time' words (italicized in the examples below, with root Hebrew words in square brackets), although not with reference to the days of creation. For example, in *Genesis 1:14*, he wrote, 'Let there be lights … *for seasons* [moed]'; in *Genesis 6:3*, 'My spirit shall not *always* [olam] strive with man'; in *Genesis 9:12* 'for *perpetual generations* [olam dor]'; in *Leviticus 24:2*, 'to burn *continually* [tamid]'; in *Numbers 24:20* 'that he perish *forever* [ad]'; in *Deuteronomy 30:20*,

'He is thy light and the *length of thy days* [*yôm orek*]'; in *Deuteronomy 32:7*, 'Remember the *days of old* [*yôm olam*]'; and so on.

Why did God not use any of these words with reference to the creation days, seeing that He used them to describe other things? Clearly it was His intention that the creation days should be regarded as being normal earth-rotation days, and it was not His intention that any longer time–frames should be inferred. Professor James Barr, professor of Hebrew at Oxford University agrees that the words used in *Genesis 1* refer to 'a series of six days which were the same as the days of 24 hours we now experience,' and he says that he knows of no professor of Hebrew at any leading university who would say otherwise.

2. Children have no problem in understanding the meaning of Genesis. The only reason why other ideas are entertained is because people apply concepts from outside the Bible, principally from evolutionary/ atheistic sources, to interpret the Bible.

3. The Bible is God's message to mankind and as such it makes authoritative statements about reality. If one removes any portion of the Bible from the realm of reality, God may still be communicating truth to us, but the reader can never be sure that he understands it as the author intended. Furthermore, if God's communication to us is outside our realm of reality, then we cannot know whether any account in the Bible means what the words actually say or whether it means something entirely different, beyond our understanding. For example, if we apply this criterion to the accounts of the resurrection of Jesus, perhaps the words could mean that Jesus did not rise from the dead physically, but in a way beyond our comprehension. When these sorts of word–games are played with the Bible, the Bible loses its authority, we lose the divine perspective on reality, and Christianity loses its life–changing power.

4. If the 'days' really weren't ordinary days, then God could be open to the charge of having seriously misled His people for thousands of years. Commentators universally understood Genesis in a straightforward way, until attempts were made to harmonize the account with longs ages and then evolution.

The Pen of Moses

In Genesis 1, God, through the 'pen' of Moses, is going out of His way to tell us that the 'days' of creation were literal earth–rotation days. To do this, He used the Hebrew word *yôm*, combined with a number and the words 'evening and morning.' If God had wanted to tell us it was an ancient creation, then there were several good ways He could have done this. If theistic evolution had been intended, then there were several constructions He could have used. If the time factor had been meant to be ambiguous, then the Hebrew language had ways of saying this. However, God chose not to use any construction which would have communicated a meaning other than a literal solar day.

Fig.11.17: Moses Wrote the Pentateuch

The only meaning which is possible from the Hebrew words used is that the 'days' of creation were 24–hour days. God could not have communicated this meaning more clearly than He did in *Genesis 1,* Figure 11.17. The divine confirmation of this, if any is needed, is *Exodus 20:9–11*, where the same word 'days' is used throughout:

'Six *days* shalt thou labor, and do all thy work: But the seventh *day* is the sabbath of the LORD thy God: in it thou shalt not do any work, thou, nor thy son, not thy daughter, thy manservant, nor thy maidservant, nor thy cattle, nor thy stranger that is within thy gates: For in six *days* the LORD made heaven and earth, the sea, and all that in them is, and rested the seventh *day*: wherefore the LORD blessed the sabbath *day*, and hallowed it.'

JOSEPHUS – GENESIS – LITERAL DAYS

Many people who compromise on the plain meaning of Genesis claim that the literal interpretation is a modern invention. Instead, they claim that most commentators in the past took a long-age view.

On the contrary, the vast majority interpreted the days of Genesis 1 as ordinary days. Furthermore, even those who did not, such as Origen and Augustine, vigorously attacked long-age ideas and affirmed that the world was only thousands of years old. Among the Jewish commentators, the first-century historian Flavius Josephus (AD 37–*ca.* 100) stands out from the rest.

Having been born in Judea and living there in his formative years, Josephus is unquestionably the most important Jewish historian outside of Scripture, Figure 11.18. Were it not for Josephus, entire periods of Jewish history would have been lost in the mists of time. Like any good Jew, Josephus recognized that one could not understand Jewish history without first understanding its religion. As Scripture defines Judaism, Josephus first explained Judaism by defining Scripture and the Jewish love of their holy books.

"For we have not an innumerable multitude of books among us, disagreeing from, and contradicting one another, [as the Greeks have], but only twenty-two books, which contain the records of all the past times; which are justly believed to be divine; and of them five belong to Moses, which contain his laws and the traditions of the origin of mankind till his death. This interval of time was little short of three thousand years; … the prophets … in thirteen books. The remaining four books contain hymns to God, and precepts for the conduct of human life."

Fig.11.18: An engraving of Flavius Josephus.

As always, Josephus cuts to the heart of the matter. No further explanation is needed to clarify his plain words. He *explicitly* states that man had been around for only 3,000 years by the time of Moses. He goes on to say that Jews hold Scripture so sacred that they would rather die than add to, subtract from, or change any of the divine doctrines of Scripture!

In the preface to *Antiquities*, easily his most important work, Josephus further explains his interpretation of Scripture. When explaining why Moses began with the creation account, Josephus records that Moses taught humanity that God blesses those who love and serve Him.

"Now when Moses was desirous to teach this lesson to his countrymen, he did not begin the establishment of his laws after the same manner that other legislators did; I mean, upon contracts and other rights between one man and another, but by raising their minds upward to regard God, and his creation of the world; and by persuading them, that we men are the most excellent of the creatures of God upon earth. Now when once he had brought them to submit to religion, he easily persuaded them to submit in all other things; … while our legislator speaks some things wisely, but enigmatically, and others under a decent allegory, but still explains such things as required a direct explanation plainly and expressly."

After explaining his methodology, Josephus launches into the Creation account. He quickly established that he considers Moses' account to be quite literal. He comments, 'And this was *indeed* the first day'[7] and 'in *just* six days the world, and all that is therein, was made.' Josephus gives no indication that he considers these words to be enigmatic or allegorical. His comments are as plain in their meaning as Moses' words in Genesis.

Josephus writes next of Eden, the Fall, and then the ten generations from Adam to Noah. Josephus allows no room for gaps between Adam and the Flood, with the 3,000 years between Moses and Adam. Several times in his discourse on the Flood (which he records as global with 'no place' uncovered), Josephus confirms the absence of gaps in the *Genesis 5* genealogies.

Throughout his writings, Josephus notes any Jewish sect that holds a different view from the mainstream position he records. Though he speaks of differences in doctrine between Sadducees, Pharisees, Essenes, and Zealots, he records not even a single dissenting Jewish voice on these key interpretations of Genesis 1–11. Clearly, for Josephus, if there were any dissent, it was not even worth mentioning, because he had shown how the meaning was unambiguous.

During his explanation of the Hebrew Scriptures, Josephus confronts opponents of Judaism who said the same things as modern opponents of Christianity. In Josephus's day, the pagan Greek historians denied the history of the Jewish people as recorded in Scripture. Similarly, in our day, uniformitarian 'scientists' deny the history of the earth and life upon it, likewise recorded in the Bible. Josephus replies to this charge in the same manner that today's church *must* respond to the opponents of a literal Genesis who claim that only secular science should speak on origins:

"And now, in the first place, I cannot but greatly wonder at those men who suppose that we must attend to none but Greeks, when we are inquiring about the most ancient facts, and must inform ourselves of their truth from them only, while we must not believe ourselves nor other men; …

"Nay, who is there that cannot easily gather from the Greek writers themselves, that they knew but little on any good foundation when they set to write, but rather wrote their histories from their own conjectures? Accordingly, they confute one another in their own books on purpose, and are not ashamed to give us the most contradictory accounts of the same things."

Josephus's writings should encourage the modern Church to stand strong on Genesis and its account of the Earth's beginning. Josephus shows that the consistent Jewish stance on Genesis in Jesus' land and time was, 'Genesis means what it says.'

Historian Josephus– Joseph bar Matthias!

Josephus's original name was Joseph bar Matthias. Although he was born a Sadducee and a friend of the Essenes, he trained as a Pharisee. Josephus also went to Rome with a diplomatic envoy from Jerusalem and later spent time in the Zealot militia.

During the Jewish Revolt of AD 66, he (unsuccessfully) defended Galilee against the Romans. He barely escaped the massacre of his garrison in AD 67, and was captured and taken to the Roman general Vespasian. Josephus shrewdly prophesied that Vespasian would become emperor. He freed Josephus when this occurred in AD 69. Seeing the hopelessness of resistance, Josephus tried to persuade the Jews to surrender Jerusalem, so he was regarded as a traitor. Instead, Jerusalem was captured violently in AD 70. Soon after, Vespasian recognized Josephus' intellect and affinities for history.

Under Imperial patronage, Josephus produced two multi-volume works on Jewish history—*Wars of the Jews* (focusing on the Maccabean revolt up to the fall of Jerusalem, *ca.* 145 BC to AD 70) and *Antiquities of the Jews* (a commentary on Jewish Scripture, tradition, and folklore covering Creation to 145 BC). In AD 100, he published *Against Apion* (a Jewish apologetic) and *The Life of Flavius Josephus* (an autobiography) under the patronage of a private citizen.

DID GOD CREATE OVER BILLIONS OF YEARS?

Is it Important?

GEOLOGIC TIME SCALE

ERA		PERIOD	EPOCH	SUCCESSION OF LIFE
GENOZOIC — recent life		QUATERNARY 0-1 Million Years Rise of Man	Recent Plaisto-cone	
		TERTIARY 62 Million Years Rise of Mammals	Pliocone Miocene Oligocene Eocene	
MESOZOIC — middle life		CRETACEOUS 72 Million Years Modern seed bearing plants, Dinosaurs		
		JURASSIC 46 Million Years First birds		
		TRIASSIC 49 Million Years Cycads, First dinosaurs		
PALEOZOIC — ancient life		PERMIAN 50 Million Years First reptiles		
	Carboniferous	PENNSYLVANIAN 30 Million Years First insects		
		MISSISSIPPIAN 35 Million Years Many crinoids		
		DEVONIAN 60 Million Years First seed plants, cartilage fish		
		SILURIAN 20 Million Years Earliest land animals		
		ORDOVICIAN 75 Million Years Early bony fish		
		CAMBRIAN 100 Million Years Invertebrate animals, Brachiopods, Trilobites		
		PRECAMBRIAN Very few fossils present (bacteria-algae-pollen?)		

***Fig.11.19:** Geologic Time Scale*

Often, people challenge biblical creationists with comments along the lines of, "I believe God created, and I don't believe in evolution, Figure 11.19, but He could have taken billions of years, so what's the big deal about the age of the earth?" Some claim that an emphasis on '6 literal days, 6,000 years ago' even keeps people away from the faith, so "Why be so dogmatic? Why emphasize something so strongly that's not a salvation issue?"

It might come as a surprise that we *agree*—to a point. The timescale *in and of itself* is not the important issue. So why do many Christians emphasize it? It's important because the issue ultimately comes down to, "Does the Bible actually mean what it plainly says?" It therefore goes to the heart of the trustworthiness of Scripture. As such, compromising with long ages also severely undermines the whole Gospel message, thus creating crises of faith for many as well as huge problems with evangelism.

PROBLEMS WITH LONG-AGE TIMESCALE

First, we need to understand where the concept of an old earth came from. The idea of millions or billions of years simply is not found anywhere in Scripture; it is a concept derived from *outside* of the Bible.

In 1830, Charles Lyell, a Scottish lawyer, released his book *Principles of Geology*. He stated that one of his aims was "To free the science [of geology] from Moses." He built his ideas upon those of another geologist, James Hutton, who advocated a uniformitarian interpretation of the world's geology. Lyell argued that the thousands of feet of sedimentary layers (laid down by water or some other moving fluid) all over the earth were the result of long, slow, gradual processes over millions or billions of years (instead of the processes of Noah's Flood).

He believed that processes observed in the present must be used to explain the geological history of the earth. So, if we currently see rivers laying down sediment at an average rate of say 1 mm (4/100th of an inch) per year, then a layer of sedimentary rock such as sandstone which is 1,000 meters (3,300 feet) thick must have taken about a million years to form. This 'present is the key to the past' assumption (and its variants) is a cornerstone of modern geology. It involves a rejection of the biblical account of a global watery cataclysm. The millions of years assigned to the various layers in the 'geological column' were adopted long before the advent of radiometric dating methods—well before radioactivity was even discovered.

Fig.11.20: Fossilized Rocks

However, here's the theological problem. Those rock layers don't just have rocks or granules in them, Figure 11.20. They contain fossils. And these fossils are indisputable evidence of death—and not just of death, but carnivory, disease and suffering. There are remains that have tooth marks in them, and even animals fossilized in the process of eating other animals. There is evidence of disease, cancers, and infection; and general suffering from wounds, broken bones, etc. Biblically, we understand these things only began to happen *after the Fall*. But because of the Bible's detailed genealogies, there's no way for the biblical Adam to exist millions of years ago, before death and suffering started happening in the uniformitarian time scale.

The implication of long-age belief is that God ordained death *before* the Fall of man, but the Bible clearly states that it was Adam's actions that brought death into the world (Romans 5:12).

OLD EARTH AND GOD

The idea that death was in creation before the Fall has major implications for the character of God. The same problem arises if one thinks that God used evolution to create.

Evolution is a random and wasteful process that requires millions of 'unfit' organisms to die. Countless transitional forms would have arisen, only to fall as casualties in the great march 'forward.' At some point, this allegedly 'good' God ordained a lottery of death that finally resulted in humans, and then God looked at His image-bearers, standing on top of layers upon layers of rocks filled with the remains of billions of dead things, and proclaimed His whole creation—along with the evidence of all the death and suffering that went into creating it—to be 'very good' (*Genesis 1:31*). Therefore, we can see that long ages don't fit in the biblical view, whether or not someone believes in evolution along with it.

To summarize, the age of the earth was derived from the rock layers, which have fossils in them, which puts death, suffering and disease before the Fall. The Bible is clear that there was no death before Adam (*Romans 5:12*).

THE GOSPEL OF AN OLD EARTH

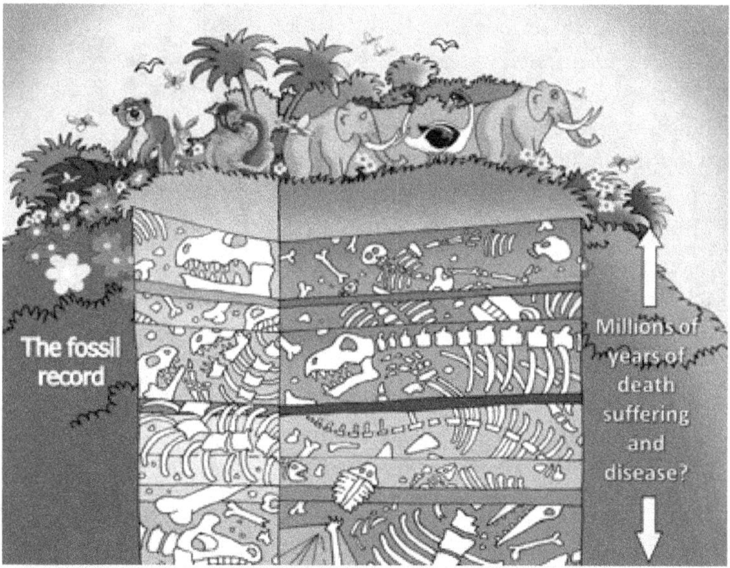

***Fig.11.21:** At the end of day 6 God pronounced his finished creation as" very good".*

If evolution were true, would Adam and Eve have been standing on a fossil graveyard of death and struggle over millions of years that God called "very good". The Bible describes death as the last enemy to be destroyed, Figure 11.21.

Some alleged 'experts' try to sidestep this 'very good' issue by saying that the Fall only caused *human* death and disease. This cannot be true. For one thing, *Romans 8:19–22* clearly teaches that the curse of death and suffering following Adam's Fall affected "the whole creation," i.e., the entire physical universe.

But even if we set that aside for the sake of argument, there is another problem, because we have human remains that are 'dated' as hundreds of thousands of years old. This is well before any possible biblical date for Adam, which places him in the Garden about 6,000 years ago. Many compromising positions see these remains as those of 'pre-Adamites'—soulless non-human animals. But these skeletons fall within the normal range of human variation. And Neandertals, for example, show signs of art, culture and even religion. And recently, the sequencing of actual Neandertal DNA shows that many of us carry Neandertal genes—i.e. we are the same created kind. To call them 'non-human animals' seems entirely contrived to salvage the long-age belief system.

Also, *Romans 5:12* states that "sin came into the world through one man, and death through sin, and so death spread to all men because all sinned". It gives no indication that the Fall caused only human death. To distort the interpretation of Romans 5 to say that death was limited to humans would mean that Adam's sin only brought a *partial* Fall to God's creation; yet *Romans 8:19–20* tells us the *whole* creation groans under the weight of sin and is subjected to futility. And *Genesis 3:17–19* tells us that the very ground was cursed so that it produced thorns and thistles. If only a partial Fall occurred, then why will God destroy all creation to bring about a new one instead of a partial restoration? Why not just restore humans if the rest of creation is still "very good"?

DEATH – LAST ENEMY

A central part of the Gospel is that death is the last enemy to be destroyed (*1 Corinthians 15:26*). Death intruded into a perfect world because of sin, and it is so serious that Jesus' victory over death cannot be entirely manifested while there is a single believer in the grave, Figure 11.22. Are we expected to believe that something the Bible authors described as an enemy was used or overseen by God for millions of years and was called "very good"?

A major part of the Gospel is the hope we have in this Resurrection and restoration of the creation to its original perfect state. The Bible is clear about the New Heavens and Earth as a place where there is no carnivory, no death, no suffering, and no sin (*Isaiah 65:17–25; Revelation 21:1–5*). But how can this be called a *restoration* if such a state never existed?

An evolutionist Anglican priest gave a good summary of what accepting death before the Fall means for Christian theology:

" *... Fossils are the remains of creatures that lived and died for over a billion years before Homo Sapiens evolved.*

Fig.11.22: Death – Last Enemy to be Defeated

Death is as old as life itself by all but a split second. Can it therefore be God's punishment for Sin? The fossil record demonstrates that some form of evil has existed throughout time.

On the large scale it is evident in natural disasters. ... On the individual scale there is ample evidence of painful, crippling disease and the activity of parasites. We see that living things have suffered in dying, with arthritis, a tumor, or simply being eaten by other creatures. From the dawn of time, the possibility of life and death, good and evil, have always existed. At no point is there any discontinuity; there was never a time when death appeared, or a moment when the evil [sic] changed the nature of the universe.

God made the world as it is ... evolution as the instrument of change and diversity. People try to tell us that Adam had a perfect relationship with God until he sinned, and all we need to do is repent and accept Jesus in order to restore that original relationship. But perfection like this never existed. There never was such a world. Trying to return to it, either in reality or spiritually, is a delusion. Unfortunately, it is still central to much evangelical preaching."

So, one can now see the slippery slope that ensues if we allow for billions of years with or without evolution, because it puts death and suffering before the Fall. Its logical corollary is that it also places evil before the Fall (which no longer exists in his view, as such, since there was nowhere to fall from). And in the process, it rules out the hope of a return to a perfect state, since there can be no return to what never was. The Gospel itself has been destroyed in the process.

Thus, what did Jesus come to save us from, if not death, suffering, sin, and separation from God? What do we do with passages like *Hebrews 9:22*, which says " ... the law requires that nearly everything be cleansed with blood, and without the shedding of blood there is no forgiveness," if death and bloodshed were occurring as 'natural' processes for millions of years before Adam? If that is the case, then the death of Christ becomes insignificant and unable to pay for our sins. And what is our hope if it is not in the Resurrection and the New Heavens and Earth?

If death is natural, why do we mourn it so? Why can we not accept death as a 'normal' part of life? This view robs the Gospel of its power and Jesus' sacrifice of its significance. Following the thought to its natural conclusion has led many people to abandon the Christian faith altogether.

CHURCH EFFECT

The widespread teaching of evolution has dire consequences for our youth, who are leaving the church in droves. Christians who 'hang in there' but accept a billions-of-years' timescale will have a much harder time defending their faith, and thus, this affects church growth. One of the major stumbling blocks to faith is the question: "**Why does a good God allow all the death and suffering in the world?**" Such believers cannot adequately explain the origin of death and suffering as a reaction to human sin.

Conversely, believers who have a biblical view of the world's history have a logical platform for introducing God to people with no scriptural background. Incidentally, this was precisely the approach that Paul used when preaching to similar Gentile audiences (*Acts 14:15–17; 17:23–31*). In Lystra, he used creation as a key identifying factor that set God apart from mere men like himself and Barnabas. And in Athens he took the stoics and other philosophers of the day 'back to Genesis' to lay a foundation to introduce them to the true God in the hope that they would repent from their useless idolatry.

If belief in the Bible as plainly written strengthens one's ability to explain the Gospel, and compromise can have such damaging effects, why would anyone compromise? Practically every Christian leader and theologian who lays out his reasons for believing in long ages rather than the biblical timescale has to admit that Genesis—when read at face value, in the Hebrew as well as the English translations—teaches a straightforward creation in six normal-length days. And that this is powerfully backed up by *Exodus 20:11*, part of the Ten Commandments, which shows the Genesis days were understood as normal-length days, with no room for millions of years or gaps in the text to insert them. Nonetheless, they unfortunately accept that science has somehow 'proved' millions of years, which is actually not the case.

INCONSISTENT CHRISTIANITY

While it is possible to be a Christian and believe in an old earth, it would indicate that one has either not thought through the consequences, or that the Bible is not the ultimate authority for one's faith. If Genesis is not real literal history, how can one know where the truth actually does begin in Scripture?

Today's 'science' also 'proves' that men don't rise from the dead. So, if we allow that same science to tell us that Jesus has not risen from the dead (which would be consistent in the compromiser's worldview) then our "preaching is in vain and your faith is in vain," as the Apostle Paul wrote (*1 Corinthians 15:14*). Placing our trust in man-made philosophies is reminiscent of the man that Jesus described in *Matthew 7:26* when He said: "But everyone who hears these words of mine and does not put them into practice is like a foolish man who built his house on sand." Conversely, in verses 25–24 He stated: "everyone who hears these words of mine and puts them into practice is like a wise man who built his house on the rock. The rain came down, the streams rose, and the winds blew and beat against that house; yet it did not fall, because it had its foundation on the rock." And because Jesus clearly believed in a literal historical Genesis, so should we.

THE BIBLE AND 6,000 YEARS CREATION

The Bible Answers Earth's Age

Many people ask, "How do we know that the earth is 6,000 years old from the Bible?" Given that the chrono-genealogies—genealogies where the age of the father at the time of the son's birth is given in an unbroken chain—end shortly after Noah, how do we get from ~1600 AM (*anno mundi* = 'year of the world') to today, which we would argue is about 6000 AM?

How Precise the Earth's Age?

The precision by which we can know the timing of historical events or ages of things is constrained by the precision of the data we're given. The timing we're given in the chrono-genealogies is accurate to within one year of the event. By this, I mean we can know that Adam was 130 years old when he fathered Seth, but we don't know if he was 130 and 3 months, or just shy of 131, for example. This is true for all the ages. Accordingly, when you add up the chrono-genealogies, we know that the Flood happened in 1656, plus up to less than 10 years, because we have 10 numbers that have less than a year of uncertainty.

If all of the numbers were recorded just shy of the next birthday (for instance, Adam was 130 and 11 months when he fathered Seth, Seth was 105 and 11 months when he fathered Enosh, and so on), the Flood could have been as late as 1665 AM. But clearly this sort of small-scale uncertainty won't give any comfort to people who want to add thousands of years to human history.

FLOOD TO PATRIARCHS

There is an unbroken chrono-genealogy from Shem to Abraham in Genesis 11, and we're given the information elsewhere in Genesis to extend the chronology until the relocation of Israel to Egypt when Jacob was 130 years old. Going by these numbers, Jacob went to Egypt in (642 + < 12 years) after the Flood, or (2298 + < 22) years AM. The chrono-genealogy ends here, with nearly 2,000 years to go until Christ. How do we extend the timeline, Table 11.1? [< means less than]

PATRIARCHS TO EXODUS

Exodus 12:40 says that Israel was in Egypt for 430 years. This harmonizes well with Genesis 15:13 where God tells Abram that his descendants will be enslaved and mistreated for 400 years (enslavement did not happen on their arrival in Egypt but sometime after Joseph died, when their number became threatening). Thus, the Exodus happened in 2728 + < 23 years AM, Table 11.1.

EXODUS TO KINGS

We know that Israel wandered in the desert for 40 years, meaning that they entered the Promised Land in 2768 + < 24 years AM. But here the chronology becomes a bit hazier for a while. This is because we don't know *exactly* how long the conquest took, or exactly how long it was before the judges started ruling Israel. We're told how long each judge ruled, and how long each period of peace lasted, but some of these clearly overlap, and some judges clearly only ruled part of Israel, while another judge was ruling another part.

Nevertheless, we have a clear statement in 1 Kings that allows us to continue a reliable chronology. *1 Kings 6:1* says "In the four hundred and eightieth year after the Israelites had come out of Egypt, in the fourth year of Solomon's reign over Israel, in the month of Ziv, the second month, he began to build the temple of the Lord."

Accordingly, if we subtract 124 years (40 each for the wandering in the desert, Saul's reign, and David's reign, and 4 for the partial reign of Solomon), we get a period of about 356 years for the judges, which fits well with the numbers in Judges if we assume a few overlaps. Therefore, Solomon began to build the Temple in 3208 + < 23 years AM. Notice that even though we're thousands of years into history at this point, the uncertainty about the dates is less than 25 years!

KINGS TO EXILE

If we go by the reigns of the kings of Judah, without assuming any co-regencies, from the Temple to the Exile of Judah would have been 429.5 years + < 21 years. But we know that there *were* co-regencies in Judah, partly by comparing the kings of Judah to the kings of Israel. If we do that, we know that from the Temple to the Exile of Judah is actually around 345 years, at around 3553 AM. At this point, it's possible to say what the date would be in our terms—and when one adjusts for the differences in calendrical systems, the vast majority consensus is 586 BC. This would mean that 1 AD would be around 4150 AM, plus or minus less than 50 years, and today we would be around 6150 AM, ± < 50 years.

THE BIBLE IS HISTORY

It's clear that from the very first verse of Genesis, the Bible is concerned with giving a factual account of how God has interacted with the earth. This means that it must give *historically accurate* details, as well as being theologically accurate. In fact, what we believe about God is based on historical claims, so if the history is inaccurate, then the theology must be as well! One of the ways the biblical authors communicated that they were giving actual history is by recording lifespans, or measuring the amount of time between certain events.

We can be confident that God's Word is accurate in its historical details as well as in what it tells us about theology, Table 11.1.

Table 11.1 : Chrono-Genealogies

Father/Event 1	Son/Event 2	Age/Length of time	Running total	Reference
Adam	Seth	130	130	Genesis 5
Seth	Enosh	105	235	Genesis 5
Enosh	Kenan	90	325	Genesis 5
Kenan	Mahalalel	70	395	Genesis 5
Mahalel	Jared	65	460	Genesis 5
Jared	Enoch	162	622	Genesis 5
Enoch	Methuselah	65	687	Genesis 5
Methuselah	Lamech	187	874	Genesis 5
Lamech	Noah	182	1056	Genesis 5
Noah	Flood	600	1656	Genesis 7:11
Flood	Arphaxad	2	1658	Genesis 11
Arphaxad	Shelah	35	1693	Genesis 11
Shelah	Eber	30	1723	Genesis 11
Eber	Peleg	34	1757	Genesis 11
Peleg	Reu	30	1787	Genesis 11
Reu	Serug	32	1819	Genesis 11
Serug	Nahor	30	1849	Genesis 11
Nahor	Terah	29	1878	Genesis 11
Terah	Abram	130	2008	Genesis 11
Abraham	Isaac	100	2108	Genesis 21:5
Isaac	Jacob	60	2168	Genesis 25:26
Jacob	Egypt	130	2298	Genesis 47:9
Jacob in Egypt	Exodus	430	2728	Exodus 12:40
Exodus	Temple begun	480	3208	1 Kings 6:1
Temple	Exile	345	3553	

"… God spoke to our fathers by the prophets, but in these last days He has spoken to us by His Son, whom He appointed the heir of all things, through whom also He created the world" (*Hebrews 1:1–2*)

"All things were made through Him, and without Him was not anything made that was made" (*John 1:3*).

"For by Him all things were created, in heaven and on earth, visible and invisible, whether thrones or dominions or rulers or authorities—all things were created through Him and for Him" (*Colossians 1:16*).

Former Chief Rabbi of the United Hebrew Congregations of the Commonwealth, Lord (Jonathan) Sacks comments: "But it [*Genesis 1*] is not a scientific account, and the proof that it isn't is that it gets through creation in 34 verses."

We have a question concerning Genesis 1 for Professor Craig, Rabbi Sacks, *et al.*: If God had wanted to convey to the Hebrew people (and to us) that He created everything in a period of six days, how could He have said it more clearly than He did in Genesis 1?

The primary reason for our faith is that there is a book, called the Holy Bible, which proclaims the fact that God is, because only divine revelation can explain the book's existence. The divine authorship of the Bible is seen in its amazing unity, its amazing preservation, its historical, scientific, and prophetic accuracy, its civilizing influence, its absolute honesty, and above all, its life changing message, which mends lives broken by sin and disbelief, which separate us from our holy Creator.

True Christian faith is based on this Word of God, the Bible, which tells us that "without faith it is impossible to please Him, for whoever would draw near to God must believe that He exists and that He rewards those who seek Him" (*Hebrews 11:6*). Similarly, it is "By faith we understand that the universe was created by the word of God, so that what is seen was not made out of things that are visible" (*Hebrews 11:3*). This does not mean, however, that Christian faith is irrational or contrary to logic and reason (*Romans 1:18–20*). Indeed, the Apostle Peter commands us: "Always be prepared to make a defense to anyone who asks you for a reason for the hope that is in you" (*1 Peter 3:15*).

STRATEGY OF THE DEVIL

In *Genesis 3* we read the account of the first temptation of Adam and Eve by Satan. The strategy of the devil was successful and led to our first parents' disregard of God's Word to them and their rebellion against God Himself. Today Satan continues to use the same tactics, which produce the same result, namely, the disregard of mankind for the Word of God, and the rebellion of mankind against the authority of God. These tactics are:

1. To doubt the Word of God

The very first temptation recorded in the Bible was for Eve to doubt the truth of something that God had said. 'And he [the serpent] said unto the woman, Yea, hath God said, Ye shall not eat of every tree of the garden?' (*Genesis 3:1*—KJV). Cf. NIV: "'Did God really say, 'You must not eat from any tree in the garden'?'"

Eve's answer to this should have been a simple repetition of what God had said to Adam, 'Of every tree of the garden thou mayst freely eat: But of the tree of the knowledge of good and evil, thou shalt not eat of it: for in the day that thou eatest thereof thou shalt surely die' (*Genesis 2:16–17*).

Instead, Eve replied, 'We may eat of the fruit of the trees of the garden: But of the fruit of the tree which is in the midst of the garden, God hath said, Ye shall not eat of it, neither shall ye touch it, lest ye die' (*Genesis 3:2–3*).

The first issue that, in his query about, 'any' tree of the garden, Satan mispresented and distorted what God had said. God had been very specific ('the tree of the knowledge of good and evil'); Satan was suitably vague ('any tree').

The second issue that Eve misquoted God no less than three times, and both diminished and added to what God had said [but see the discussion in the later article *Did Eve lie before the Fall?*—:

Fig.11.23: Strategy of the Devil

1. She undervalued her privileges by misquoting the divine permission. God had said they could freely eat of every tree (except one); she reduced this to, 'We may eat of the 'trees.' Figure 11.23.
2. She exaggerated the restrictions by misquoting the divine prohibition. God had said nothing against touching; she included this in God's command.

3. She underrated her obligations by misquoting the divine penalty. God had said they would 'surely die'; she changed this to 'lest you die'.

Dr Henry Morris, commenting on this, says 'It is always dangerous to alter God's Word, either by addition (as do modern cultists) or by deletion (as do modern liberals). God, being omniscient, can always be trusted to say exactly, and only, what He means (*Deuteronomy 4:2; Proverbs 30:5; Revelation 22:18–19*); and finite man is inexcusable when he seeks to change God's Word. Such will lead either to divine reproof (*Proverbs 30:6*) or death (*Revelation 22:19*).'

2. To deny the Word of God

Having misrepresented what God had said, and sown the seeds of doubt in Eve's mind, Satan proceeded to outright denial of the truth of what God had said. 'And the serpent said unto the woman, Ye shall not surely die' (*Genesis 3:4*).

Note the progression: Satan distorted the Word of God, which caused the woman to doubt the Word of God, and finally he denied the Word of God.

3. To disregard the judgment of God

That sin has consequences is something that we probably all learn only from experience, and Eve as yet had had none in this area. What she did have was a solemn forewarning of the judgment of God—the death penalty that would follow disobedience. Any thought of such judgment was quickly expunged from her mind by Satan's suggestion that 'in the day ye eat thereof, then your eyes shall be opened, and ye shall be as gods [Hebrew: 'God'], knowing good and evil' (*Genesis 3:5b*).

To be as God was the same desire that had led to Satan's own downfall (*Isaiah 14:13–14*); he now sought to infect Eve with the same desire and she found the prospect irresistible. However, the benefit offered was spurious. From now on, she (and Adam) would know good by the loss of it, and evil by bitter experience. Furthermore, they would know good without having the power to consistently do it, and they would know evil without having the power to refrain from doing it. Far from 'being as God', from now on they would be slaves to Satan. Their eyes were opened not to superior wisdom, but to guilt expressed in shame and fear.

4. To defame the character of God

Satan's suggestion that if Eve ate the fruit she would be 'as God' was put in the context that God knows that you will obtain this advantage (*Genesis 3:5*). The goodness of God was thus impugned, as if He was unfairly and arbitrarily withholding something to which Eve was entitled and from which she would greatly profit; as if through meanness He was keeping her in a state of ignorance and dependence on Himself.

Once Eve had allowed her mind to travel thus far down the road of resentment against God, it was but a small step to total rebellion:

1. She saw that the tree was 'good for food' (i.e., appealing to her bodily appetite), 'pleasant to the eyes' (i.e., appealing to her senses and emotions, 'desired to make one wise' (i.e., appealing to her intellect).

2. 'She took of the fruit and ate thereof.' It was her own act and deed.

3. 'She gave also unto her husband with her; and he did eat' (*Genesis 3:6*). Having fallen into sin herself, she desired that Adam should join her, and he willingly did so.

Thus, the forbidden fruit was looked upon, desired, taken, eaten, and given to another. In the process the Word of God was rejected, the will of God was resisted, and the way of God was repudiated.

The essence of all sin is the desire in the heart to be independent of God. This results in the choosing of self-interests rather than God's interests, and the gratifying of self as the chief end rather than obeying God. In the case of Adam and Eve, the final act was an expression of the sin that had already been committed in the heart and mind.

In the past one-and-a-half centuries the rise of Darwinism and the associated acceptance of theistic evolution and liberal theology by many church leaders have done more to cast doubt on and to deny the truth of the Word of God than any other cause. The effect has been that Western society not only does not believe in the judgment of God—it does not even believe in the existence of a God whose chief attribute is holiness. Satan's strategy, 'which worked with Eve, has proved to be no less effective with modern man.

GOD IMPOSED DEATH FOR SIN

The question is sometimes asked: Why did God impose the death penalty, not only on Adam and Eve, but also on the whole human race, just because they ate a piece of fruit in the Garden of Eden? Surely that was a very small thing to incur such a huge penalty? The answer lies in our having a correct understanding of what was involved in this event, as well as of the holiness of God, Figure 11.24.

Fig.11.24: God's Justice

On probation

In the Garden of Eden, God gave Adam everything needed for his enjoyment, with one significant restraint. 'And the LORD God commanded the man, saying, Of every tree of the garden thou mayest freely eat: But of the tree of the knowledge of good and evil, thou shalt not eat of it: for in the day that thou eatest thereof thou shalt surely die' (*Genesis 2:16–17*).

Why this one restraint?

It is not hard for us to see this as a test of man's love for God. There was just one thing that Adam was forbidden to do—to partake of one tree, and that because God purposed to make a simple test of man's obedience. Man was thus given a choice, which showed that he was a free moral agent before God, and which would also show, when exercised, whether he was prepared to trust and obey God and reciprocate His love, or go his own way. In short, Adam was on probation.

Was it Fair of God to do This?

Yes, indeed it was. The Creator has every right to set the rules, not only for Adam and Eve, but also for us as well. God chose to make them free moral beings but, if they had been able to do as they pleased in every respect and had not had right and wrong laid down for them, they could not have been free moral agents. In their case they were told not to eat of 'the tree of the knowledge of good and evil'. Notice that it was not called 'the tree of good and evil'; nor was God the author of evil. But it was called 'the tree of the knowledge of good and evil'. Our first parents were not to *experiment* with evil to see what it was like. They were clearly warned of the penalty if they did so. They had the power to make a choice. This choice was otherwise unlimited and so they were not short of food. And they should have trusted that God, who made them and all things around them, would know better than the tempter what was in their best interests. God's best is always only available through faith and obedience.

But was it just for such a very small offence to have carried such a huge penalty?

Yes! So small a command presented the best test of their obedience, and it showed that God had the right to make demands of Adam and to expect to be obeyed. Had the sin been some base crime, this would have drawn our attention away from the act of disobedience to the terrible character of the sin. 'Big' sins could then be regarded as the only sins. And the penalty attached to the command shows that Adam was not left in ignorance of its meaning or importance (*Genesis 2:17*). Furthermore, if God ordained such a severe penalty for what some may say was a minor offence or a mere peccadillo, how seriously then does it show that God regards all sins, including those that we label 'big' ones?

The motive for Adam and Eve's disobedience was not appetite, but the ambition to be as God (*Genesis 3:5*). It is clear, therefore, that sin is essentially rebellion against God's revealed will.

A matter of the will more than of the hand, sin is an act of rebellion, revolution, and anarchy against God's righteous government. As such it is an affront to the holiness of God. The measure of God's wrath against sin is the measure of His holiness. And the measure of the penalty—death—is the measure of the enormity of the offence.

THE CONSEQUENCES

In the event, Adam and Eve (to whom Adam must have communicated God's command) chose to disobey God. They ate the forbidden fruit and the consequences followed. There were at least three:

1. **Guilt.** The first consequence of their disobedience was a feeling of guilt: 'And the eyes of them both were opened, and they knew that they were naked; and they sewed fig leaves together, and made themselves aprons. And they heard the voice of the LORD God walking in the garden in the cool of the day: and Adam and his wife hid themselves from the presence of the LORD God amongst the trees of the garden' (*Genesis 3:7–8*).

    Their guilt expressed itself in two ways—shame and fear. As a result of their sin, Adam and Eve experienced the accusations of a newly awakened faculty—conscience. They had wanted to be as gods, knowing good and evil (*Genesis 3:5*), but what they experienced was not the glory of deity but the shame of fallen humanity. When Satan promised them enlightenment, he had left out a vital part of the truth: that they would be reminded of good without having the power to do it, and that they would know evil without having the power to avoid it.

    Why did this shame manifest itself in their desire to cover their nakedness with fig leaves?

    Before they ate the forbidden fruit, Adam and Eve had committed no sin and so there was nothing to cause them shame, either regarding their bodies or anything else. Indeed, it is just possible that in their innocent state their bodies exhibited a beauty of holiness or radiance like that of Moses' face when he communed

Fig.11.25: God's Justice Collides with God's Mercy – Curtsy – wikipedia.org

with God on Mount Sinai. Now, however, the innocence and the beauty of holiness which they once had were gone and their awareness of their physical nakedness was also evidence of their awareness of their awful spiritual nakedness before God. God had wanted them to 'be fruitful and multiply', but now 'their children would all be contaminated with the seed of rebellion, so that their feeling of guilt centered specially on their own procreative organs. The result was that they suddenly desired to hide these from each other, and from God.'

The second consequence of their guilt was fear of God, whom they now dreaded to meet. As they thought about God and His holiness, His glory which once had been their delight must now have seemed to them more like a fire which they could neither endure nor escape from.

God's two questions to Adam, 'Where art thou?' (*Genesis 3:9*) and 'Hast thou eaten of the tree …?' (*Genesis 3:11*), might have been for the purpose of encouraging Adam and Eve to confess their sin and repent of it, as God in His omniscience already knew the answers. Instead Adam blamed Eve and in the process managed to suggest that it was God who was responsible(!)—'The woman whom thou gavest to be with me, she gave me of the tree, and I did eat' (*Genesis 3:12*). Eve, in turn, blamed the serpent— 'The serpent beguiled me, and I did eat' (*Genesis 3:13*). Both were guilty, but both tried to shift the responsibility for their sin to someone else.

2. Then came **condemnation**. First the serpent was cursed. Eve was then condemned to sorrow and subjection. Last of all Adam was dealt with—the ground was cursed for his sake—with the result that his life from then on would be one of sorrow, hardship, and toil. Spiritually he died immediately. Ultimately, he would die physically, too.

3. Finally there was **separation** of Adam and Eve from the garden and from fellowship with God.

We should not think that these consequences were arbitrary or something that God thought up in a fit of angry disappointment. Rather, they spell out what it means for man to put himself out of fellowship with God. Consider a deep-sea diver who feels frustrated at the restriction imposed on his movements by the hose that connects him to the air pump in the boat at the surface of the water and cuts the line … So, Adam cut his spiritual lifeline to the One who is the source of all peace and goodness and life, and found in the process that all of creation was against him. Sin is always a perversion of God's best and brings all of life under judgment— the judgment of futility and death.

DYING YOU SHALL DIE

How do these consequences compare with the penalty that God had previously announced: '… in the day that thou eatest thereof thou shalt surely die' (*Genesis 2:17*)?

It is sad that some Christians have made shipwreck of their faith in the reliability of God's Word and the faithful interpretation of Genesis in particular, because of the wording of this verse as it appears in most versions of the Bible, including the KJV. These people point out that Adam did not die on the day he ate the fruit, but lived for 930 years (*Genesis 5:3–5*). So was God speaking the truth when He said they would die, or was Satan when he said they would not (*Genesis 3:4*)?

Two things need to be said. First, a literal translation of the Hebrew of *Genesis 2:17* is, 'in the day you eat thereof, dying you shall die'. The second is that in the Bible the concept of 'death' in the spiritual sense has the meaning of separation from God rather than of annihilation.

In the Garden of Eden, on the day that they sinned, Adam and Eve were no longer innocent and holy. They now had a sinful nature. Their former fellowship with God was broken. There was a very real separation of their souls from God, and because of this, on that day, spiritually they died. They continued to live physically, but from that day on their human bodies began to die—a process which continued until the day that there was a separation between their souls and their bodies in physical death.

Therefore, on the day that they ate the fruit, literally, 'dying they died'!

GOOD NEWS AND BAD NEWS

The Apostle Paul, writing about this event, tells us that there is both good news and bad news! The bad news is that Adam's sin was imputed, reckoned and imparted to every member of the human race (*Romans 5:12*). Adam stood as the representative of the race; indeed, he was the human race, and all future generations were in him. The result is that we have all sinned and are under the judgment of God (*Romans 3:19,23*). We were born into the world spiritually dead and, as Jesus said to Nicodemus, we must be born again (*John 3:3–6*).

The good news is that if we repent and believe the Gospel, Christ's death on the cross fully paid the penalty for our sin and appeased God's holy wrath against our sin. Christ's righteousness is imputed to us and we are reconciled to God (*Romans 5:17–21*). God's holiness is thus not compromised in our salvation. He can justly forgive us our sins because the penalty for them has been paid in full by Jesus. As the Apostle John affirms, 'If we confess our sins, He is faithful and just to forgive us our sins, and to cleanse us from all unrighteousness' (*1 John 1:9*).

Notice that word 'just'

This then leads us to one other reason why God imposed the death penalty for sin on the human race.

If mankind was immortal, we would all be cut off from God for eternity. However, because of Christ's death on the cross and His resurrection, if we repent and have faith in the atoning work of Christ for us, our physical death then ushers us into the glorious presence of God in Heaven to be united with him for eternity. How wonderful of God that death, which was the ultimate penalty for sin, should be the very means whereby believers are restored to God and to beautiful holy perfection forever!

If Adam were to have been some form of ape-man which God zapped to form the first human when He breathed into him the breath of life, as some theistic evolutionists require, several faith-destroying conclusions would follow.

If the text of Genesis cannot be trusted to mean what it says, then the Genesis account of sin is a mere human myth, and the trustworthiness of the Bible is invalidated right at the beginning. If that is the case, then how are we to know that the account of the redemption unfolded in it later on is not a myth too? Again, how did sin originate, if not according to the Genesis narrative? Likewise, the restriction on Adam's eating the fruit of a certain tree would then appear to be ridiculous in view of the probability that Adam (a great survivor) could well have spent his life up until then in killing and eating his fellow hominoids. If Adam were to have been the product of millions of years of death and bloodshed in the survival of the fittest, human death as the penalty for disobeying God would have no meaning. If human death is not the penalty for human sin, then Christ's death on the cross cannot have been the atonement for human sin (*1 Corinthians 15:21*).

The historical and faithful view of the Genesis story is the only one that shows the true meaning and purpose of God's plan for mankind. It explains the need for and basis of our salvation, it exalts the Word of God, and it glorifies the Name and Being of God.

God is further revealed in Jesus Christ. The Gospels record that Jesus did things that only God can do, such as raise the dead to life (*John 11:17 44*), calm storms (*Matthew 8:23 27*), and forgive sins (*Mark 2:1 12*). He claimed to send prophets (*Matthew 23:34*) and the Holy Spirit (*Luke 24:49*), and He accepted worship (*Matthew 14:33*). He was much more than just a prophet or a good man. The converted atheist, C.S. Lewis, said that there are three options (after accepting the first point, that the Bible reported Jesus accurately): Jesus was either a liar, a lunatic, or Lord (God). His life and Resurrection proved that He was indeed Lord. No wonder that soon after His death, the New Testament authors recognized Him as God and Creator (*John 1:1 3, Colossians 1:15 20; Hebrews 1:3*).

DANGERS OF THEISTIC EVOLUTION

Churches taking a public stand against atheistic evolutionary indoctrination send a strong message to theistic evolutionists in their own ranks.

The atheistic formula for evolution is:

Evolution = matter + evolutionary factors (chance and necessity + mutation + selection + isolation + death) + very long time periods.

In the theistic evolutionary view, **God** is added:

Theistic Evolution = matter + evolutionary factors (chance and necessity + mutation + selection + isolation + death) + very long time periods + **God**.

In this system God is not the omnipotent Lord of all things, whose Word has to be taken seriously by all men, but He is integrated into the evolutionary philosophy. This leads to substantial Dangers for Christians.

I. Misrepresentation of the Nature of God

The Bible reveals God to us as our Father in Heaven, Figure 11.26, who is absolutely perfect (*Matthew 5:48*), holy (*Isaiah 6:3*), and omnipotent (*Jeremiah 32:17*). The Apostle John tells us that 'God is love', 'light', and 'life' (*1 John 4:16; 1:5; 1:1–2*). When this God creates something, His work is described as 'very good' (*Genesis 1:31*) and 'perfect' (*Deuteronomy 32:4*).

Theistic evolution gives a false representation of the nature of God because death and ghastliness are ascribed to the Creator as principles of creation. (Progressive creationism, likewise, allows for millions of years of death and horror before sin.)

Fig.11.26: The Penalty of Death is Paid on Full

II. God Becomes a God of the Gaps

The Bible states that God is the Prime Cause of all things. 'But to us there is but one God, the Father, of whom are all things … and one Lord Jesus Christ, by whom are all things, and we by Him' (*1 Corinthians 8:6*).

However, in theistic evolution the only workspace allotted to God is that part of nature which evolution cannot 'explain' with the means presently at its disposal. In this way He is reduced to being a 'god of the gaps' for those phenomena about which there are doubts. This leads to the view that 'God is therefore not absolute, but He Himself has evolved—He is evolution'.

III. Denial of Central Biblical Teachings

The entire Bible bears witness that we are dealing with a source of truth authored by God (*2 Timothy 3:16*), with the Old Testament as the indispensable 'ramp' leading to the New Testament, like an access road leads to a motor freeway (*John 5:39*). The biblical creation account should not be regarded as a myth, a parable, or an allegory, but as a historical report, because:

a. Biological, astronomical and anthropological facts are given in didactic [teaching] form.

b. In the Ten Commandments God bases the six working days and one day of rest on the same time-span as that described in the creation account (*Exodus 20:8–11*).

c. In the New Testament Jesus referred to facts of the creation (e.g., *Matthew 19:4–5*).

d. Nowhere in the Bible are there any indications that the creation account should be understood in any other way than as a factual report.

The doctrine of theistic evolution undermines this basic way of reading the Bible, as vouched for by Jesus, the prophets and the Apostles. Events reported in the Bible are reduced to mythical imagery, and an understanding of the message of the Bible as being true in word and meaning is lost.

IV. Loss of the Way for Finding God

The Bible describes man as being completely ensnared by sin after Adam's fall (*Romans 7:18–19*). Only those persons who realize that they are sinful and lost will seek the Savior who 'came to save that which was lost' (*Luke 19:10*).

However, evolution knows no sin in the biblical sense of missing one's purpose (in relation to God). Sin is made meaningless, and that is exactly the opposite of what the Holy Spirit does—He declares sin to be sinful. If sin is seen as a harmless evolutionary factor, then one has lost the key for finding God, which is not resolved by adding 'God' to the evolutionary scenario.

V. The Doctrine of God's Incarnation is Undermined

The incarnation of God through His Son Jesus Christ is one of the basic teachings of the Bible. The Bible states that 'The Word was made flesh and dwelt among us' (*John 1:14*), 'Christ Jesus … was made in the likeness of men' (*Philippians 2:5–7*).

The idea of evolution undermines the foundation of our salvation. Evolutionist "Hoimar von Ditfurth" discusses the incompatibility of Jesus' incarnation with evolutionary thought: "Consideration of evolution inevitably forces us to a critical review … of Christian formulations. This clearly holds for the central Christian concept of the 'incarnation' of God … The absoluteness with which the event in Bethlehem has up to now been regarded in Christian philosophy, is contrary to the identification of this man who personifies this event (= Jesus), with man having the nature of homo sapiens."

VI. The Biblical Basis of Jesus' Work of Redemption Is Mythologized

The Bible teaches that the first man's fall into sin was a real event and that this was the direct cause of sin in the world. 'Wherefore, as by one man sin entered into the world, and death by sin; therefore, death passed upon all men, for that all have sinned' (*Romans 5:12*).

Theistic evolution does not acknowledge Adam as the first man, nor that he was created directly from 'the dust of the ground' by God (*Genesis 2:7*). Most theistic evolutionists regard the creation account as being merely a mythical tale, albeit with some spiritual significance.

However, the sinner Adam and the Savior Jesus are linked together in the Bible—*Romans 5:16–18*. Thus, any theological view which mythologizes Adam undermines the biblical basis of Jesus' work of redemption.

VII. Loss of Biblical Chronology

The Bible provides us with a time-scale for history and this underlies a proper understanding of the Bible. This time-scale includes:

a. The time-scale cannot be extended indefinitely into the past, nor into the future. There is a well-defined beginning in *Genesis 1:1*, as well as a moment when physical time will end (*Matthew 24:14*).

b. The total duration of creation was six days (*Exodus 20:11*).

c.   The age of the universe may be estimated in terms of the genealogies recorded in the Bible (but note that it cannot be calculated exactly, Table 11.1). It is of the order of several thousand years, not billions.

d.   *Galatians 4:4* points out the most outstanding event in the world's history: 'But when the fulness of the time was come, God sent forth His Son.' This happened nearly 2,150 years ago, Table 11.1.

e.   The return of Christ in power and glory is the greatest expected future event.

Supporters of theistic evolution (and progressive creation) disregard the biblically given measures of time in favor of evolutionist time-scales involving billions of years both past and future (for which there are no convincing physical grounds). This can lead to two errors:

a.   *Not all statements of the Bible are to be taken seriously.*

b.   *Vigilance concerning the second coming of Jesus may be lost.*

VIII. Loss of Creation Concepts

Certain essential creation concepts are taught in the Bible. These include:

1.   God created matter without using any available material.

2.   God created the earth first, and on the fourth day He added the moon, the solar system, our local galaxy, and all other star systems. This sequence conflicts with all ideas of 'cosmic evolution', such as the 'big bang' cosmology.

Theistic evolution ignores all such biblical creation principles and replaces them with evolutionary notions, thereby contradicting and opposing God's omnipotent acts of creation.

IX. Misrepresentation of Reality

The Bible carries the seal of truth, and all its pronouncements are authoritative—whether they deal with questions of faith and salvation, daily living, or matters of scientific importance.

Evolutionists brush all this aside, e.g., "Richard Dawkins" says, 'Nearly all peoples have developed their own creation myth, and the Genesis story is just the one that happened to have been adopted by one particular tribe of Middle Eastern herders. It has no more special status than the belief of a particular West African tribe that the world was created from the excrement of ants.'

If evolution is false, then numerous sciences have embraced false testimony. Whenever these sciences conform to evolutionary views, they misrepresent reality. How much more than a theology which departs from what the Bible says and embraces evolution!

X. Missing the Purpose

In no other historical book do we find so many and such valuable statements of purpose for man, as in the Bible. For example:

A.   Man is God's purpose in creation (Genesis 1:27–28).

B.   Man is the purpose of God's plan of redemption (Isaiah 53:5).

C.   Man is the purpose of the mission of God's Son (1 John 4:9).

D.   We are the purpose of God's inheritance (Titus 3:7).

E.   Heaven is our destination (1 Peter 1:4).

However, the very thought of purposefulness is anathema to evolutionists. 'Evolutionary adaptations never follow a purposeful program; they thus cannot be regarded as teleonomical.' Thus, a belief system such as theistic evolution that marries purposefulness with non-purposefulness is a contradiction in terms.

THEISTIC EVOLUTION INVOLVEMENT

The following evolutionary assumptions are generally applicable to theistic evolution:

a. The basic principle, evolution, is taken for granted.

b. It is believed that evolution is a universal principle.

c. As far as scientific laws are concerned, there is no difference between the origin of the earth and all life and their subsequent development (the principle of uniformity).

d. Evolution relies on processes that allow increases in organization from the simple to the complex, from non-life to life, and from lower to higher forms of life.

e. The driving forces of evolution are mutation, selection, isolation, and mixing. Chance and necessity, long time epochs, ecological changes, and death are additional indispensable factors.

f. The time line is so prolonged that anyone can have as much time as he/she likes for the process of evolution.

g. The present is the key to the past.

h. There was a smooth transition from non-life to life.

i. Evolution will persist into the distant future.

In addition to these evolutionary assumptions, three additional beliefs apply to theistic evolution:

A. God used evolution as a means of creating.

B. The Bible contains no usable or relevant ideas which can be applied in present-day origins science.

C. Evolutionistic pronouncements have priority over biblical statements. The Bible must be reinterpreted when and wherever it contradicts the present evolutionary worldview.

The doctrines of creation and evolution are so strongly divergent that reconciliation is totally impossible. Theistic evolutionists attempt to integrate the two doctrines; however such syncretism reduces the message of the Bible to insignificance. The conclusion is inevitable: There is no support for theistic evolution in the Bible.

THE CREATOR GOD JESUS CHRIST

The Bible affirms in several places that Jesus Christ is the Creator God, Figure 11.27. For example, 'All things were made by him [the Word, in Greek ὁ λόγος, = Jesus Christ]' (*John 1:1,3*), and 'For by him [Jesus Christ] were all things created' (*Colossians 1:16*).

Fig.11.27: The Resurrected Jesus is God

If this is true, we should expect to see some parallelism between what happened at creation and the works of Jesus during his ministry on earth. What do we find?

I. First let us consider what kind of evidence we are looking for.

Some of the essential and distinctive elements of creation, as revealed in Genesis chapter 1, as well as elsewhere in the Bible, are:

i. Creation involved the act of God in bringing into being immediately and instantaneously matter, which did not previously exist, without the use of pre-existing materials or secondary causes; for example, in the creation of the heavens and the earth, as recorded in *Genesis 1:1*. Creation also involved the shaping, combining, or transforming of existing materials, as when God created Adam from the dust of the ground (*Genesis 2:7*), and Eve from Adam's rib (*Genesis 2:21–22*).

ii. Creation involved the imparting of life to otherwise lifeless matter.

iii. The mechanism of creation, or the means whereby the above aspects were accomplished, was by the Word of the Lord, that is, God said (= God willed it to happen) … and it happened.

iv. The purpose or motive of God in creating was to display His glory, to make known His power, His wisdom, His will, and His holy name, and that He might receive glory from His created beings.

We should not expect to find exact parallels between the miracles of Jesus and what happened at Creation, as Jesus did not come to re-create the universe, but 'to seek and to save that which was lost,' and 'to give his life a ransom for many.' With this in mind, let us compare these four aspects of creation with the works of Jesus.

I. CREATION OUT OF NOTHING AND/OR FROM EXISTING MATERIALS

Several of Jesus' miracles involved the creation of new material. Whether this was out of nothing or from existing materials is not spelt out by the Gospel writers, as they major on the fact of the miracles and the effects they produced (John emphasizes the teaching that Jesus drew from them), rather than on any analyses of the *modus operandi*.

Water to Wine

Jesus' first *miraculous sign to His disciples* involved the creation of wine (His first *miracle* recorded in the Gospels is actually the creation of the universe (*John 1:3*), as mentioned above). At a wedding breakfast, Jesus instructed the waiters to fill six stone water-pots with water, and then to take them to the master of ceremonies of the wedding banquet. When they arrived, the water had been turned into wine, that is, there had been the instantaneous creation of the carbon atoms and chemical molecules that made up the grape sugar, carbon dioxide, coloring matter, etc., of the wine, Figure 11.28.

Fig.11.28: Jesus Turned Water to Wine

Bread and Fish

Other examples are the two times when Jesus fed a multitude: on the first occasion more than 5,000 people from five loaves and two fish, and on the second occasion more than 4,000 people from seven loaves and a few little fish.

Fig.11.29: Feeding Multitudes of People with 5 Loaves and 2 Fish

Original Bread and Fish – New Bread and Fish

Here there were bread and fish to begin with on both occasions. Jesus either caused these original items to multiply, or He may have dispensed all the original food and then created new loaves and fishes until everyone was fed. Either way, Jesus created sufficient extra bread and fish, not only to feed many thousands of people, but also to provide 12 basketfuls of leftovers on the first occasion and seven basketfuls of leftovers on the second, Figure 11.29. This involved not just the creation of the appropriate carbohydrate, protein and other molecules, but their immediate arrangement into the complex forms and structures needed to make baked bread and fish (albeit dead and cooked).

Miracle Healer

Lepers, Blinds and Paralytics

Some of Jesus' miracles of healing, for example, of lepers, the blind, and paralytics, involved the instant repair of tissues, nerves, muscles, etc., and the instantaneous growth or regrowth of healthy cells. The net result was the creation of healthy functioning parts of the body to replace diseased, non-functioning or atrophied parts, Figure 11.30.

II. THE GIVING OF LIFE

Jesus gave life to the dead on three occasions:

- to a widow's son,
- to Jairus' daughter, and
- to his friend Lazarus, Figure 11.31.

Fig.11.30: Jesus Healing the Blind

In the case of Lazarus, the body had been in the grave for four days, and Martha's words are recorded for us: '...by this time there is a bad odor, for he has been there four days.'

Dead Resurrected

This illustrates that the process of decomposition whereby a dead body eventually becomes dust had already begun. Therefore, here we have a parallel with what happened on the sixth day of creation when God formed Adam from the dust of the ground and breathed into his nostrils the breath of life, and Adam became a living being. Jesus called Lazarus back to life, and the molecules of matter that were in the process of becoming dust became, again, a living human being.

Fig.11.31: Jesus Brought Back Lazarus to Life from the Dead

In the case of the widow's son and of Jairus' daughter, death was more recent, that is, probably on the same day that Jesus gave life to their dead bodies. The principle still applies.

III. THE METHODOLOGY JESUS USED

Jesus appeared to use a variety of means in performing His miracles. These included touching lepers, the blind, and the deaf; the use of saliva to heal a deaf mute and a blindman; the use of clay (with instructions to wash) to heal a blind man; and the word of command to heal, to raise the dead, and to exorcise demons.

What happened in these and in all of Jesus' miracles was that Jesus willed the event to happen and it did.

However, what happened in these and in all of Jesus' miracles was that Jesus willed the event to happen and it did. This is nowhere better illustrated than in the healing of the nobleman's son. Jesus was at Cana in Galilee and a certain royal official asked Him to travel to Capernaum to heal his son who was close to death. The Apostle John records what happened, as follows:

'So, He, came again to Cana in Galilee, where He had made the water wine. And at Capernaum there was an official whose son was ill.'

'When this man heard that Jesus had come from Judea to Galilee, he went to Him and asked Him to come down and heal his son, for he was at the point of death.'

'So, Jesus said to him, "Unless you see signs and wonders you will not believe."'

'The official said to Him, "Sir, come down before my child dies."'

'Jesus said to him, "Go; your son will live." The man believed the word that Jesus spoke to him and went on his way.'

'As he was going down, his servants met him and told him that his son was recovering.'

'So, he asked them the hour when he began to get better, and they said to him, "Yesterday at the seventh hour the fever left him."'

'The father knew that was the hour when Jesus had said to him, "Your son will live." And he himself believed, Figure 11.32, and all his household.' (*John 4:46–53*).

Capernaum was about 27 kilometers (17 miles) from Cana as the crow flies, which means there was no way that the sick son, or anyone else in Capernaum, could have heard Jesus or been influenced by His physical presence in Cana.

Jesus willed the sick boy to recover, at a distance of 27 kilometers, and he did so. Similarly, Jesus willed the water to become wine, as it was being taken into the wedding feast in Cana, and it did so. He willed the bread and fish to form and they did, and He willed the 10 lepers to become well after they had left Him and were on their way to the priests, and they were healed.

It is interesting that a Gentile centurion recognized this authority of Jesus. The centurion had sent servants to request Jesus to come and heal his servant, as Luke records:

Fig.11.32: Jesus Healed the Centurian Son

'And Jesus went with them. When He was not far from the house, the centurion sent friends, saying to Him, "Lord, do not trouble yourself, for I am not worthy to have you come under my roof.'

'Therefore, I did not presume to come to you. But say the word, and let my servant be healed.'

'For I too am a man set under authority, with soldiers under me: and I say to one, 'Go,' and he goes; and to another, 'Come,' and he comes; and to my servant, 'Do this,' and he does it.'"

'When Jesus heard these things, he marveled at him, and turning to the crowd that followed him, said, "I tell you, not even in Israel have I found such faith."'

'And when those who had been sent returned to the house, they found the servant well.' (*Luke 7:6–10*)

The centurion recognized that the voice of Jesus could not be heard by his sick servant, but the result, brought about by the exercise of Jesus' authority, would be no less effective because of this.

IV. JESUS' GLORY SEEN IN HIS MIRACLES

After narrating Jesus' first miraculous sign to His disciples—the turning of water into wine—the Apostle John says, He 'manifested forth his glory; and his disciples believed on him.' When Jesus heard that Lazarus was sick, He said, 'This sickness is not unto death, but for the glory of God, that the Son of God might be glorified.' And then, after Lazarus had died and before Jesus raised him to life, He said to Martha, 'Said I not unto thee, that, if thou wouldest believe, thou shouldest see the glory of God?'

John calls Jesus' miracles' signs and in his Gospel John shows which way the signs point: 'these are written, that ye might believe that Jesus is the **Christ, the Son of God**; and that believing ye might have life through his name.'

Jesus Christ is the Creator God

Jesus Christ is the Creator God. Not only does Scripture affirm it, but during His earthly life and ministry He did the very things we would expect the Creator God to do. He did them in the way that we would expect the Creator God to do them—by His word of authority and the exercise of His will. And the doing of them displayed His glory, Figure 11.33.

This is a source of praise and inspiration for those who believe the Word of God, and at the same time it is a ***reproof of the doctrine of theistic evolution***. The thought that Jesus might have used evolutionary chance random processes to heal the sick or give life to the dead is as unsustainable as the idea that He used such processes to create and give life to all things 'in the beginning'.

Fig.11.33: *Jesus Ascended to Heaven*

APPENDIX

Theistic Evolution

- This article has been adapted from chapter 8 'The Consequences of Theistic Evolution', from Prof. Dr Werner Gitt's book, *Did God use Evolution?*, Christliche Literatur-Verbreitung e.V., Postfach 110135, 33661 Bielefeld, Germany.
- E. Jantsch, *Die Selbstorganisation des Universums,* München, 1979, p. 412.
- Hoimar von Ditfurth, *Wir sind nicht nur von dieser Welt*, München, 1984, pp. 21–22.
- Richard Dawkins, *The Blind Watchmaker,* Penguin Books, London, 1986, p. 316.
- H. Penzlin, Das Teleologie-Problem in der Biologie, *Biologische Rundschau,* 25 (1987), S.7–26, p. 19.

Jesus Christ is the Creator God

- Psalm 19:1.
- Exodus 9:16: cf. Romans 9:17, 22–24: Ephesians 1:5–10; 3:9–11.
- 1 Chronicles 16:29: Psalm 29:1, Revelation 4:11.
- Luke 19:10.
- Matthew 20:28.
- John 2:1–11.
- It is interesting that this is the only miracle which all four Gospel writers record: Matthew 14:15–21; Mark 6:35–44; Luke 9:12–17; John 6:5–14.

- Matthew 15:32–38; Mark 8:1–9.
- Luke 5:12–13; Luke 17:11–19.
- Matthew 9:27–30: Mark 8:22–25; John 9:1–41.
- Luke 5:17–26: Luke 6:6–10.
- Luke 7:11–16.
- Luke 8:41–42 and 49–55.
- John 11:1–44.
- John 11:39.
- Genesis 2:7.
- Mark 7:31–35.
- Mark 8:22–25.
- John 9:1–41.
- Possibly to increase the sense of expectancy on the part of those who would relate in a particular way to touch—the blind, the deaf, and lepers.
- Luke 17:11–19.
- John 2:11. See also Luke 17:15,18; John 11:4,40.
- John 11:4,40.
- Miracles per se are not necessarily evidence of deity; rather they are evidence of supernatural power. Others in the Bible, from Pharaoh's magicians (Exodus 7:22) to the false prophet (Revelation 19:20), are said to perform miracles.
- John does this by showing the occasion, the teaching that Jesus drew from them (for example, 'I am the Bread of Life' after the feeding of the 5,000; 'I am the Resurrection and the Life' after the raising of Lazarus), the increased faith of those who were willing to receive truth, and the increased spiritual blindness of those who rejected Christ's claims.
- John 20:31.
- John 1:3; 1 Corinthians 8:6; Ephesians 3:9; Colossians 1:16; Hebrews 1:2.

  i. Hebrew 'any'. The question is a little ambiguous and could mean either, 'Has God indeed enjoined that you should not dare to touch any tree, or, 'Have you not then the liberty granted you of eating promiscuously from whatever tree you please?'—John Calvin, *Genesis*, Banner of Truth Trust, 1965, p. 148.

  ii. Compare the modern theistic-evolutionary jibe, 'Did God really say, "Six days"?'.

  iii. Adapted from WH. Griffith Thomas, *Genesis*, Eerdmans, Michigan, 1946, p. 48.

  iv. Dr Henry M. Morris, *The Genesis Record*, Master Books, El Cajon, California, 1976, p. 111.

  v. The Apostle Paul wrote, concerning the strategy of Satan in the world, that '… we are not ignorant of his devices' (*2 Corinthians 2:11b*). However, he also wrote to the same church at Corinth, 'There hath no temptation taken you but such as is common to man: but God is faithful, who will with the temptation also make a way to escape, that ye may be able to bear it' (*1 Corinthians 10:13*).

  vi. to David C.C. Watson, dated 23 April 1984. Note that Prof. Barr does not say that he believes that Genesis is historically true; he is just telling us what, in the unanimous opinion of the world's leading Hebrew-language professors (including himself), the Hebrew words used were intended to convey.

  vii. The standard model of the big bang now seems to demand that the universe is about 5% ordinary matter, which is observed through telescopes; 22.5% is dark matter, which is not observed; and the remaining is a mysterious dark energy, 72.5%. These are mysterious unknown forms of matter and energy, called 'dark' because their existence is not directly observed.

viii. A 'light year' is the *distance* light travels in one year, about 6,000,000,000,000 yesrs.

ix. Dark matter is required to hold the galaxies together during all the supposed time the universe has existed. This is because in deep time most galaxies would have flown apart if their visible stars provided the only sources of gravity.

REFERENCES

1. Barrow, John and Tipler, Frank, *The Anthropic Cosmological Principle*, Clarendon Press, 1986.

2. William Lane Craig, Barrow and Tipler on the Anthropic Principle vs. Divine Design, 2005; Critical review of *The Anthropic Cosmological Principle, International Philosophical Quarterly* 27:437–47, 1987.

3. Barnes, L.A., The Fine-Tuning of the Universe for Intelligent Life, arxiv.org, 2 June 2012.

4. Barnes, L.A., Christmas Tripe—A Fine-Tuned Critique of Richard Carrier (Part 3), Letters to Nature blogspot, 23 December 2013; bold/italics in original.

5. Lewis, G.F. and Barnes, L.A., *A Fortunate Universe: Life in a finely tuned cosmos*, Cambridge University Press, UK, 2016.

6. See review, Statham, D., A naturalist's nightmare (review of *A Fortunate Universe*), *J. Creation* 32(1):48–53, April 2018; creation.com/fortunate-universe.

7. M. Saebo, in his *Theological Dictionary of the Old Testament* 6:22, says that *yôm* is: 'the fundamental word for the division of time according to the fixed natural alternation of day and night, on which are based all the other units of time (as well as the calendar).' Cited from Ref. 1, p. 72.

8. For a further discussion of the meaning of *yôm*, see Charles Taylor, *The first 100 words*, The Good Book Co, Gosford, NSW, Australia, p. 21, 1996.

The Hebrew words, anglicized spellings, and biblical references are cited from *Young's Analytical Concordance to the Bible.*

9. Interestingly, the fossil record contains thorns. A conventional interpretation of the fossil record (which denies the global Flood) places them at 'hundreds of 'millions' of years before any human being. See W.N. Stewart and G.W. Rothwell, *Paleobotany and the Evolution of Plants* (Cambridge, UK: Cambridge University Press, 1993), p. 176–172. Return to text.

10. Tom Ambrose, 'Just a pile of old bones', *The Church of England Newspaper*, A Current Affairs section, 21 October 1994.

11. Published in 2013 by Search for Truth Enterprises Ltd, Scotland. For more details see comments in our article *Exploring the God Question: 1. The Cosmos, Part 1 (The big bang).*

12. As advocated by the Narrator in the second DVD program *Life and Evolution, Part 1*, where he says: "For atheists, there appears to be no better weapon to derail belief in God than evolution." Note also that atheist William B. Provine has said: "Evolution is the greatest engine of atheism ever invented." (Source: Slide from W.B. Provine's 1998 "Darwin Day" address, "Darwin Day" website, University of Tennessee Knoxville TN, 1998.

13. Boëthius, *The Consolation of Philosophy (De consolatione philosophiae)* 2(7):3–7, ad 524.

14. Lewontin, R., Billions and billions of demons, *The New York Review*, January 9, 1997, p. 31.

15. Little wonder that the American atheist activist Eugenie Scott said: "I have found the most effective allies for evolution are people of the faith community. One clergyman with a backward collar is worth two

biologists at a school board meeting any day!" (*Research News Opportunities in Science and Theology* (later renamed *Science & Theology News*) 2(8):2, April 2002.

16. Adapted from Williams, A., and Hartnett, J., *Dismantling the Big Bang*, Master Books, Arkansas, 2013, pp. 13–14.

17. Genesis 5 goes from Adam to Noah and his sons; Genesis 11:10ff goes from Shem to Abram; Genesis 21:5 states that Abraham was 100 when Isaac was born; Genesis 25:26 states that Isaac was 60 when Jacob and Esau were born, and Genesis 47:9 says that Jacob was 130 when he went to Egypt. Some argue for gaps in the Genesis 5 and 11 genealogies. For the reasons to take them as unbroken genealogies, see Sarfati, J. Biblical chronogenealogies, *J. Creation* 17(3):14–18, December 2003, creation.com/chrono-genealogies.

18. For more detail about the challenges of interpreting the chronology of the kings of Israel and Judah, see Kaiser W., *A History of Israel: From the Bronze Age Through the Jewish Wars* (Broadville & Holman: Nashville, TN, 1998), p. 292–300.

19. The idea expounded by some popular preachers that sin is a lack of self-esteem of 'not having faith in yourself', and that Jesus: died 'to change us from negative thinking to positive thinking' is clearly contrary to the plain teaching of the Bible.

20. 'Moses wist not that his face shone while he talked with Him' (Exodus 34:29).

21. Dr Henry Morris, The Genesis Record. Master Books. El Cajon, California, 1976, p. 115.

22. Cf. Hebrews 12:29; Revelation 6:15–16; 20:15.

23. Sullivan, W., Fred Hoyle dies at 86; opposed 'big bang' but named it, <www.nytimes.com/2001/08/22/obituaries/22HOYL.html>, August 22, 2001.

24. Clayton, J., excerpt from *The Source*, <http://howardpublishing.com/Books/Books/Chapters/thesource.htm>, August 22, 2001.

25. Hoyle, F., *Home Is Where the Wind Blows*, University Science Books, Mill Valley, California, 414, 1994, as reported in *The Skeptic*, 16(1):52.

26. Arp, H.C., Burbidge, G., Hoyle, F., Narlikar, J.V. and Wickramasinghe, N.C., The extragalactic universe: an alternative view, *Nature* 346:807–812, August 30, 1990.

27. Hoyle, F., The big bang in astronomy, *New Scientist* 92(1280):527, November 19, 1981.

28. Hoyle on evolution, *Nature* 294(5837):105, November 12, 1981.

29. Hoyle F., The universe: past and present reflections; in: *Engineering and Science*, p. 12, November 1981.

30. Burbidge, E.M., Burbidge, G.R., Fowler, W.A. and Hoyle, F., Synthesis of the *Elements in Stars, Revs. Mod. Physics* 29:547–650, 1957, often referred to as the B^2FH paper after their initials. This is an attempt to explain the evolutionary origin of the chemical elements in the stars, but rejecting evolution does not entail rejecting helium fusion.

12

DIVINE DESIGN

DARWIN AND DIVINE DESIGN

While Charles Darwin, Figure 12.1, in his writing, made it very clear that he did not accept the Genesis account of creation, the picture we have of Darwin's views about the existence of a Creator is, at best, confusing. In the 1st edition of his Origin of Species, Darwin wrote, in his Conclusion:

" … I should infer from analogy that probably all the organic beings which have ever lived on this earth have descended from someone primordial form, into which life was first breathed." [emphasis added]

Fig.12.1: *Darwin (1809–1882) Curtsy - Wikipedia. com*

That almost sounds like biblical language (Genesis 2:7). However, Darwin cannot be referring to the God of the Bible, as the biblical Creator breathed life into the first human directly—Adam was not "descended from someone primordial form" along with "all the organic beings which have ever lived on this earth." Nevertheless, it does seem that Darwin was leaving the reader with the impression that he believed in some kind of a creator. In fact, in the 2nd edition, Darwin added "by the Creator" to the end of the sentence.

WHY THE ADDITION?

Perhaps science historian James Strick's observations about Darwin are pertinent here. He writes that Darwin's public writing was framed so as to not alienate people who, while taking a liberal view of the Bible, nevertheless believed in a Creator:

"Darwin went out of his way, even misrepresenting his own views on life's origin, to use language that gave these readers some breathing room."

If so, given the controversy that erupted with the publication of the 1ˢᵗ edition of *Origin*, Darwin might well have considered that adding "***by the Creator***" would be strategically prudent.

Public vs Private

Darwin's private dealings present a different picture. His famous advocate Thomas Huxley, often dubbed "Darwin's bulldog," was a man "anxious to banish from science all supernatural explanations for the origin of life." According to Strick:

"Judging by his private correspondence, Darwin seems largely to have concurred with Huxley's version of a naturalistic origin of life. … But Darwin never aired his thoughts on the subject in public."

Perhaps some of today's evolution-accepting church leaders would be well-advised to read some of that private correspondence. For example, in a letter to the American biologist Asa Gray, Darwin wrote:

"Your question what would convince me of Design is a poser. If I saw an angel come down to teach us good, and I was convinced from others seeing him that I was not mad, I should believe in design. If I could be convinced thoroughly that life and mind was in an unknown way a function of other imponderable force, I should be convinced. If man was made of brass or iron and no way connected with any other organism which had ever lived, I should perhaps be convinced."

In light of that, the polite concession made sometimes by creationists today that, "If Darwin had known what we know today about biology and genetics, he might not have become a Darwinist", appears overly generous.

In contrast, *Romans 1:20* is unquestionably hard-hitting: anyone who denies the existence of God, given the evidence of what has been made, is "without excuse." And that applied just as much in Darwin's time as it does today.

MIRACULOUS DESIGN

Eggshell Nanostructure - Purposeful Construction

The major issue of whether the chicken or the egg came first has been asked for thousands of years. However, any reader of Genesis will readily see that the answer is the chicken (or more accurately the land fowl kind), created on Day 5, which then laid an egg.

Fig.12.2: Chicken Egg-Shells

Renowned evolutionist "David Attenborough," who denies any role for God in the egg's design, has nonetheless described eggs as "miracles of nature" and the egg as "an excellent life support system," Figure 12.2. The eggshell not only protects the chick developing inside, it acts as a semipermeable membrane that lets air and moisture pass through about 7,000 pores in a controlled fashion. This allows the chick inside to breathe, while protecting it from drying out due to a net loss of water.

Now details of the first ever nanostructure examination of fully formed eggshells of the domestic chicken, *Gallus gallus domesticus*, have been published showing superb design in their formation, function and dissolution.

Researchers from McGill University were able to accurately cut thin slices of the eggshell and "found that a factor determining shell strength is the presence of nanostructured mineral associated with osteopontin, an eggshell protein." Osteopontin was discovered to be a binding agent, helping to form the superstructure of the eggshell, guiding the framework and controlling the arrangement of the calcium carbonate in the shell. In the outer layer of the eggshell, there were high levels of osteopontin, meaning that the structure is more closely and densely formed. This keeps the hard-shell protective on the outside while the chick is getting ready to hatch. However, in the inner layer of the eggshell, there were lower levels of osteopontin. This means the nanostructure was larger and more loosely arranged, which made the calcium carbonate more accessible and thus allowed the inner layers to dissolve more readily.

Eggshells – Tiny Teeth

Adding to these marvels of the eggshell's purposeful design is the ingenious tool the tiny chick needs to break out to freedom—the 'egg tooth.' Figure 12.3 This is a horny protrusion on the tip of its upper beak that starts to develop on day 7 of its gestation.

Some three days before hatching, the growing chick is finding it hard to get enough oxygen through the pores in the shell, so it uses its 'toothed' beak to slice a hole through the membrane into the air sac at the shell's flatter end. This stored air gives it just enough extra oxygen to cope with the coming task of 'breakout'.

At the right time, a muscle behind the chick's neck begins spasming, encouraging it to 'pip' through the outer membrane and shell with its tiny 'tooth' tool. Thousands of times it chips the shell, rotating counter clockwise at its flatter end. This mammoth task requires hours of rest between bursts of activity.

Finally, it happens! Fresh air. Success! Figure 12.4. With one huge kick, the chick is born—exhausted, wet and sticky. The 'tooth' gradually shrivels and falls away, culminating this unique and purposeful process, the information for which was programmed into its DNA all along.

Fig.12.3: Toothed Beak Chick – Curtsy - Wikipedia

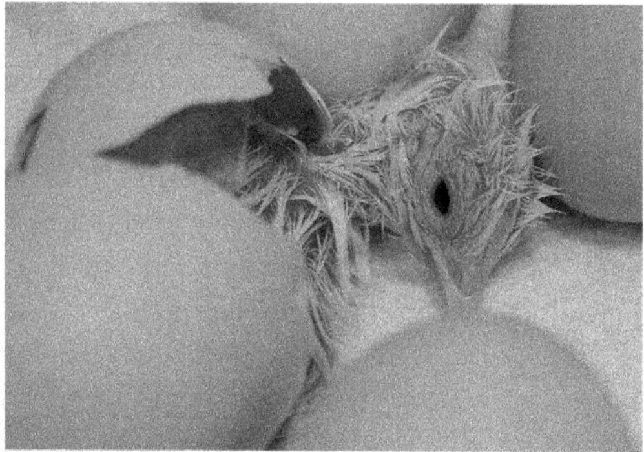

Fig.12.4: Baby Chick is Being Hatched

REASONS FOR DISSOLUTION

Hatching is the climax of the egg's experience, and the nanostructure of the egg demonstrates its wonderful dual-function design allowing the chick to hatch. The inner layer of the eggshell changes as the embryo grows and develops inside. The developing chick requires calcium to help form its bones, and it obtains this through dissolving the innermost layer of the eggshell.

In addition to helping the chick develop its bones, this dissolution also weakens the shell from inside, allowing the chick to be able to hatch when it comes to maturity, normally at 21 days.

Studying this process at the nanostructure level has made this design feature more fully understood. The research team highlighted that, "Such a process would allow retention of overall shell layer structure but with some thinning and compromised strength, a feature ultimately necessary for successful chick pipping to puncture/break the shell during hatching".

Potential Benefits

These latest nanostructure findings have been acclaimed as useful in the design and strengthening of bioinspired material, as well as helping to understand controlled solubility in biostructures. The finding will also be of particular importance to the agri-food industry as "eggshell quality is a major concern to the poultry industry since the percentage of broken or cracked (with possible microorganism invasion) eggshells can range from 13% to 20%." One of the paper's authors, Dr "Marc McKee," explained that, "Understanding how mineral nanostructure contributes to shell strength will allow for selection of genetic traits in laying hens to produce consistently stronger eggs for enhanced food safety."

The Evolutionary Setback

Despite the rather obvious design features of the chicken egg, and the potential applications of copying some of these, the researchers paid the usual homage to blind evolution. McKee said: "When you think about it, we should be making materials that are inspired by nature and by biology because, boy, it is really hard to beat hundreds of millions of years of evolution in perfecting something." Yet the paper never even hinted at explaining how such a wonderful biomineralized life-supporting chamber could actually arise through hundreds of millions of years of a bit-by-bit evolutionary process. Like the waving of a magic wand, it's as if just writing the words makes it so!

However, there are many difficulties in this idea of the egg being perfected by selection of random changes over millions of years. Achieving this rather complicated balance of structural and mechanical properties needs to happen in different ways at different times of development. If the shell were not sufficiently strong outside, then it would not protect the chick. If it did not dissolve inside, then there would not be enough calcium for the bones to form, nor would the chick be able to break through and hatch. And if the thinning of the shell happened too soon, then it would compromise its protective function. How did the ancestors of today's chickens reproduce for millions of years before that process was allegedly perfected to the point of ensuring the next generation of birds? It is undoubtedly, living matters tuned exceptionally well to bring forward a living creature for the benefits of all human beings,

The research discussed here only scratches the surface of the many mechanisms involved in egg design and function, including other proteins, which are still poorly understood. The credit for this 'excellent' design does not belong to evolution, but rather to the One who is "worthy … to receive glory and honor and power, for you created all things, and by your will they existed and were created" (*Revelation 4:11*).

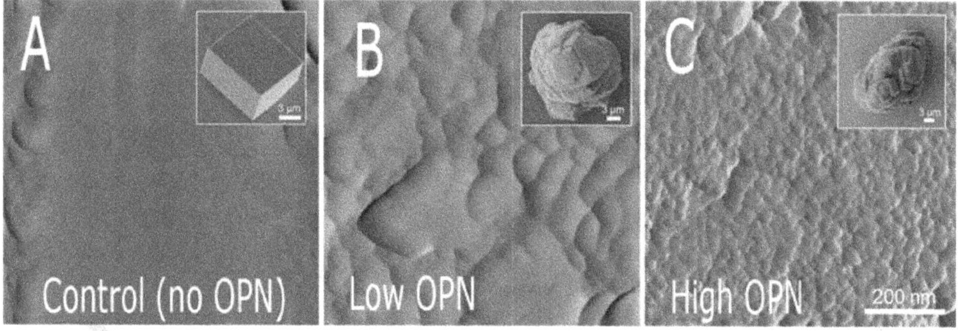

Fig.12.5: *Three types of Osteopontin Curtsy - CC-BY-NC-4.0 Athanasiadou, D. et al.*

To demonstrate how the osteopontin (OPN) affected nanostructure in synthetic calcium carbonate, calcite crystals were grown in its presence. Pictures A, B and C show no, low and high concentrations of osteopontin added to the synthetic calcium carbonate respectively, Figure 12.5. Notably, the measured nanostructure size from the synthetic calcite grown at the low osteopontin concentration was similar to the size found in the inner region of the eggshell, whereas the higher osteopontin concentration produced a nanostructure size similar to the outer part of the eggshell.

Design – Tall-Neck Giraffes

Fig.12.6: *Curtsy - ©iStockPhoto.cpm/paulbanton*

The giraffe is an animal that certainly stands 'head and shoulders' above every other animal, Figure 12.6. An adult male can reach a height of 3 meters (10 feet) at the shoulder, with a neck that can extend for a further 2.5 meters (8 feet). Its front legs are about 10% longer than the hind legs.

There are few more iconic images of Africa than a group of these magnificent creatures silhouetted against the warm oranges of the setting sun. Their uniquely long necks and stilt-like legs give the appearance of slowness to their graceful, almost casual way of moving. Yet, an adult giraffe can give most other creatures 'a run for their money,' with a top speed of around 55 km/h (34 mph).

The giraffe (*Giraffa camelopardalis*) is an even-toed ungulate (hoofed animal). It is also the world's largest ruminant (animals that partly digest their food and then regurgitate it to chew as 'cud'). The giraffe is placed in the family Giraffidae, a group that contains only two animals— the other being the Okapi. This is a curious animal in its own right, with a giraffe-like head, zebra-striped legs and hindquarters, and a body shape much like that of a large gazelle.

Far more than just a beautiful and impressive animal to look at, the giraffe has a whole range of interesting design features. These mostly either involve supporting its amazing neck or are in some way related to it. Long and powerful, this 225-kilogram (500 pound) structure enables the giraffe to reach foliage that other species can only dream of.

Yet, despite its impressive size, the giraffe's neck still contains just seven cervical vertebrae (neck bones). This is the same as most other mammals—but the giraffe's vertebrae are of course longer (up to 25 cm = 10 in) and they are bound together with ball-and-socket joints. This is the same kind of joint that links our arm to our shoulder, giving a 360-degree range of motion. Therefore, the bony structure of the giraffe's neck demonstrates an excellent balance of weight, flexibility and durability. Indeed, substantially durable is a giraffe's neck, that adult males will enthusiastically club one another with them in order to win mates.

To further assist in supporting the neck, the vertebrae over the shoulders have long vertical extensions, which allow for a very large ligament (the nuchal ligament, which runs from the back of the skull all the way down to the base of the tail, but it is at its thickest just over the shoulders). This ligament helps counteract the weight of the giraffe's head and neck, acting like a giant rubber band, pulling the neck up. This means very little muscular energy is required to hold the head up.

Having one's head perched approximately 5.5 meters (18 ft) in the air may provide an excellent view and premium browsing options, but it does pose the problem of how to get the blood all the way up there. The higher you need to push a liquid up a pipe against gravity, the more powerful a pump you need.

Thankfully, the Creator knew about this and furnished the giraffe with a suitably large heart (up to 60 cm (2 ft) long in an adult male), which generates a blood pressure about twice that of a human or other large mammal. The artery walls have extra elasticity to ensure that they can handle this high pressure close to the heart. Furthermore, to prevent the blood from rushing too quickly back down the neck again, the jugular veins in the neck partially contract to restrict return flow.

Now this is all very well for when the giraffe is walking around with its head up, but what about when it wants to take a drink? When it lowers its head, all that high pressure blood would likely rush downhill (further assisted by gravity) and blow out the delicate blood vessels in the brain and eyes—if it weren't for a series of clever mechanisms working in concert with one another. When the head is lowered, special shunts in the arteries supplying the head restrict blood flow to the brain, diverting it into a web of small blood vessels (the *rete mirabile* or 'marvelous net').

This network of vessels near the brain gently expands to accommodate the increased local blood pressure. Valves in the jugular veins also prevent returning blood from flowing backward while the head is lowered.

All of this is controlled by a complex series of mechanisms that constantly monitor the pressure in the blood vessels and make whatever adjustments are needed to ensure that the proper pressure is maintained in all situations. This means that even if the giraffe lifts its head up quickly mid-drink (perhaps in response to a nearby lion), proper blood supply is maintained to the brain, so that the giraffe doesn't faint (probably much to the lion's disappointment—a giraffe can kill a lion with its powerful kick).

NASA GRAVITY-SUITS

Of course, this high blood pressure, combined with the effect of gravity on such a tall body, would also be a problem for the giraffe's legs. The animal would bleed profusely from any cut, and there is a very real danger of blood pooling in the lower extremities. To combat this, the skin on the giraffe's legs is extremely tough, and tightly fitted by way of a firm inner fascia to prevent blood pooling. (This has been studied by NASA scientists developing the special 'gravity-suits' worn by astronauts to help maintain correct circulation while in space.) To prevent excess bleeding, the blood vessels in the giraffe's legs run deep (away from the skin's surface), and those capillaries that do reach the surface are very narrow, with blood cells only 1/3 the size of ours. Additionally, these smaller blood cells allow for faster absorption of oxygen, ensuring a good supply to the extremities of such a large animal.

Many have asked how the giraffe got all these interesting features. Some suggest that you can start with a non-giraffe and, through successive small changes, end up with a giraffe. However, the fossil evidence of giraffes in the past shows them to be much the same as the ones we see in Africa today. Fossil evidence of transitional forms, or 'not-quite-giraffes,' is "completely lacking". The selective advantage of a long neck for reaching higher leaves in a drought is often discussed, but this does not account for the survival of baby giraffes (who are unable to reach this food supply). And female giraffes would have a selective disadvantage because they are shorter. In addition, giraffes spend a good portion of their time, legs splayed, browsing grass or low-lying shrubs.

In any case, the idea that the neck became elongated stepwise over successive generations under environmental/selection pressures is now shown to be a great deal more complex than previously thought, with a whole assortment of structures and systems that need to be in place to accommodate the long neck. Many of these features involve and affect parts of the body seemingly unrelated to the neck, but nonetheless connected through the necessities of function or support. It illustrates the point that an organism is a finely balanced collection of interconnected (and often interdependent), systems. And the only One who can achieve such a delicate balancing act is the creative Genius who designed it in the first place.

Giraffe Gems

The giraffe's scientific name (*Giraffa camelopardalis*), is similar to its older English name of camelopard. It refers to its irregular patches of color on a light background, Figure 12.7, which bear a token resemblance to a leopard's spots, and its face, which is similar to that of a camel.

With their long legs, giraffes walk by moving both legs on one side of their body forward at the same time—known as 'pacing' (other quadrupeds usually walk by moving diagonally opposite limbs forward at the same time). This allows a longer stride, thus fewer steps and less energy used.

The irregular brown markings that cover most of its body are unique to each giraffe, like fingerprints in humans. They were often thought to be for camouflage, but giraffes show no interest in hiding—fairly pointless for such a towering beast anyway. Instead, the markings are used as thermal windows to regulate temperature. Each one has a large blood vessel around its border; by directing blood flow to or away from the smaller vessels branching off to the center, the giraffe can radiate heat away or retain it as appropriate.

The collective name for a group of giraffes (as in 'pride of lions' or 'gaggle of geese') is a 'journey of giraffes.'

Fig.12.7: Curtsy ©iStockPhoto. cpm/sburel

Coping with Extreme Height: Giraffes and Sauropods

Some of their design features have unfortunately not fossilized. But they must have had them to survive, as can be inferred by comparison with the giraffe. Because it is so tall (6 m), a giraffe must have a way of pumping blood uphill to its brain. Accordingly, its heart is about 60 cm long, with walls as much as 7 cm thick, and a mass of 11 kg. This is needed to deliver blood pressure twice as much as humans: about 180/220, to deliver blood to its brain at 70/110.

For comparison, normal human blood pressure is 120/80, while 140/90 is high blood pressure (hypertension). Even worse is a pressure at 180/120 or more—this is a hypertensive crisis that often requires hospitalization to prevent an imminent hemorrhagic stroke.

Also, humans with hypertension often develop thickened hearts that become stiffer because of too much connective tissue (fibrosis). The stiffening means the heart can't refill with as much blood after every beat, a disease called diastolic heart failure. This causes shortness of breath and fatigue. But giraffes have large hearts without fibrosis.

A giraffe heart also has a different rhythm from hearts of 'ordinary' animals. It delays the pump of the ventricles longer so that they have a chance to fill with more blood. This is why giraffes have no problem running fast and pumping lots of blood, even with a larger heart.

But what would happen if the giraffe bent down to drink—wouldn't that high pressure, no longer working against gravity, cause a hemorrhagic stroke?

For a long time, the explanation involved shunting off blood into the *rete mirabile* or 'marvelous net'—a network of smaller blood vessels. However, recent research shows that the blood pressure drops only about 0.5% in this network, so it wouldn't protect the brain.

Ingenious Automated Pressure Regulating System

The likely answer is that when the head is down, the big veins in the neck store over a liter of blood. This is blood that doesn't reach the heart. Thus, the heart cannot generate its usual high blood pressure, so the brain is not overwhelmed. As soon as the giraffe lifts its head, the blood rushes back to the heart, which can then generate the high pressure needed to pump blood to the raised brain. That way it doesn't have a dizzy spell like humans sometimes experience when they get up too quickly. This is an ingenious automated mechanical pressure regulating system.

And there are also design features needed for their long legs—to avoid both blood pooling and hemorrhage from a cut because of such high pressure assisted by gravity on top.

Humans with hypertension often have swelling (edema) in their ankles because the pressure forces some water out of blood vessels into tissues. Sometimes they wear elastic support stockings or pressure bandages to compress tissues to keep water out.

Giraffes have natural pressure bandages comprising both tight skin and tight fascia, the connective tissue surrounding and stabilizing muscles. This causes the tissue pressure in the lower legs to be about five times higher than in the neck, which resists swelling. Massive bleeding is prevented by having very deep veins, and very narrow capillaries supplying blood near the surface.

Since a giraffe requires very high blood pressure, *a fortiori* (even more so), the even taller *Brachiosaurus* would need an even higher blood pressure with a neck several times longer than a giraffe's. The same must be true for the Diplodocid dinosaur *Barosaurus,* if it lived as it is often pictured or displayed rearing up, with its neck reaching very high. The larger sauropod hearts must have been much bigger even than a giraffe's, and able to pump blood with pressure two or three times as high. And given the giraffe's feedback mechanism to regulate blood pressure to the brain, its pressure bandages in the legs, we can only wonder at the design features that God must have programmed into the sauropods.

Some proposals have not won wide acceptance. For example, did long-necked sauropods have multiple hearts in the neck? However, no known vertebrate animal has more than one heart, although cephalopods such as squid do. Or, the larger animals floated on the surface of the water and used their long necks to feed on underwater plants. But as above, sauropods have many features of land-walking creatures.

Design Perfection in Human Embryo

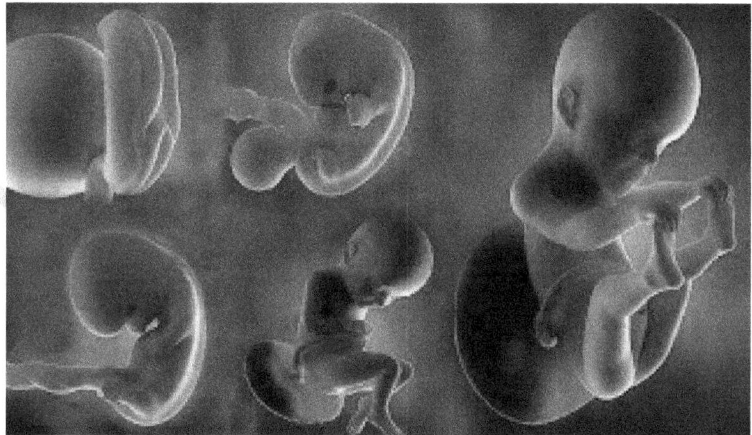

Fig.12.8: Baby Marvelous Development – Curtsy American Pregnancy Association

One of the most marvelous types of evidence of creative design and organization is the astonishing process of how a human being develops in their mother's womb, Figure 12.8. But right at this point, evolutionists come on with one of their strongest arguments. They say, in effect, "Look, if you're talking about creation, then surely the Creator must not be very good at it, or else there wouldn't be all those mistakes in human embryonic development."

Figure 12.9 shows an early stage in human development. Consider it your first "baby picture." You start off as a little, round ball of "unformed substance." Then gradually arms, legs, eyes, and all your other parts appear.

At one month, you're not quite as charming as you're going to be, and here's where the evolutionist says, "There's no evidence of creation in the human embryo. Otherwise, why would a human being have a yolk sac like a chicken does and a tail like a lizard does? Why would a human being have gill slits like a fish does? An intelligent Creator should have known that human beings don't need those things." Well, there they are, "yolk sac, gill slits, and a tail." Why are they there?

What might a creation scientist say?

Figure 12.9. The marvelous development of the human embryo should make everyone a creationist, it seems to me— but evolutionists say that the so-called "gill slits, yolk sac, and tail" are useless evolutionary leftovers (vestiges) that virtually "prove" we evolved from fish and reptiles. How does the creationist respond?

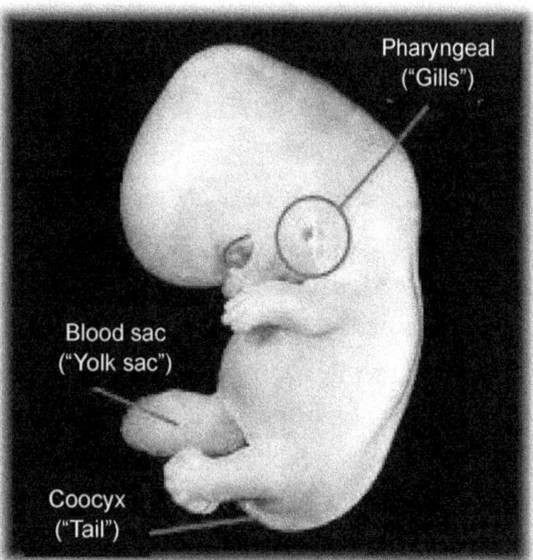

Fig.12.9: Baby Gills-Slits and Tail – Curtsy Christian Heritage Fellowship

The evolutionist believes these structures are there only as useless leftovers, or "vestiges" of our evolutionary ancestry—remainders of the times when our ancestors were only fish and reptiles. Actually, the evolutionary idea of vestigial organs slowed down scientific research for many years. If you believe something is a useless, non-functional leftover of evolution, then you don't bother to find out what it does.

Fortunately, other scientists didn't take that view. They felt instead that such structures probably had a function important for human development.

Sure enough, studies have shown that at least 178 of 180 organs once listed as evolutionary vestiges have quite important functions in human beings. Take the yolk sac, for instance. In chickens, the yolk contains much of the food that the chick depends on for growth.

However, we, on the other hand, grow attached to our mothers and they nourish us. Does that mean the yolk sac can be cut off from the human embryo because it isn't needed? Not at all.

The so-called "yolk sac" is the source of the human embryo's first blood cells, and death would result without it. Now here's an engineering problem for you. In the adult, you want to have the blood cells formed inside the bone marrow. That makes good sense, because the blood cells are very sensitive to radiation damage and bone would offer them some protection.

But you need blood in order to form the bone marrow that later on is going to form blood. So, where do you get the blood first?

Why not use a structure similar to the yolk sac in chickens? The DNA and protein for making it are "common stock" building materials. And, since it lies conveniently outside the embryo, it can easily be discarded after it's served its temporary—but vital—function. Notice, this is exactly what we would expect as evidence of good creative design and engineering practice.

Suppose you were in the bridge-building business, and you were interviewing a couple of engineers to determine whom you wanted to hire. One fellow says, "Each bridge I build will be entirely different from all others". Proudly he tells you "Each bridge will be made using different materials and different processes so that no one will ever be able to see any similarity between the bridges I build". How does that sound?

Now the next fellow comes in and says, "Well, out back in your yard I saw a supply of I-beams and various sizes of heavy bolts and cables. We can use those to span either a river or the San Francisco Bay. I can adapt the same parts and processes to meet a wide variety of needs. You'll be able to see a theme and a variation in my bridge building, and others can see the stamp of authorship in our work." Which fellow would you hire?

As A.E. Wilder-Smith points out (1980), we normally recognize in human engineers the principles of creative economy and variations on a theme. That's what we see in human embryonic development. The same kind of structure that can provide food and blood cells to a chicken embryo can be used to supply blood cells (all that's needed) for a human embryo. Rather than reflecting time and chance, adapting similar structures to a variety of needs seems to reflect plan and purpose. The same is true of the so-called "gill slits".

In the human embryo at one month, there are wrinkles in the skin where the "throat pouches" grow out. Once in a while, one of these pouches will break through, and a child will be born with a small hole in the neck. That's when we find out for sure that these structures are not gill slits. If the opening were really part of a gill, if it really were a "throw-back to the fish stage" then there would be blood vessels all around it, as if it were going to absorb oxygen from water as a gill does.

But there is no such structure. We simply don't have the DNA instructions for forming gills.

Unfortunately, some babies are born with three eyes or one eye. That doesn't mean, of course, that we evolved from something with one eye or three eyes. It's simply a mistake in the normal program for human development, and it emphasizes how perfect our design features and operation must be for life to continue.

The throat (or pharyngeal) grooves and pouches, falsely called "gill slits" are not mistakes in human development. They develop into absolutely essential parts of human anatomy.

The middle-ear canals come from the second pouches, and the parathyroid and thymus glands come from the third and fourth. Without a thymus, we would lose half our immune systems. Without the parathyroids, we would be unable to regulate calcium balance and could not even survive. Another pouch, at one time thought to be vestigial by evolutionists, becomes a gland that assists in calcium balance.

Far from being useless evolutionary vestiges, then, these so-called "gill slits" are quite essential for distinctively human development.

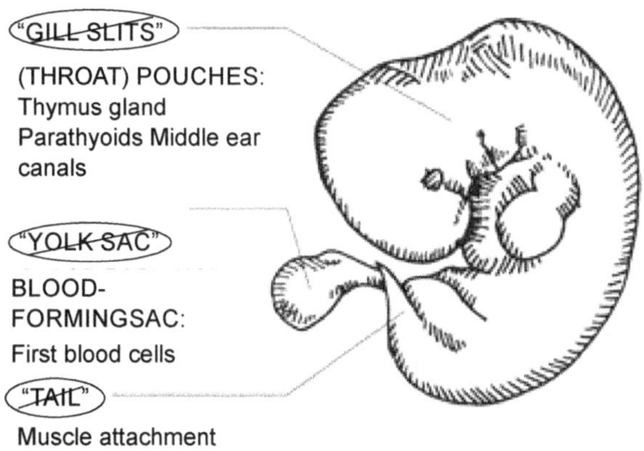

Fig.12.10: *Baby Development of Muscle – Blood - Throat*

Figure 12.10, far from being "useless evolutionary leftovers," the mis-named structures above are absolutely essential for normal human development. Similar structures are used for different functions in other embryos—and we normally consider variation on a theme and multiple uses for a part as evidence of good creative design and engineering skill.

As with yolk sacs, "gill slit" formation represents an ingenious and adaptable solution to a difficult engineering problem. How can a small, round egg cell be turned into an animal or human being with a digestive tube and various organs inside a body cavity?

The answer is to have the little ball (or flat sheet in some organisms) "swallow itself", forming a tube which then "buds off" other tubes and pouches. The anterior pituitary, lungs, urinary bladder, and parts of the liver and pancreas develop in this way. In fish, gills develop from such processes, and in human beings, the ear canals, parathyroid, and thymus glands develop. Following DNA instructions in their respective egg cells, fish and human beings each use a similar process to develop their distinctive features.

What about the "tail"?

Some of you have heard that man has a "tail bone" (also called a coccyx), and that the only reason we have it is to remind us that our ancestors had tails.

You can test this idea yourself, although I don't recommend it. If you think the coccyx is useless, fall down the stairs and land on it. Some of you may have actually done that—unintentionally. What happens? You can't stand up; you can't sit down; you can't lie down; you can't roll over. You can hardly move without pain. In one sense, the coccyx is one of the most important bones in the whole body. It's an important point of muscle attachment required for our distinctive upright human posture (and also for defecation, but I'll say no more about that).

So again, far from being a useless evolutionary leftover, the coccyx is quite important in human development.

True, the end of the spine sticks out noticeably in a one-month embryo, but that's because muscles and limbs don't develop until stimulated by the spine, Figure 2. As the legs develop, they surround and envelop the coccyx, and it winds up inside the body. Once in a great while a child will be born with a "tail."

But, is it really a tail?

No, it's not even the coccyx. It doesn't have any bones in it; it doesn't have any nerve cord either. The nervous system starts stretched out open on the back. During development, it rises up in ridges and rolls shut. It starts to "zipper" shut in the middle first, then it zippers toward either end.

Once in a while it doesn't go far enough, and that produces a serious defect called spina bifida. Sometimes it rolls a little too far. Then the baby will be born—not with a tail, but with a fatty tumor. It's just skin and a little fatty tissue, so the doctor can just cut it off. It's not at all like the tail of a cat that has muscle, bones, and nerve, so cutting it off is not complicated. (So far as I know, no one claims that proves we evolved from an animal with a fatty tumor at the end of its spine.)

AMAZING HUMAN DEVELOPMENT

The details of human development are truly amazing. We really ought to stop, take a good look at each other, and congratulate each other that we turned out as well as we did!

Evolutionists used to say that human embryonic development retraced stages in our supposed evolutionary history. That idea, the now defunct "biogenetic law," was summarized in the pithy phrase, "ontogeny recapitulates Phylogeny." The phrase means that the development of the embryo is supposed to retrace the evolution of its group.

The fertilized egg, for example, would represent our one-celled ancestors, sort of the "amoeba stage." Figure 12.11.

Sure enough, we start as small, round structures looking somewhat like single cells. But notice how superficial that argument is. The evolutionists were just looking at the outside appearance of the egg cell. If we look just on outside appearance, then maybe we're related to a marble, a beebee (air-gun projectile), or a ball bearing—they're small, round things! An

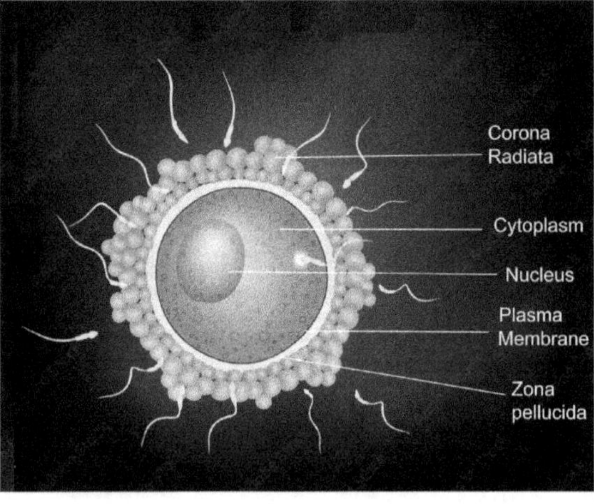

Fig.12.11: Fertilized Human Egg – Curtsy – Adobe Stock

evolutionist (or anyone else) would respond, of course, "That's crazy. Those things are totally different on the inside from a human egg cell."

But, that's exactly the point. If you take a look on the inside, the "dot" we each start from is totally different from the first cell of every other kind of life.

A mouse, an elephant, and a human being are identical in size and shape at the moment of conception. Yet in terms of DNA and protein, right at conception each of these kinds of life is as totally different chemically as each will ever be structurally. Even by mistake, a human being can't produce yolk or gills or a tail, because we just don't have, and never had, those DNA instructions.

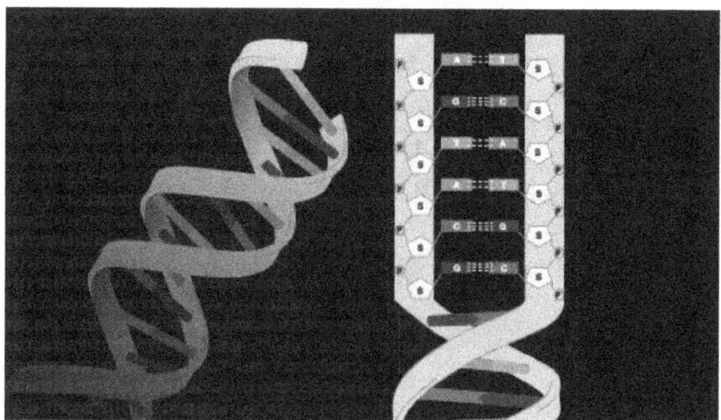

Fig.12.12: Human DNA – Curtsy National Human Genome

The human egg cell, furthermore, is not just human, but also a special individual. Eye color, general body size, and perhaps even temperament are already present in DNA, ready to come to visible expression, Figure 12.12.

KNIT TOGETHER IN MOTHER'S WOMB

Fearfully and Wonderfully Made

Perhaps the most eloquent summary of embryonic development was written over 3,000 years ago by the Psalmist David (*Psalm 139*). He talks about his "unformed substance," and that's just how we all begin—as unformed substance. Then he talks about being "knit together" in his mother's womb, and that's similar to what we observe in embryonic development. The substance is there to start with, like yarn in a basket. And the plan is there, written out in DNA like a knitter's plans. Instructions are creatively conceived and spelled out ahead of time. As the sleeves of a sweater take shape from yarn in the skillful knitter's hands, so the human embryo takes shape according to the plan of his or her Creator.

We are indeed "fearfully and wonderfully made."

Innovations Biometric Design

Fig.12.13: Lobsters Miraculous Eyes – Curtsy - © iStockphoto/JackJelly

The exquisite designs of the organisms on Earth have served as inspiration to countless engineers and scientists. Important inventions, such as Velcro or the airplane, owe their inceptions to those who first observed these qualities in God's handiwork and then sought to mimic these abilities. Although an ancient practice, biomimetics has grown in modern times as a form of reverse engineering. In practice, designs or processes in nature are studied for the purpose of finding practical applications and/or with the hope of designing artificial imitations. Recently the U.S. Department of Homeland Security (DHS) has funded biomimetic research that led to the development of an imaging device that can literally see through walls.

LOBSTER EYE DESIGN

The unique design of the eye of the lobster, Figure 12.13, is one such example that has been intensely studied to help understand how it allows some organisms to see in low light and murky waters. Rather than bending (refracting) the light to focus the image on the retina, several of the long-bodied decapod crustaceans (shrimps, prawns, crayfish and lobsters) possess *reflecting* compound lobster eye design incorporates square facets that are arranged radially to form an optic array with a 180° field of view. The geometric assemblage of facets has all of the hallmarks of intelligent design and defies attempts to explain it through natural mechanisms.

Science & Technology Directorate, Department of Homeland Security

Simply put, these facets are tiny square-shaped tubes with walls that act as mirrors to reflect the incoming light. The walls of each facet are perfectly aligned so that the reflected light is flawlessly focused toward the receptor layer so that they all merge at the same point, Figure 12.14. The design creates an intensified, super-positioned image because the light from many facets combines to form a single image. As many as 3,000 reflective facets are found in some species such as the Norway lobster *(Nephrops norvegicus)*, and increases in sensitivity up to 1,000-fold above that of the more common apposition type eye (where light remains within a single facet/ommatidium). Truly amazing!

Ears that hear and eyes that see—the Lord has made them both. *(Proverbs 20:12)*

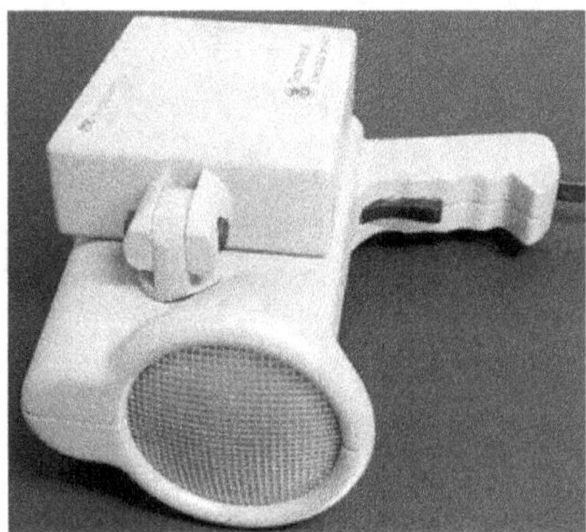

Fig.12.14: LEXID Lobster Eye Imaging Device.

BIOMIMETIC APPLICATIONS

The decapod's eye has the ability to intensify a low brightness image that is captured from a broad field of view using the technique of reflective super-positioning, Figure 12.15. Developing a system similar to that possessed by the lobster has intrigued engineers since the mechanism was first made known, Figure 12.14.

In a 2006 press release, UK researchers at the "University of Leicester" announced that they were developing an X-ray telescope that draws from the design features of the lobster eye. The "Lobster All-Sky X-ray Monitor," which was originally proposed by "Roger Angel" of the "University of Arizona" in 1977, replicates the eye's ability to focus images from all around without

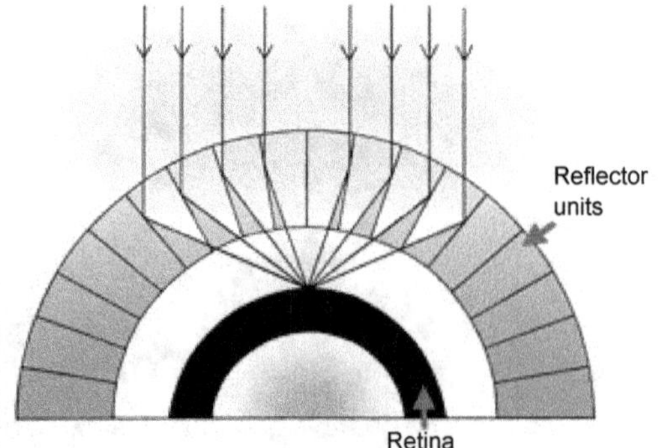

Fig.12.15:

turning. Dr "Nigel Bannister," University of Leicester, stated: "The great advantage of the Lobster design is an almost unlimited field of view." The device may be used aboard the International Space Station or perhaps mounted on a free-flying satellite.

Illustration of the decapod reflective compound eye. Adapted from Denton, M., *Nature's Destiny: How the laws of biology reveal purpose in the universe*, ch. 15, The Free Press, New York/London, 1998.

More recently, the Physical Optics Corporation in Torrance, CA, operating under the U.S. Department of Homeland Security Science and Technology (S&T) Directorate has implemented the design of the lobster's eye to create an imaging device. Known as LEXID ("lobster eye x-ray imaging device,") Figure 12.14, the new handheld imaging system can see through walls of various thicknesses and materials, and identify contents. The marvelous potential of the device has sparked interest from the U.S. Coast Guard, the U.S. Customs and Border Protection, and the Transportation Security Administration, which are organizations responsible for scrutinizing what is coming into the country.

LEXID works by emitting low-level X-rays, which the lobster eye optics focuses into a collector. It then produces an interpretation of the returning X-rays on a small liquid crystal display, which is currently clear enough to reveal weapons or the presence of humans behind concrete walls.

Although still in the developmental and testing phase, the prototype produced with just under one million dollars of Homeland Security money is expected to be ready for on-the-job DHS testing in the near future. In addition to DHS-intended applications, the inventors also envision a virtually unlimited number of alternative uses for the device, in fields ranging from construction to archaeology. Who would think that research revealing the incredible design of the lobster eye might someday safeguard nations or lead to other potential discoveries of historical significance?

EVOLUTION OR DESIGN

The compound eye is one of the most complex and diverse organs. Crustaceans can be found with nine of the ten types of compound eyes, and four distinct types are present in the decapod subgroup. Apposition eyes (where each lens element contributes a unique part of each total image) are more common in crustaceans, but the reflecting superposition type discussed here is typical in the decapods (crayfish, lobsters, etc.).

Evolutionists have attempted to construct phylogenies (evolutionary family trees) by comparing the types of compound eyes present in existing groups. They generally assume that the apposition eyes evolved first, since they are the most common type of compound eye. They are also present in larval stages of all decapods, and possessed by all 'lower crustaceans', such as the trilobite. The advanced reflecting superposition optics that inspired the LEXID are assumed to have developed by Darwinian processes in an ancient common ancestor of the decapods. However, no specific mechanisms for such a development have yet been put forth and experts admit that the overall structure of the eye would have to be radically transformed at once, or non-functioning intermediates would result.

According to "Edward Gaten," University of Leicester:

"The evolution of superposition eyes from the apposition eyes found in primitive crustaceans poses a particular problem. The apposition eye produces multiple inverted images whereas in the superposition eye a single erect image is present. To make this transition without going via non-functioning intermediates requires a continuing correction of the focusing properties of the dioptric apparatus so that light leaving the crystalline cone is either a focal or is focused onto the rhabdom layer."

Instead of recognizing the rather obvious implication, that a masterful intellect is responsible for the engineering wonders possessed by the organisms on Earth, the materialistic worldview causes many to naïvely accept that blind natural processes have given these amazing creatures the "appearance of having been designed for a purpose." Nevertheless, the discipline of biomimetics speaks loudly against such materialistically derived fallacies.

Engineers attempt to copy biological structures and processes because their designs are superior to those devised by the human mind. While LEXID and other devices based on the lobster eye are indeed ingenious technological innovations, they are but crude copies of the real thing.

CREATIVITY OF THE ONE WHO CREATED ALL

Designs offer clear testimony of the existence and creativity of the One who created them, and the biological realm contains countless novelties that scientists struggle to comprehend, and in many cases, they remain a mystery. Although we may never fully understand certain aspects of the creation, the qualities of the Creator are revealed through what is made, thereby leaving those who would deny His existence without any excuse (*Romans 1:20*).

INVERTED RETINA – BAD DESIGN?

The 'inverted' arrangement of the vertebrate retina, in which light has to pass through several inner layers of its neural apparatus before reaching the photoreceptors, has long been the butt of derision by evolutionists who claim that it is inefficient, and therefore evidence against design.

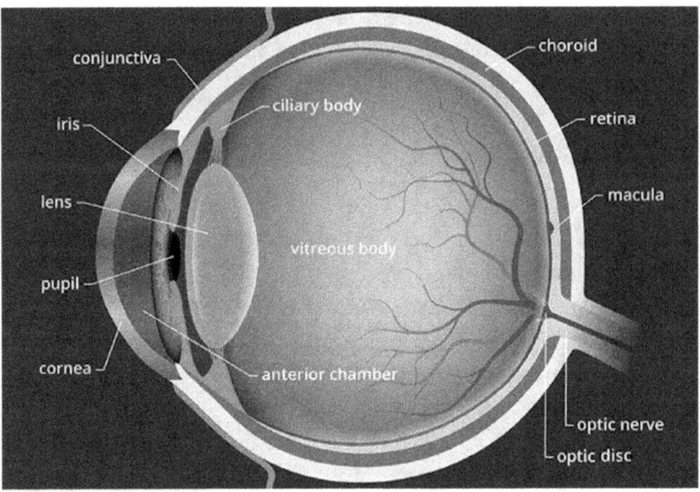

Fig.12.16: Human Eyes – Inverted Retina – Curtsy All About Vision

The lens of the eye, also called the crystalline lens, is an important part of the eye's anatomy that allows the eye to focus on objects at varying distances. It is located behind the *iris* and in front of the vitreous body.

Evolutionist claimed the reasons for our having the inverted retina and why the opposite arrangement (the verted retina), in which the photoreceptors are innermost and the first layer to receive incident light, would be liable to fail in creatures who have inverted retinas, Figure 12.16.

Creation scientists suggest that the need for protection of the retina against the injurious effects of light, particularly with the shorter wavelengths, and of the heat generated by focused light necessitates the inverted configuration of the retina in creatures possessing it.

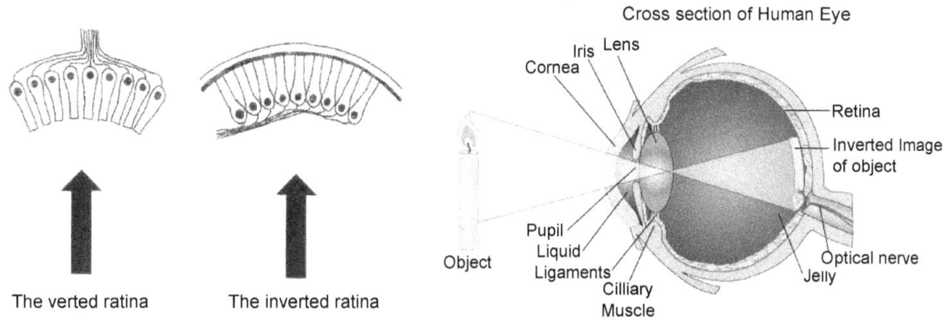

Fig.12.17: *Verted Vs Inverted Retina – Curtsy Sematic Scholar*

Figure 12.17 the two arrangements of photoreceptors. The arrows indicate the direction of incident light.

Evolutionists frequently maintain that the vertebrate retina exhibits a feature which indicates that it was not designed because its organization appears to be less than ideal. They refer to the fact that for light to reach the photoreceptors it has to pass through the bulk of the retina's neural apparatus, and presume that consequent degradation of the image formed at the level of the photoreceptors occurs. In biological terms this arrangement of the retina is said to be *inverted* because the visual cells are oriented so that their sensory ends are directed away from incident light, Figure 12.17. It is typical of vertebrates but rare among invertebrates, being seen in a few mollusks and arachnids.

The opposite arrangement, which is more commonly seen in the invertebrates, is said to be verted, i.e., the receptive elements face the surface. This is thought to be more efficient by many evolutionists who ridicule the inverted arrangement as being 'back-to-front' or 'inside-out'. However, Dawkins, a leading atheistic evolutionist, while admitting that light traversing the inverted retina is not disturbed significantly during its passage to the photoreceptors, writes as follows:

'Any engineer would naturally assume that the photocells would point towards the light, with their wires leading backwards towards the brain. He would laugh at any suggestion that the photocells might point away, from the light, with their wires departing on the side nearest the light. Yet this is exactly what happens in all vertebrate retinas. Each photocell is, in effect, wired in backwards, with its wire sticking out on the side nearest the light. The wire has to travel over the surface of the retina to a point where it dives through a hole in the retina (the so-called 'blind spot') to join the optic nerve. This means that the light, instead of being granted an unrestricted passage to the photocells, has to pass through a forest of connecting wires, presumably suffering at least some attenuation and distortion (actually, probably not much but, still, it is the principle of the thing that would offend any tidy-minded engineer). I don't know the exact explanation for this strange state of affairs. The relevant period of evolution is so long ago.'

Before considering the validity or otherwise of this evolutionary viewpoint, it may be helpful to review briefly some basic ocular anatomy and terminology, Figure 12.18.

Light enters the human eye via the transparent *cornea,* the eye's front window, which acts as a powerful convex lens. After passing through the pupil (the aperture in the iris diaphragm) light is further refracted by the *crystalline lens.* An image of the external environment is thus focused on the *retina* which transduces light into neural signals and is the innermost (relative to the geometric center of the eyeball) of the three tunics of the eye's posterior segment. The other two tunics of the eye's posterior segment are the white tough fibrous *sclera* which is outermost and continuous with the cornea anteriorly, and the *choroid*, a pigmented and highly vascular layer which lies sandwiched between the retina and sclera.

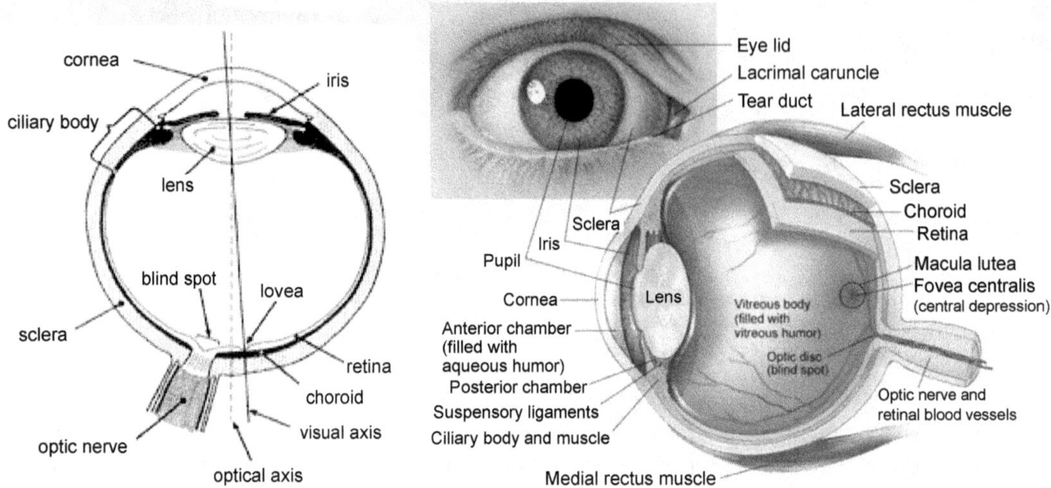

Fig.12.18: Diagram of Eye Cross Section.

The retina consists of ten layers, Figure 12.18, of which the outermost is the dark *retinal pigment epithelium* (RPE) which because of its melanin pigment is opaque to light. The RPE cells have fine hair-like projections on their inner surface called *microvilli* which lie between and ensheath the tips of the photoreceptor outer segments.

There is thus a potential plane of cleavage between the RPE and the photoreceptors which is manifested when the neurosensory retina becomes separated from the RPE, e.g., as a result of injury, a condition known as retinal detachment.

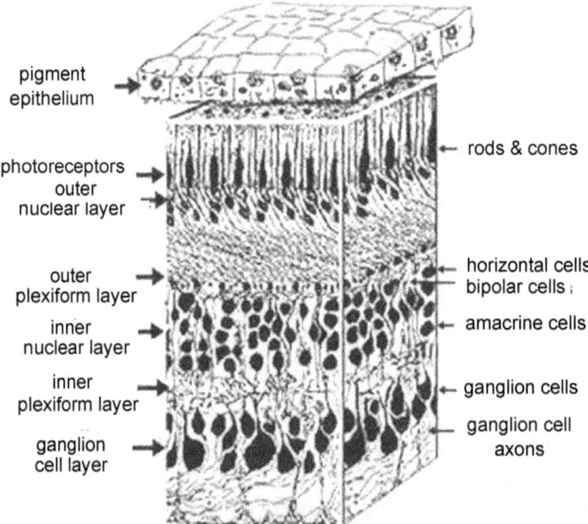

Fig.12.19: Diagram showing the Layers of the Retina – Curtsy Holga Kolb Webvision

Each photoreceptor, whether rod or cone, consists of an inner and an outer segment, the former having organelles (intracellular apparatus) for manufacturing the visual pigment present in the latter. The rod and cone layer and all eight layers internal to it constitute (in distinction from the RPE) what is known as the *neurosensory retina* which is virtually transparent to light. By means of many complex nerve connections within the neurosensory retina, electrical impulses generated by light reaching the photoreceptors are processed and transmitted to the retina's nerve fiber layer and thence pass up the optic nerve to the brain.

In many species for whom vision in very low levels of illumination is important, a layer of reflective crystalline material, the *tapetum* (Latin: carpet) is incorporated in the RPE or choroid. Acting as a mirror, the tapetum reflects light which has passed between the photoreceptors, so augmenting the light bombarding the photoreceptors. Hence the proverbial 'cat's eyes' when caught by a beam of light in the dark.

THE RETINAL PIGMENT EPITHELIUM

Fundamental to understanding the inverted retina is the crucial role played by the RPE. Many of its important functions are now well known. Each RPE cell is in intimate contact with the tips of 20 or more photoreceptor outer segments which number over 130 million. Without the RPE the photoreceptors and the rest of the neurosensory retina cannot function normally and ultimately atrophy. Thus, if the neurosensory retina becomes separated ('detached') from the RPE the vision of the affected area of retina will deteriorate or be extinguished in time, Figure 12.19.

The outer segment of a photoreceptor consists of a stack of discs containing light-sensitive photopigment. These discs are being continually formed by the inner segment from where they move in succession outwards in the outer segment towards the RPE which phagocytoses (Greek: φάγω (*phagō*) = eat) them and recycles their chemical components, Figure 12.20.

The RPE stores vitamin A, a precursor of the photopigments, and thus participates in their regeneration. There are four photopigments which are all bleached on exposure to light: rhodopsin (found in the rods, for night vision) and one for each of the three different types of cones (one for each of the primary colors). It synthesizes glycosaminoglycans for the interphotoreceptor matrix, i.e., the material lying between and separating the photoreceptors.

Besides oxygen, the RPE selectively transports nutrients from the choroid to supply the outer third of the retina and removes the waste products of photoreceptor metabolism to be cleared by the choroidal circulation. By selective pumping of metabolites and the presence of its tight intercellular junctions, the RPE acts as a barrier, called the *blood-retinal barrier*, preventing access of larger or harmful chemicals to retinal tissue, thereby contributing to the maintenance of a stable and optimal retinal environment.

The RPE has complex mechanisms for dealing with toxic molecules and free radicals produced by the action of light. Specific enzymes such as the superoxide dismutase, catalases, and peroxidases are present to catalyze the breakdown of potentially harmful molecules such as superoxide and hydrogen peroxide. Antioxidants such as a-tocopherol (vitamin E) and ascorbic acid (vitamin C) are available to reduce oxidative damage.

Our photoreceptors thus continually synthesize new outer segment discs with their specific photopigments, recycling materials from used discs digested by the RPE. This prompts the question, 'Why have such a complicated process?' The answer must be that it is an example of biological renewal, by means of which tissues exposed to damaging chemicals, radiation, mechanical trauma, etc., are able to survive.

Without self-renewal, tissues such as the skin, the lining of the gut, blood cells etc., would quickly accumulate fatal defects. In the same way, by the continual replacing of their discs the photoreceptors counter the relentless process of disintegration accelerated by toxic agents, particularly short wavelength light.

The RPE cells contain granules of the pigment melanin which absorbs scattered and excess light, so improving visual acuity (VA—i.e., the measure of sharpness of vision or of the ability to perceive as distinct two points close together). Between 25 and 33% of all light entering the eye is absorbed by pigment granules in the RPE and the choroid. The spectral absorption of melanin increases with decreasing wavelength and it is the shorter wavelengths which have the greater energy. It thus also protects the photoreceptors from photic injury and is thought to function as a suppressor of photosensitized molecules, including singlet oxygen, as well as a quencher of free radicals. Besides this, the melanin pigment of the RPE and of the choroid screens the photoreceptors from light entering the eye through the sclera.

Such intensive metabolic activity in the RPE requires a good blood supply which, as will now be considered, is provided more than adequately by the choroid lying in contact with it.

THE CHOROIDAL HEAT SINK

It has been observed that the damage to photoreceptors in an experimental model is strongly related to temperature, and other studies have confirmed that heat exacerbates photochemical injury. Any system designed to protect against the latter should also protect against the former.

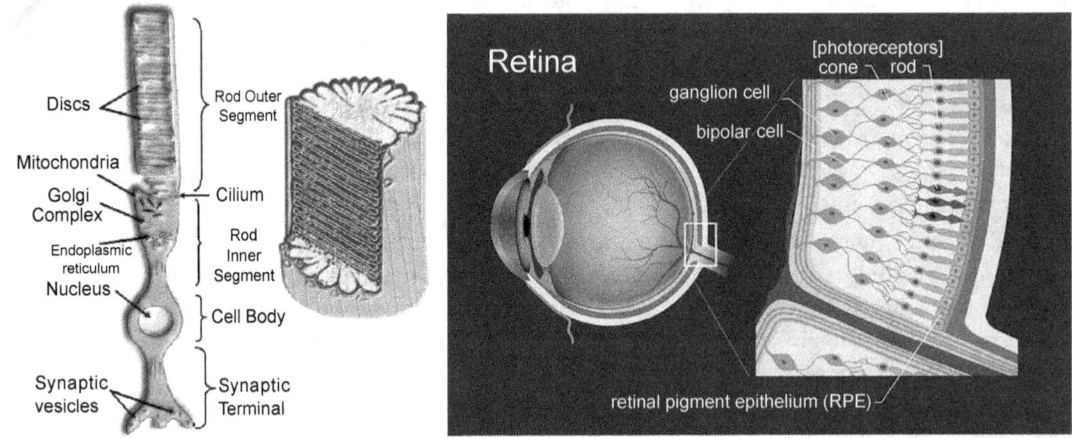

Fig.12.20: *Diagram of a Rod Photoreceptor*

In 1980, a paper was published which explained for the first time something already known about the choroid. It is very high rate of blood flow which far exceeds the nutritional needs of the retina, despite the latter being highly active metabolically, as indicated, Figure 12.20. The paper referred to earlier experiments showing that much less light energy was required to cause a retinal burn in dead animals than in living animals.

It went on to describe experiments with animals which demonstrated that reducing the choroidal blood flow rendered the retina more susceptible to light-induced thermal injury. The choroid takes 85% of the ocular blood flow, and the choroid is remarkable for having the highest blood flow per gram of tissue of all tissues in the body, four times greater even than that of the renal (kidney) cortex. The authors also noted that little oxygen is extracted from blood flowing through the choroid.

The choroidal capillaries (the *choriocapillaris*) form a rich plexus lying immediately external to the RPE, predominantly its central area, and separated from it by only a very thin membrane (Bruch's). The absorption of excess light by the RPE produces heat in the outer retina which has to be dissipated if thermal damage to the delicate and complex biological machinery, its own and that of its neighborhood, is to be avoided.

The authors of this study cogently argue that an important function of the choroid with its torrential blood flow (in local terms) and its close proximity to the RPE, is to act as a heat sink and cooling device. Still more fascinating are the results of further studies by the same workers indicating that there are central (via the brain), light-mediated nervous reflexes regulating choroidal blood flow, increasing the blood flow with increased illumination.

It is evident therefore that for the human retina to function, the presence of both the RPE and the choroid are essential. But both structures are opaque, the RPE because of its melanin and the choroid because of its blood and melanin. It follows that for light to reach the photoreceptors, both RPE and choroid have to be located external to the neurosensory retina; hence we can conclude that there are sound reasons for the inverted configuration of the human and vertebrate retina. Two other design features related to the inverted configuration should be considered.

THE FOVEOLA

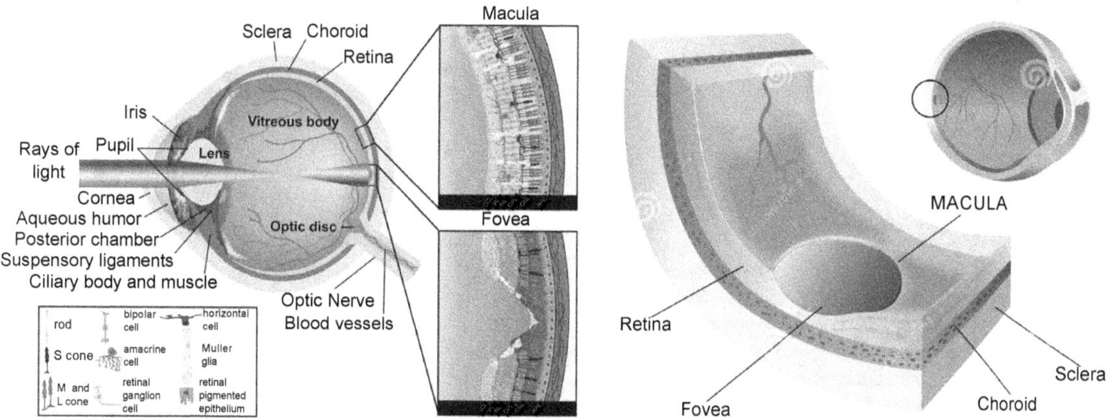

Fig.12.21: *Human Fovea*

Although the neurosensory retina is virtually transparent apart from the blood in its very slender blood vessels, there is an additional refinement of its structure in its central region called the macula. The retina and the occipital cerebral cortex (called the visual cortex) of the brain, to which the former transmits visual information, are so organized that the VA is maximal in the visual axis, line of sight—Figure 12.18. The visual axis passes through the *foveola* which forms the floor of a circular pit with a sloping wall, the *fovea* (Latin: pit) at the center of the macula (Figures 2 & 5). Away from the fovea the VA diminishes progressively towards the periphery of the retina. Thus, the color photoreceptors—the cones for red, green and possibly also blue—have their greatest density of 150,000 per square mm at the foveola, which measures only 300–330 μm across.

Moreover, the foveolar cones differ from those elsewhere in being taller, slenderer, perfectly straight and accurately oriented to be axial with respect to incident light, for maximal VA and sensitivity. In this area, blood vessels are absent and the retina is much thinner, being reduced to only photoreceptors (cones) with minimal supporting tissue. The inner neural elements of the neurosensory retina are displaced from the foveola radially to allow unimpeded access of light and elimination of what little scattering of light occurs elsewhere. The central retina serves primarily color and form perception while the peripheral retina is more concerned with light and motion detection and with night vision. And consistent with these functions, there is a low ratio of receptors to ganglion cells (1:1.2 or more) at the fovea where optical aberrations are also minimal. By contrast, elsewhere in the retina the ratio is reversed with 4–6 cones or up to 100 rods to each ganglion cell, Figure 12.21.

XANTHOPHYLL PIGMENT

The optical system of the human eye is such that ambient light tends to fall with peak intensity on the macular area of the retina with much less on the retinal periphery. It must be significant therefore that not only is melanin more abundant in the macular region because its RPE cells are taller and more numerous per unit area than elsewhere but there is also in the retina's central area the yellow pigment *xanthophyll* (Greek: ξάνθος *xanthos*, yellow). In this region of the retina, xanthophyll permeates all layers of the neurosensory retina between its two limiting membranes and is concentrated in the retinal cells, both the neurons and the supporting tissue cells. Recently attention has been drawn to the presence of a collection of retinal supporting tissue cells (called Müller cells after the person who first described them) over the internal surface of the fovea and forming a cone whose apex plugs the foveolar depression (Figure 5). Besides providing structural support for the fovea, it is thought that the Müller cells, particularly in this location, act as a reservoir of xanthophyll.

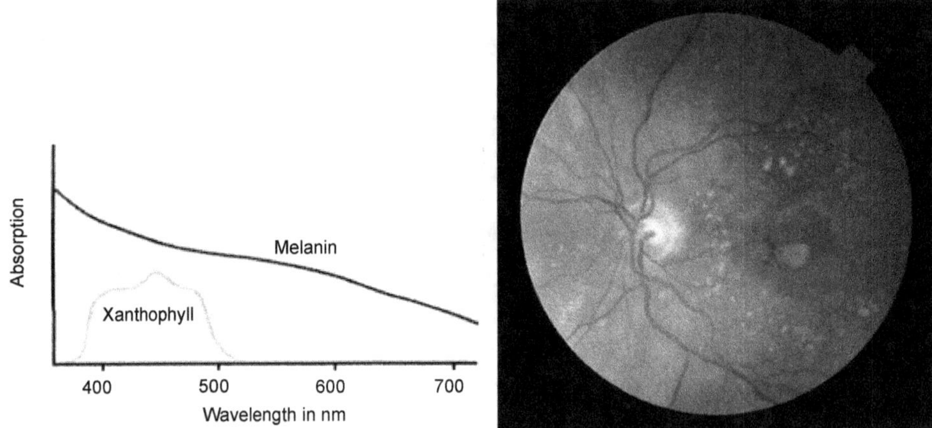

Fig.12.22: *Ocular Pigment Spectral Absorptions – Curtsy Springer Link*

Retinal xanthophyll is a carotenoid, chemically related to vitamin A, whose absorption spectrum peaks at about 460 nm and ranges from 480 nm down to 390 nm, Figure 12.22. It helps to protect the neurosensory retina by absorbing much of the potentially damaging shorter wavelength visible light, i.e. blue and violet, which is more scattered by small molecules and structures. Studies have shown that the retina's sensitivity to photic damage increases exponentially with decreasing wavelength, being six times more sensitive to ultraviolet radiation (UVR) than to blue light. However, almost all of the non-ionizing radiation with wavelengths shorter than 400 nm is blocked by the combination of cornea and lens, leaving a remaining dangerous band of wavelengths, 420–450 nm in the blue part of the spectrum, against which xanthophyll is an effective shield.

The presence of this pigment was well demonstrated in practice when ophthalmologists used to use the argon laser in which the emitted blue light was at first not subtracted; xanthophyll would absorb the blue light and produce an unwanted burn in the neurosensory retina. Modern ophthalmic argon lasers are made to emit only green light for this reason.

THE BLIND SPOT

Because of the retina's inverted arrangement, the axons (nerve fibers) transmitting data to the brain pass under cover of the retina's inner surface to converge to a small area which is the optic nerve head, where they all exit the eye together as the optic nerve. The optic nerve head has no photoreceptors and so is blind, thereby producing a small blind spot in the visual field. This also has been a target of criticism by evolutionists who suggest that it can significantly disadvantage creatures that have it. As Williams puts it:

'Our retinal blind spots rarely cause any difficulty, but rarely is not the same as never. As I momentarily cover one eye to ward off an insect, an important event might be focused on the blind spot of the other.'

Notwithstanding, this issue has to be viewed in perspective: the blind spot is centered at 15° away from the visual axis (3.7 mm from the foveola) and is very small in relation the visual field of an eye, occupying less than 0.25%. As mentioned above, the further away a point in the retina is from the foveola, the less will be its VA and its sensitivity. The retina surrounding the optic nerve head, in the light-adapted state, has a VA of only about 15% of that at the foveola. We can safely infer that the theoretical risk referred to by Williams arising from the blind spot in a one-eyed person, is negligible; and, in keeping with this, it is considered safe for a one-eyed person to drive a private motor car, i.e. for non-vocational purposes.

Because the two visual fields overlap to a large degree, the blind spot of one eye is covered by the other eye's visual field. It is true that occlusion or loss of one eye is a handicap, but this is not because of the blind spot of the seeing eye for the reasons given above. Rather, and much more important to survival in a threatening situation when sight may be vital, it is the loss of stereopsis (binocular distance or depth perception) together with a moderate reduction of *peripheral* visual field which would constitute a handicap.

However, suppose the Creator had sought to avoid the situation described by Williams by creating a very large single eye with two identical optical systems capable of converging their visual axes and two foveae far enough apart to achieve the level of stereopsis we enjoy. The result would be manifestly impractical. Indeed, such a hypothetical one-eyed creature would be the more vulnerable should its only eye be injured or covered.

HUMAN AND ANIMAL VISION

The human eye's visual acuity, while good, is not as high as that of some animals, such as birds, whose retinae are also inverted. But variations in performance are related to the demands of the individual creature's way of life.

The human visual system cannot register motion as accurately and sensitively as that of a fly, but if it did, we would see all fluorescent lighting and television flickering continually. We cannot see at night as well as a cat, but we surpass it in some other areas. For example, cats have no color vision. The human eye represents an excellent balance between versatility and performance, which has enabled man's astonishing technological achievements in antiquity. Latterly, man's capacity to construct devices to see the far distant, the microscopic, and to see in the darkest night, has augmented the practical scope of our vision to exceed that of any other creature.

THE VERTED RETINAE OF INVERTEBRATES

Some evolutionists claim that the verted retinae of cephalopods, such as squids and octopuses, are more efficient than the inverted retinae found in vertebrates. But this presupposes that the inverted retina is inefficient in the first place. As shown above, evolutionists have failed to demonstrate that the inverted retina is a bad design, and that it functions poorly; they ignore the many good reasons for it.

Also, they have never shown that cephalopods actually see better. On the contrary, their eyes merely 'approach some of the lower vertebrate eyes in efficiency' and they are probably color blind. Moreover, the cephalopod retina, besides being 'verted', is actually much simpler than the 'inverted' retina of vertebrates; as Budelmann states, 'The structure of the [cephalopod] retina is much simpler than in the vertebrate eye, with only two neural components, the receptor cells and efferent fibers.' It is an undulating structure with 'long cylindrical photoreceptor cells with rhabdomeres consisting of microvilli,' so that the cephalopod eye has been described as a 'compound eye with a single lens.' The *rhabdomeres* act as light guides, and their *microvilli* are arranged such that the animal can detect the direction of polarized light—this foils camouflage based on reflection.

Finally, in their natural environment cephalopods are exposed to a much lower light intensity than are most vertebrates and they generally live only two or three years at the most. Nothing is known about the lifespan of the giant squid; in any case it is believed to frequent great depths at which there is little light. Thus for cephalopods there is less need for protection against photic damage. Being differently designed for a different environment, the cephalopod eye can function well with a 'verted' retina.

Although it would appear at first sight that the inverted arrangement of the retina has disadvantages and is inefficient, in reality these objections amount to little. Even evolutionists concede that the inverted retina serves those creatures that possess it, very well; it affords them superb visual acuity. We have reviewed the necessity for this arrangement which turns on the nature of the photoreceptors.

Light at various wavelengths is capable of very damaging effects on biological machinery. The retina, besides being an extremely sophisticated transducer and image processor, is clearly designed to withstand the toxic and heating effects of light. The eye is well equipped to protect the retina against radiation we normally encounter in everyday life. Besides the almost complete exclusion of ultraviolet radiation by the cornea and the lens together, the retina itself is endowed with a number of additional mechanisms to protect against such damage:

- The retinal pigment epithelium produces substances which combat the damaging chemical by-products of light radiation.
- The retinal pigment epithelium plays an essential part sustaining the photoreceptors. This includes recycling and metabolizing their products, thereby renewing them in the face of continual wear from light bombardment.
- The central retina is permeated with xanthophyll pigment which filters and absorbs short-wavelength visible light.

The photoreceptors thus need to be in intimate contact with the retinal pigment epithelium, which is opaque. The retinal pigment epithelium, in turn, needs to be in intimate contact with the choroid (also opaque) both to satisfy its nutritional requirements and to prevent (by means of the heat sink effect of its massive blood flow) overheating of the retina from focused light.

If the human retina were 'wired' the other way around (the verted configuration), as evolutionists such as Dawkins propose,[2] these two opaque layers would have to be interposed in the path of light to the photoreceptors which would leave them in darkness!

Thus, I suggest that the need for protection against light-induced damage, which a verted retina in our natural environment could not provide to the same degree, is a major, if not *the* major reason for the existence of the inverted configuration of the retina.

Recently, other researchers adduced from analyzing baby zebrafish eyes, "evidence that the inverted retina actually is a superior space-saving solution, especially in small eyes." And the Müller glial cells in front of the photoreceptors act as a fiber-optic plate, funneling more light, increasing image sharpness, and reducing distortion from stray light, so is "an optimal structure." This structure also helps separate the colors, so that green and especially red light is directed to the cones that are more sensitive to these wavelengths, while the high energy blue photons are directed to the rods that respond more strongly to them, which optimizes our color vision.

MADE IN THE IMAGE OF GOD

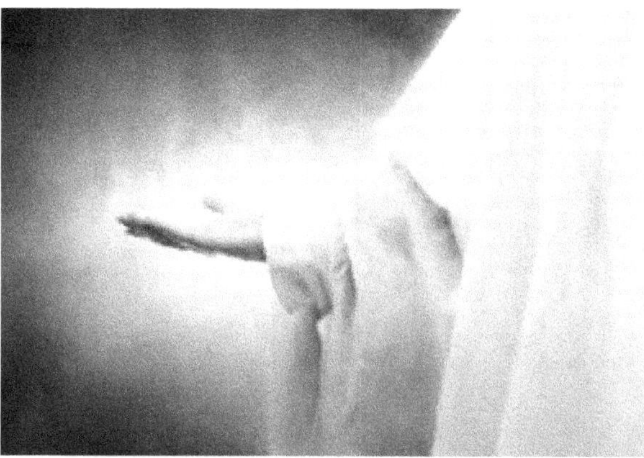

Fig.12.23: Christ is the visible image of the invisible God. He existed before anything was created and is supreme over all creation, for through him God created everything in the heavenly realms and on earth. He made the things we can see and the things we can't see—such as thrones, kingdoms, rulers, and authorities in the unseen world. Everything was created through him and for him. Colossians 1:15-16

The historical biblical account of the creation of Adam and Eve (*Genesis 1:26-27*) states that God made the first man and woman '***in His own image.***' What does this mean? And why is it important?

The occasion was the sixth day of Creation Week, after God had prepared the earth as a habitation, and after He had created the fish, the birds, and the other animals. These were all created by divine fiat, which means that God commanded (or willed) each event to happen and it was done, Figure 12.23. In the case of the creation of man, there is a difference. The inspired record reads: 'And God said, let us make man in our image, after our likeness.... So, God created man in His own image, Figure 12.24, in the image of God created He him; male and female created He them'(*Genesis 1:26-27*).

Fig.12.24: God Created Man in His Own Image

A DIVINE COUNCIL

In this intriguing verse God appears to be talking to someone. "Calvin" said, 'This is the language of one apparently deliberating... he enters into consultation.' Many other commentators refer to it as 'a council'. However, if this is so, with whom is God consulting? And why? Does the Bible give us any hint?

Since the Lord needs no other counsellor, any such consultation must have taken place within the Godhead—between God the Father and God the Son, and God the Holy Spirit. *Hebrews 1:2* tells us that when God made the worlds (KJV) or universe (NIV), He did so by His Son.

God could easily have commanded the creation of man by His own Word, as He had done in the case of the animals (*Genesis 1:20,24*) and the plants (*Genesis 1:11*), but He chose not to. Man is not a close cousin of the animals, nor a distant relative of primitive plant life, nor a product of slime. Rather, he is someone great, wonderful and different, the most excellent of all God's works, and a special expression of the divine nature, created by God's own personal activity. God introduces him with solemnity, dignity, and the honor of an intimate deliberation on the part of the Godhead.

Fig.12.25: So God created human beings in his own image. In the image of God he created them; male and female he created them. (Genesis 1:26).

Although man was formed from the dust of the ground, God personally 'breathed into his nostrils the breath of life; and man became a living soul' (*Genesis 2:7*). Man's life is thus not the result of spontaneous reorganization of molecules within his body, nor is it derived by evolution from any animal or 'lower hominid' (as theistic evolutionists teach), but is a direct gift from God. This is further emphasized in the Bible by Luke's genealogy of Adam, where he designates Adam as being not the son of an anthropoid ape, but 'the son of God' (*Luke 3:38*).

What was there for God to discuss with the Son? *Revelation 13:8* tells us that the Lamb of God (i.e., Jesus) 'was slain from the foundation of the world'. Nonetheless, the sixth day of Creation Week the foundation of the world had been well and truly laid, but with the creation of man, God was about to initiate that chain of events which would lead inexorably to the future crucifixion and death of the Lord Jesus Christ on the Cross as an atonement for the sin of mankind. In His earthly life, Jesus affirmed His willingness to do the will of the Father in this regard, in the Garden of Gethsemane (*Matthew 26:39, 42*). Did He perhaps also do so in the heavenly council, immediately before God created Adam?

IN THE IMAGE OF GOD

When God created man in His own image, He purposed that mankind (both man and woman), Figure 12.25 would resemble God in certain ways, and share certain of the divine prerogatives. Concerning this we note:

I. It was not a Physical Likeness

Although God is spirit (*John 4:24*) and does not have a body like a man, when He appeared visibly to men according to the Old Testament record, He did so in the form of a human body (e.g. *Genesis 18:1-2; 32:24, 28,30*). Dr :Henry Morris" writes: 'There is something about the human body therefore, which is uniquely appropriate to God's manifestation of Himself, and (since God knows all His works from the beginning of the world—*Acts 15:18*), He must have designed man's body with this in mind. Accordingly, He designed it, not like the animals, but with an erect posture, with an upward gazing countenance, capable of facial expressions corresponding with emotional feelings, and with a brain and tongue capable of articulate, symbolic speech.'

Furthermore, the human body was the form in which God the Son would be incarnated or 'made in the likeness of men' (*Philippians 2:7*). Thus, God made man in that bodily form which He Himself would one day assume—the form in which He wished to reveal Himself.

II. It was a Mental Likeness

God endowed man with intellectual ability which was and is far superior to that of any animal. Thus, man was given a mind capable of hearing and understanding God's communication with him, emotions capable of responding to God in love and devotion, and a will which enabled him to choose whether or not to obey God. Man was thus equipped, not only to 'love God and obey Him forever', but also to do God's work on earth—to be His regent and govern the creation in co-operation with his Creator.

This is seen in God's command to Adam and Eve that they exercise dominion over the earth and its animals (*Genesis 1:26,28*), in Adam's task of cultivating the garden (*Genesis 2:15*), and in the statement that Adam gave names to certain of the animals on the earth (*Genesis 2:19-20*).

Man's intellectual gifts are further seen in his ability to design things and then make them, to appreciate beauty, to compose glorious music, to paint pictures, to write, to count to large numbers and do mathematics, to control and use energy for his own benefit (e.g., fire, electricity, nuclear power), to organize, to reason, to make decisions, to be self-conscious, to laugh at himself, and to think abstractly. All this behavior is non-instinctive, as distinct from animal behavior, and as such it is of unlimited variety.

III. It was a Moral Likeness

Man only, of all God's creatures, has a spirit or God-consciousness, that is, a capacity for knowing God and holding spiritual communion with Him through prayer, praise, and worship. Since the Fall (Genesis chapter 3), man has had inborn moral awareness of good and evil, or conscience, which he perceives in his spirit.

Man was made not only negatively innocent (that is, without sin), but positively holy, otherwise Adam could not have had communion with God, who cannot look upon iniquity (*Habakkuk 1:13*). This is further confirmed by Genesis 1:31, when God affirms that everything He had made (including man) was 'very good', which would not have been true if man had been morally imperfect.

IV. It was a Social Likeness.

God's social nature and intrinsic love is seen in the doctrine of the Trinity. God—who *is* love—created man with a social nature and a need for love. The statement in *Genesis 3:8* that 'they heard the voice of the LORD God walking in the garden in the cool of the day' suggests that Adam and Eve enjoyed fellowship and communion with God, perhaps on a daily basis.

God also provided for human fellowship and love in a very special and intimate way. Before He created Eve He said, 'It is not good that the man should be alone; I will make him a help meet for him' (*Genesis 2:18*). He then made Eve out of a bone taken from Adam (*Genesis 2:21-24*), a fact which Jesus used in His debate with the Pharisees to uphold the sanctity of marriage and the intimacy of love within the marriage relationship (*Matthew 19:4-6; Mark 10:6-8*).

When God created the vegetation and the animals, He made them all 'after its/their kind' (the phrase occurs ten times in *Genesis 1:11-25*). When He created Adam, He made him after the God-kind — in the image and likeness of God (cf. *Acts 17:28*). After the Fall, man is still said to be in God's image (*Genesis 9:6; 1 Corinthians 11:7*) and likeness (*James 3:9*). However, this image was defiled by man's rebellion at the Fall, and all aspects of God's image were tarnished. Nevertheless, these aspects were perfect in the Lord Jesus Christ, who was and is 'the image of the invisible God' (*Colossians 1:15*), and 'the express image' of God (*Hebrews 1:3*), both in His life on earth and in Heaven.

The Apostle Paul says that we are transformed or renewed into the image of God by the Gospel, and that this image is then 'in righteousness and true holiness' (*Colossians 3:10; Ephesians 4:24*). This is not something that the natural man can bring about by his own efforts, but is the result of our 'receiving Christ' in faith and repentance (*John 1:12; Galatians 2:20*). It is accomplished by the Holy Spirit (*Titus 3:5; Romans 8:28-29*), who takes up His abode within God' children (*1 Corinthians 3:1; 6:19*). 'God is long-suffering towards us, not willing that any should perish, but that all should come to repentance' (*2 Peter 3:9*).

EVOLUTION OF THOUGHT

Unique Bacterium – No Missing Link

The discovery of a unique bacterium which acts as a natural fertilizer has been claimed by the "National Science Foundation" as a possible descendant of an ancient 'missing link.' The bacterium was found in rice paddies in Thailand by an Indiana University research team headed by professor of biology "Howard Gest." The National Science Foundation says this unique bacterium, Helio-bacillus mobilis, Figure 12.26, may be the descendant of the 'missing link,' which was the immediate forerunner of the cyanobacteria that supposedly released enough oxygen to form earth's life-sustaining atmosphere.

Absurdity, the biggest 'missing link' is the absence of the correct theory.

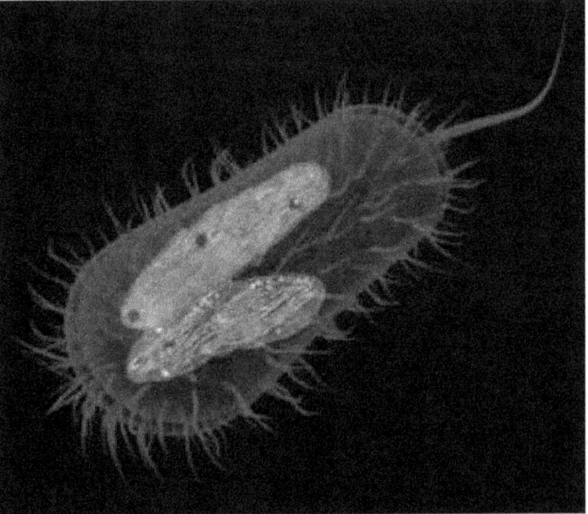

Fig.12.26: Unique Bacterium, Helio-Bacillus Mobilis - Curtrsy - iGem,org

SHAPE OF AN APE – NO MISSING LINK!

The analysis of bones found in Tanzania in 1986 has led scientists to admit that a fossil which they thought was human-like is really more like an ape.

The fossil was said to comprise more than 300 bones from a 40-year-old female who died 1.8 million years ago. The discovery was made by "Donald Johanson" (discoverer of the famous '**Lucy**' fossil), Figure 12.27, and others, and consisted of skull fragments and limb bones from a single individual. It was assigned to the species Homo habilis, which would make it human.

However, analysis of the bones is now said to indicate that this Homo habilis stood only about one meter high, her upper arm bone was almost as large as her thigh bone, her hands almost came down to her knees, and her long powerful arms had curved bones in the hands features which were very ape-like.

The features described are so apelike we could safely conclude that the creature was a type of ape - not human at all.

BAT-MAN THEORY

A biologist who last year proposed a theory that fruit bats, Figure 12.28 and humans shared the same ancestry has defended his theory against alternative proposals.

Professor "Jack Pettigrew," from the University of Queensland in Australia, believes that several lines of evidence, including DNA studies, now support his 'flying primate' theory. He believes that fruit bats, primates and humans once shared the same branch of the evolutionary tree.

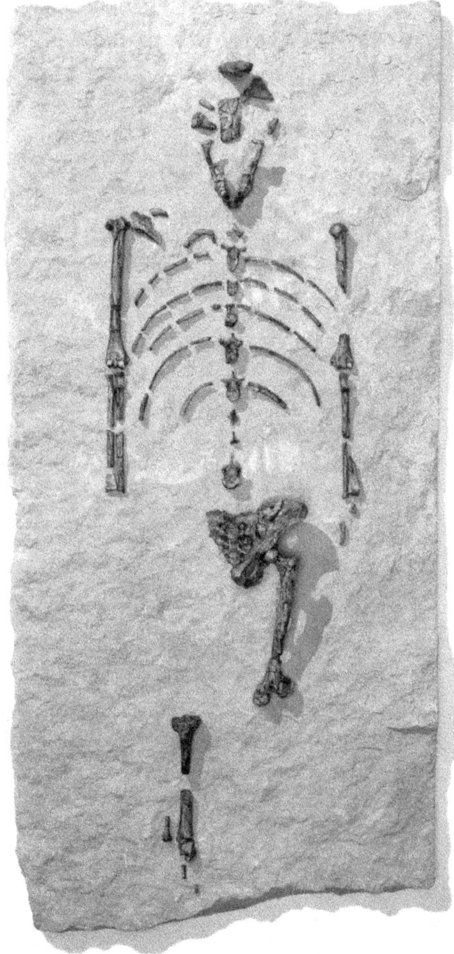

Fig.12.27: *The Famous 'Lucy' Fossil*

Fig.12.28: *Fruit Bats – Curtsy Subphoto.com/Shutterstock*

Professor "Pettigrew" has defended his idea against a study by researchers at the University of New South Wales which claimed that cows are more closely related to humans than are fruit bats.

Christian scientists will agree with Professor Pettigrew that the University of New South Wales study is totally wrong, and with the University of New South Wales researchers that Professor "Pettigrew" is wrong.

TO MARS AND BACK SAFELY

A team of United States space managers plans to send an unmanned spacecraft to Mars for a year to collect rocks and other samples and bring them back to the Earth, Figure 12.29. They hope to bore below the Martian surface to discover some of the red planet's history and possibly whether Mars has or had life.

Fig.12.29: Mission to March – Curtsy NASA

"Michael Carr," chairman of the project's science committee, said samples from Mars could allow scientists to piece together the planet's history. Many scientists are eager to discover if Mars has supported life. Mr. Carr said most scientists believe there is no life on the planet, but it may have existed in the past. 'Life may have got started on Mars, but could not sustain itself', he said. It is felt that if amino acids are found in the permafrost below the surface, there would be reason to believe that life had tried to start there.

If amino acids are found on Mars it wouldn't prove anything about life. Amino acids are not alive. They are only building blocks of proteins.

UNIVERSE AGE CHANGES

A new study by Dutch astronomer suggests that scientists have overestimated the age of the universe. Professor "Harvey Butcher," from the University of Groningen in The Netherlands, estimates the universe as being 'only' II billion years old. His study conflicts with suggestions that some stars are 16 billion to 20 billion years old, but agrees with some earlier estimates.

Cambridge University astronomer "Gerry Gilmore" said that if Professor Butcher's conclusions were true, 'both the accepted physics of stellar evolution and the age of the universe require substantial revision.'

It should be noted that over the past seven years the universe has variously been dated at 15 billion years, 12 billion, 19 billion, 8 billion, 20 billion and now 11 billion (see Creation Ex Nihilo, January 1983, p. 5, and August 1984, p. 44). As each scientist believes the others' estimates are wrong, we can agree with them all that all the others are wrong.

Dinosaur Deaths Due to Acid Rain?

Just as a consensus was growing around the idea that dinosaurs became extinct after a comet hit earth, with effects like a nuclear war, the notion has started to lose ground. A different explanation is now gaining support from evolutionists: that volcanoes spewed out Sulphur, which eroded the ozone layer, turned to acid rain, cooled the climate, then wiped out the dinosaurs, Figure 12.30.

Fig.12.30*: Dinosaur Deaths Curtsy – Curtsy Museum of Natural History*

A group of scientists believes this idea can explain all the things that the comet impact theory can, and more. They say the gradual and selective nature of extinctions seems consistent with an eruption, but not a collision. And they believe it is wrong to concentrate solely on the dinosaurs. They say the really spectacular change was the disappearance of so many marine creatures, and they claim that plankton were killed by acidification.

Did Sunburn Kill the Dinosaurs?

The dinosaurs were wiped out not only by acid rain, but by lethal sunburn, according to scientists who have concluded that the impact theory of dinosaur extinction is wrong. British scientists "Charles Officer," "Anthony Hallam" and "Charles Drake," and American volcanologist "Joseph Devine," paint a picture of rapid decline from Eden-like conditions to an environmental disaster. They imply that the sky was veiled by fine volcanic dust that blocked out light and heat, clouds were raining dilute Sulphuric and hydrochloric acid, and landscapes were bathed in deadly ultra-violet radiation after the destruction of the ozone layer.

The group says the final factor that killed the dinosaurs may not have been starvation but the destruction of the earth's protective ozone. This would have allowed lethal levels of ultra-violet radiation to reach earth's surface, possibly causing the dinosaurs to die of sunburn.

But if the dinosaurs couldn't survive, how did anything else?

Are Parasites Part of Creation?

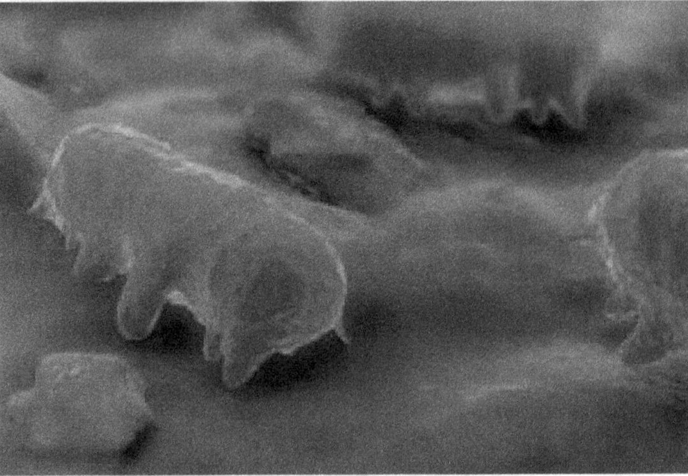

Fig.12.31: *Schistosoma–haematobium. Curtsy - Image stockxpert*

If God created everything 'very good', why are there harmful parasites?

When one of my colleagues worked as a doctor in a bush hospital in Tanzania, East Africa, he treated many patients for a disease called Bilharzia, or Schistosomiasis. This is a parasitic disease caused by several species of fluke, or flatworm, of the genus *Schistosoma*. The particular form of the disease which I treated was the one caused by the species *Schistosoma haematobium*.

This worm, Figure 12,31, like so many other parasites, has a bizarre and complex life-cycle which defies an evolutionary explanation.

It assumes seven different forms as it progresses through its life-cycle, and part of that cycle takes place within an intermediate host, a freshwater snail. The complexity of such organisms' points to a Creator; but the existence of disease-causing parasites and microbes does also raise questions concerning God's "very good" creation and the global Flood of Noah's day.

EVIDENCE FOR CREATION

As far as their physical structure is concerned, all living creatures—including the most basic and 'simple' forms of life (one-celled organisms)—are stupendously complex, sophisticated, computerized machines. Molecular biologist "Michael Denton" has estimated that if we knew how to build a machine as complex as the cell, it would take at least *one million years* to build one cell—working day and night, and churning out the parts on a mass production basis.

One kind of complexity is the way in which some animals exist in several completely different forms during their life-cycles. The butterfly, for example, exists as an egg, a caterpillar, a chrysalis and a butterfly, Figure 12.32. Within the chrysalis, the internal organs of the caterpillar dissolve into a soup, and then

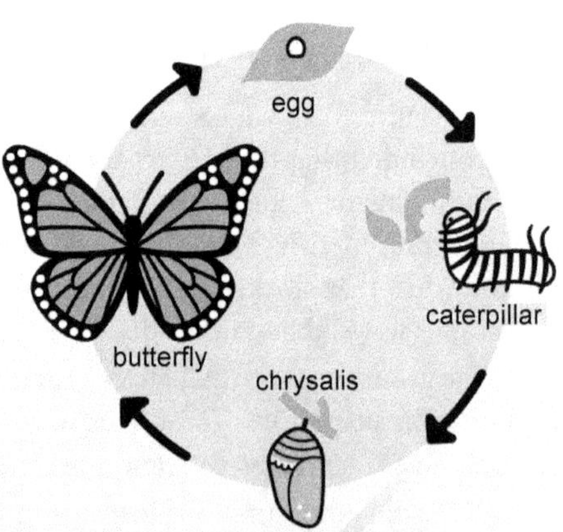

Fig.12.32: *Butterfly Natural Life Cycle – Curtsy Arizona State University*

this soup organizes itself into a beautiful, incredibly complex, adult butterfly! The whole process, from egg to butterfly, is coded in the genes, and is a marvel which defies any kind of evolutionary explanation.

The humble *Schistosoma* may not excite our admiration in the same way; but during its life-cycle it exists in *seven different forms* (including the egg)! Again, it is a marvel which defies an evolutionary explanation.

A God of love?

Schistosoma parasites, Figure 12.33, like every other living creature, are evidence for creation; but they also raise certain questions. Sir David Attenborough has often been asked why he does not give the credit to God when he describes the wonders of creation. This was his reply on one occasion:

"When Creation scientists talk about God creating every individual species as a separate act, they always instance hummingbirds, or orchids, sunflowers and beautiful things.

"But I tend to think instead of a parasitic worm that is boring through the eye of a boy sitting on the bank of a river in West Africa, [a worm] that's going to make him blind.

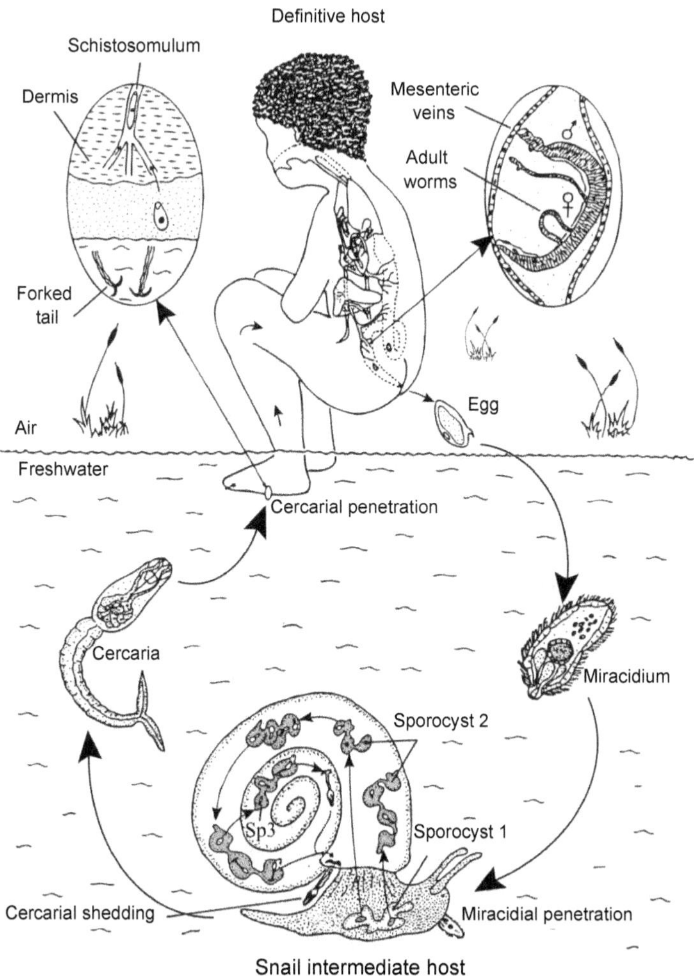

Fig.12.23: *Schistosoma Parasites Life Cycle – Curtsy Global Water Pathogen*

"And [I ask them], 'Are you telling me that the God you believe in, who you also say is an all-merciful God, who cares for each one of us individually, are you saying that God created this worm that can live in no other way than in an innocent child's eyeball? Because that doesn't seem to me to coincide with a God who's full of mercy.'"

Observing carefully, you find that Attenborough did not give a *scientific* reason for rejecting belief in a Creator. He gave a *theological* reason for rejecting belief in the good and loving God of the Bible. The first question, therefore, is "Why did God, if He is good and loving, create disease-causing parasites—not to mention all the other disease, violence, suffering and death which we see around us in nature?"

The straightforward, biblical answer is that these evils did not exist in the original creation. They came into the world only after the Fall, when Adam and Eve disobeyed God (*Genesis 3:14–24; Romans 8:18–25*). They were part of the "Curse" (*Revelation 22:3*). Six times we are told that God saw that His creation was "good", and the seventh time He saw that it was "very good" (*Genesis 1:4–31*).

Seven is the "perfect number", and the sevenfold repetition of the phrase "it was good / very good" signifies the absolute goodness and perfection of the original creation. The presence of any kind of evil, moral or physical, is absolutely denied. Disease is clearly a physical evil. It was not present in the original creation, and it will not be present in the new creation either, when God creates a new heaven and a new earth (*Isaiah 11:6–9; Isaiah 65:17–25; Acts 3:21; Romans 8:18–25; 2 Peter 3:1–13; Revelation 21:1–5; Revelation 22:3*).

Theistic evolutionists and other Christians who accept the secular belief in "millions of years" cannot give this answer. They believe that God's method of creation was a process of millions of years of death, disease, violence, suffering and waste.

The fossils (which they believe represent millions of years

Fig.12.34: *Fossilized Cancer Tumor – Curtsy Luisville Fossile and Beyond*

of life and death on earth before the Fall) show clear evidence of all these evils, Figure 12.34. For example, some of the fossils show evidence of cancer. Can cancer be described as "very good"? It is hardly surprising that many non-Christians say that such a god (if he were to exist) would be cruel or incompetent, and would be unworthy of their worship.

THE ORIGIN OF HARMFUL PARASITES

The second question is this: If disease-causing parasites did not exist in the original creation, how did they come to be here now? We do not know the answer for certain, but we can reasonably speculate. I think it is unlikely that they were specially created after the Fall, *as Genesis 2:1–3* says that God *finished* his work of creation "by the seventh day" (cf. *Exodus 20:11*).

If that is correct, these parasites must have been benign and beneficial in their original form. Perhaps some were independent and free-living, and others had beneficial symbiotic relationships with animals or humans. But when Adam and Eve sinned, things began to go wrong. These once-harmless creatures degenerated, and became parasitic and harmful. The Fall and subsequent Curse were unique events which took place soon after the beginning of human history. So, there is no compelling reason why God should not have caused these organisms to become pathogenic (disease-causing) miraculously. However, there are other methods He could have used.

Perhaps some became parasitic as a result of mutations. They degenerated, lost their ability to live independently or symbiotically, and became harmfully dependent on their hosts. Note that mutations overwhelmingly cause *loss* of genetic information. Such degenerative changes are evident in disease-causing microbes like a *Mycoplasma* that causes a type of pneumonia and the germs that cause leprosy and cholera. Additionally, the sin of Adam and Eve caused the Earth to be cursed too, consequently, the microbes and parasites have carried the curse of deformity and harm to breed the same kind and fill the Earth.

Other kinds of genetic change may have been involved too. For example, microbes can swap genes. The bacterium that causes bubonic plague probably resulted from more than one kind of change.

In many cases, however, the life-cycle of the parasite is so complex that new genetic information may have been needed. Mutations do not provide new genetic information; so, the information may have been there from the beginning. However, it was in a 'switched off' mode before the Fall, and was not 'switched on' until after the Fall. God could have included this genetic information because of his *foreknowledge* that Adam and Eve would disobey him.

Another possibility is that some parasites had symbiotic relationships with animals, and they became parasitic and harmful only when they invaded humans. Also, degeneration in the human immune system could have contributed to our susceptibility to parasitism by formerly benign organisms.

PARASITES ON THE ARK

The third question is this: If the eight people on board the Ark were the only human survivors of the global Flood, how could they have carried all the disease-causing parasites and microbes which can survive only in humans? Biblical creation scientists including the author are ridiculed about this, not only by non-Christians, but also by theistic evolutionists and other old-earth creationists, who believe that Noah's Flood was local. Possible answers are not difficult to find, however, if one is prepared to look for them.

The most obvious answer is that most, if not all, present-day pathogenic organisms that can dwell only in humans did not exist in their present form at the time of the Flood. There has been plenty of time for them to develop *since* the Flood. Many pathogenic organisms which existed in the pre-Flood world may have perished in the Flood, but others have taken their place, Figure 12.35.

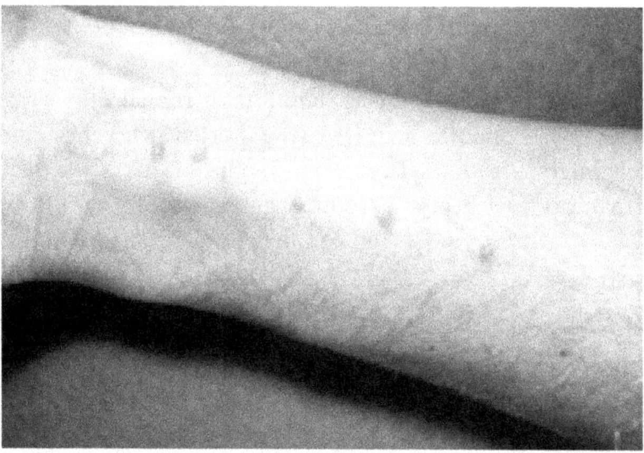

Fig.12.35: Skin Vesicles on the Forearm, Penetration by Schistosoma – Curtsy Wikipedia

One suggestion is that some pathogens which can survive only in humans today were much less specialized at the time of the Flood. At that time, they were able to flourish in a wide variety of animals; but since then, they have become entirely dependent on human hosts. Common viral diseases of humans today may well have derived from animal diseases. A *New Scientist* report states:

"Just as historians such as "William McNeill," of the University of Chicago, and other researchers trace smallpox back to cowpox, so measles probably evolved from rinderpest or canine distemper, and influenza from hog diseases."

Some of Dr Wieland's suggestions involve organisms which were *already* pathogenic (usually in animals) when the Flood began. Asymptomatic carriers could have carried some diseases on the Ark. All the organisms which became pathogenic after the Fall, however, had to be derived from organisms which were entirely benign.

The *Schistosoma* parasite, like so many other pathogenic organisms, is a thoroughly unpleasant creature. It speaks to us of a creation which has been spoiled because of man's rebellion against God. In spite of this, its complexity and remarkable life-cycle are a witness to the power and divinity of God (*Romans 1:20*).

The existence of such pathogenic organisms raises certain questions. To answer these questions, we could invoke miraculous intervention by God—and in fact it would have been quite reasonable for Him to intervene miraculously in the unique circumstances of the Fall and the Flood—but we do not have to do this. We just have to be *willing* to look for reasonable answers, and take the trouble to do so!

The parasite begins life as an egg, which is shed into the ureter or the bladder of the human host. It is passed out of the body with the urine, and on contact with fresh water, it hatches to release a free-swimming "miracidium". This is a ciliated organism, swimming by means of its many hair-like cilia. It finds a fresh-water snail and penetrates the snail's foot.

Here it transforms into a "primary sporocyst". Germ cells within the primary sporocyst then begin to divide and produce "secondary sporocysts". These migrate to the snail's hepatopancreas. Germ cells within each secondary sporocyst then begin to divide again, this time producing thousands of new parasites, called "cercariae", which emerge daily from the snail host.

These have tails, and are highly mobile. They are the larvae capable of infecting humans. Penetration of the human skin occurs after the cercaria has attached to and explored the skin. The parasite secretes enzymes that break down the skin's protein to enable penetration of the cercarial head through the skin. As it penetrates the skin, it loses its tail and transforms into the migrating "schistosomulum" stage.

The schistosomulum enters the blood stream and travels to the lungs, where it undergoes further developmental changes before migrating to the liver.

In the liver, it begins to feed on red blood cells, and it is there that the nearly mature worms' pair, the longer female residing in the 'gynaecophoric channel' of the shorter male. The adult worms are about 10 mm long. The worms ultimately migrate from the liver to the venous plexus of the bladder, ureters and kidneys. When the parasites reach maturity, they begin to produce eggs, and these pass through the ureteral or bladder wall into the urine.

INEXPLICABLE DESIGN – INSECT METAMORPHOSIS

Darwinists who want to make a strong case for evolution will routinely avoid certain biological topics—chiefly because those topics resist all gradualist explanations. One is the unique and complex method by which insects grow.

Metamorphosis (from Greek words meaning 'change of form'), describes how most insects change from juveniles to adults, often developing adult body structures and ways of life completely different from those of their youth, Figure 12.36. While the juvenile of a particular species may look like a glorified worm, the adult might have five-centimeter (2-inch) wings and no functioning jaws. Let's examine the life of a moth.

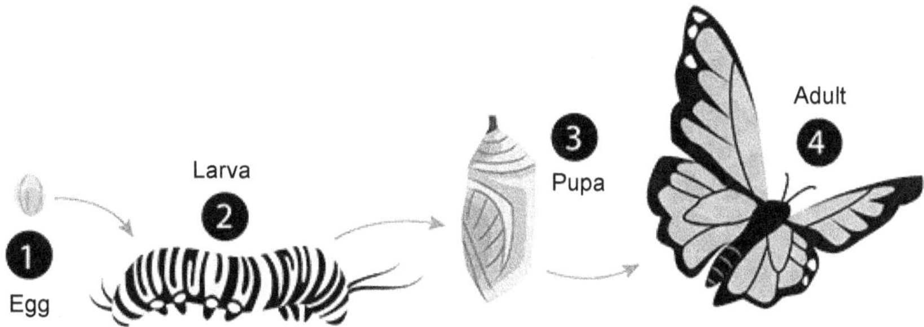

Fig.12.36: *Insect Metamorphis – Curtsy Photo by Stephen Atkins*

METAMORPHOSIS OF THE MOTH

In order for a tiny, newly hatched caterpillar to grow into a brilliant moth or butterfly, it must first become fat—*very* fat. In fact, the caterpillar seems to have only two preoccupations in life: eating and molting.

Although mammals, fish, birds, and reptiles have skeletons that support their body, arthropods—including insects—do not. Instead, God designed them with a hard skin or shell called an exoskeleton. A caterpillar's exoskeleton may seem soft, but it gives the caterpillar its entire shape (which sometimes includes fleshy spikes or horns). Yet it remains flexible enough to allow for gymnastic stretches toward high leaves. Further, the exoskeleton doesn't grow; a larger exoskeleton forms folded beneath the smaller one.

When the time is right, the tight, old skin splits open and the caterpillar wriggles carefully out, ready to try out its new exterior. After each molt, the caterpillar is bigger and may be a slightly different shape or color, Figure 12.37.

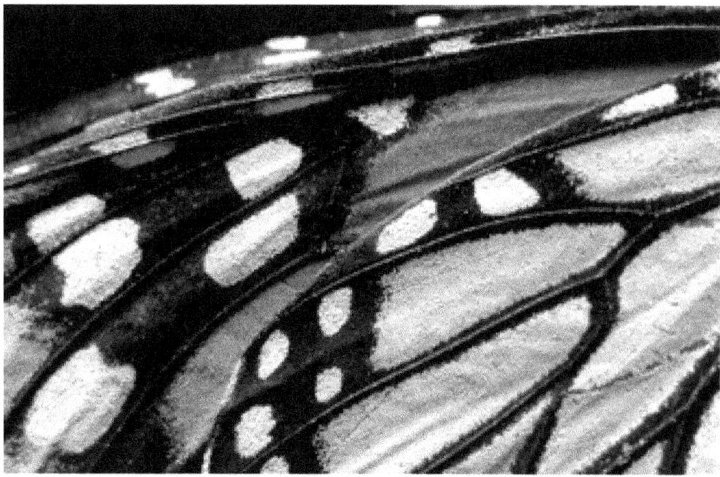

Fig.12.37: *Moth – Curtsy Photo stock.xchng*

Meanwhile, deep within the caterpillar's body are clusters of cells—*imaginal discs.* These are positioned to grow into wings, jointed legs, and compound eyes. After the caterpillar has molted into its largest body size, it prepares to become a pupa by spinning a cocoon, burrowing underground, or in the case of most butterflies, forming a chrysalis.

As it lies there very still, hormones from within the caterpillar's brain signal the body to develop into the adult stage. These cause the imaginal discs to burst into action, forming antennae, scaled wings, reproductive organs, and every other body part needed by the adult.

Even the muscular system must be reorganized to accommodate the wings. Some muscles are destroyed, some are 'reconstructed,' and others are formed brand new. When the adult moth or butterfly emerges, it looks nothing like the squirmy worm it once was.

COMPLETE VS INCOMPLETE

The type of metamorphosis described above, which insects like moths, butterflies, bees, flies and ants undergo, is known as 'complete' metamorphosis and involves four stages, Figure 12.38:

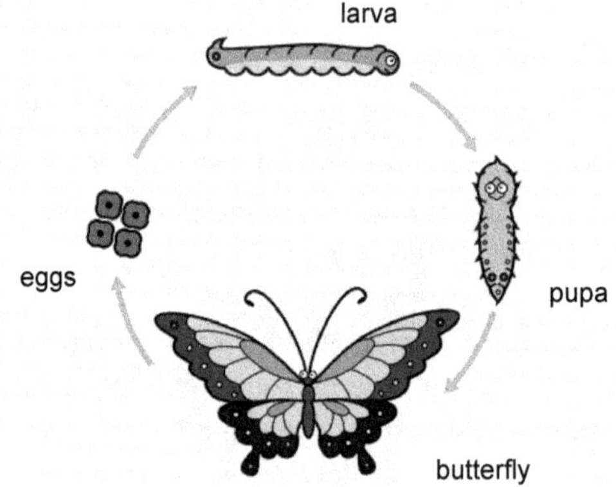

Fig.12.38: Four Stages of Metamorphis

a. egg,   b. larva,   c. pupa and   d. adult.

In the case of a moth, the caterpillar is the larva, the cocoon is the pupa, and the colorful winged creature you might catch in a net is the adult.

However, a second type of metamorphosis is used by insects such as termites, grasshoppers, crickets, cicadas and aphids. This involves changing from:

a. egg to   b. nymph to   c. adult —

Only three stages. This 'incomplete' metamorphosis does not involve a pupa. The nymph simply looks like a miniature adult, and as it molts it grows progressively larger, until it reaches its adult form with fully developed wings and reproductive organs.

Each stage in the insect's life is crucial. Darwinists face colossal problems when they attempt to explain the origin of metamorphosis in terms of random mutation and natural selection, because any gap or error in the cycle normally kills the insect or prevents reproduction. If a caterpillar can't squeeze out of its old exoskeleton, if it isn't able to form a cocoon or chrysalis, or if it fails to rearrange muscles and grow body parts as a pupa, it dies. It never becomes an adult, and therefore it does not reproduce itself.

REMARKABLE CHANGE

Evolutionary theory also fails to reasonably account for the radical diversity of insect growth. Consider the lengths of time different species use to develop. Cicada nymphs burrow underground and spend as many as 17 years feeding off tree roots before they emerge to molt into their adult stage. Certain fly larvae (maggots) can become pupae one week after hatching, and some aphids hatch, molt, reproduce and die in a little over ten days.

The aquatic larvae of such insects as mayflies and mosquitoes are especially troublesome for evolutionists. Mayflies are falsely called 'the most primitive winged insects,' but they are not primitive at all. Rather, they live two or three years underwater breathing with gills like fish, then emerge as adults into the air. Then they live only one day to swarm, mate, and lay eggs.

Mosquitoes also spend their larval and pupal stages underwater, before emerging as adults to fly around, suck blood (at least many females do), and reproduce. The female of one species of aquatic moth spends her whole life underwater, while the adult male flies around freely, returning to the water only to mate. Aquatic insects are simply too diverse for evolutionary theory to predict—so that Darwinists are forced to admit that aquatic living 'has been developed by totally unrelated species.'

MIRACULOUS – SOCIAL METAMORPHOSIS

If all this weren't enough, we can't forget the metamorphosis of social insects such as ants, Figure 12.39. Ant larvae have no legs and are incapable of finding food, moving, or cleaning themselves. They are totally dependent on the 24-hour care of adult worker ants, without which they would quickly die. Young ants even need the adult workers to cut them out of their cocoons at the end of the pupa stage.

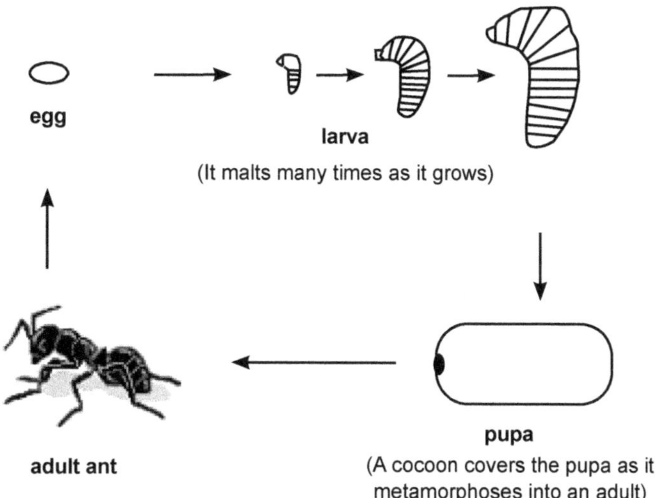

Fig.12.39: Ant Social Metamorphis

Unless ants had a complex, interdependent society from the very beginning, how did 'ancient' ant larvae survive?

The Bible has the only plausible answer:

God created them to be social from the start, just like He created caterpillars to gorge, mayfly nymphs to swim, and moths to fly and reproduce. The characteristics of the stages of metamorphosis did not evolve randomly over time. From egg to adult, insects live out the unique plans designed for them by God.

THE MONARCH BUTTERFLY

Fig.12.40: Monarch Butterfly - Curtsy Photos by Bob Moul, <www.Pbase.com/rcm1840>.

1. *Monarch caterpillar (larva),* feeding on milkweed plant leaves. The larva hatches from the egg, and remains in this stage for about two weeks, although this depends on temperature, Figure 12.40 -1.

2. *Monarch Pupa (Chrysalis):* The caterpillars then attach themselves head down to a convenient twig. They shed their outer skin and make a cocoon, to transform into a pupa (chrysalis). Here the most amazing transformation occurs, where most of the caterpillar tissues dissolve and are re-formed into a butterfly—under its genetic program. At first it looks like a waxy jade vase, Figure 12.40 -2.

3. *Monarch Pupa* moments before butterfly emerges: As the metamorphosis progresses, the chrysalis becomes increasingly transparent, so the colors of the butterfly can be seen. Just after this photo was taken, the butterfly emerged, Figure 12.40 -3.

4. *Freshly Emerged Monarch* with chrysalis shell. After 15–9 days, the butterfly finally emerges. It inflates its wings by pumping blood from a pool of blood it has stored in its abdomen into the wing veins (below left). The butterfly waits until its wings stiffen and dry before it flies away to start the cycle of life all over again, Figure 12. -4.

5. *Adult Female Monarch*: Thicker veins and the lack of the nodes on the upper hind wing veins distinguish the female from male Monarch Figure 12.40 -5.

EVOLUTION DOES NOT EXPLAIN COMPLETE METAMORPHOSIS

In a 1999 issue of *Nature*, two scientists, "James Truman" and "Lynn Riddiford" presented their hypothesis of how complete insect metamorphosis evolved. In the article the authors tried to explain the evolution of four-stage metamorphosis from three-stage metamorphosis by proposing that the latter actually contains four stages. They called this arbitrarily-defined fourth stage the 'pro-nymph,' and described it to be a period which always precedes the first molt, and which varies in duration among different species, sometimes ending as soon as the insect hatches out of its egg. This 'pro-nymph,' they argued, evolved into our modern larva.

In plainer terms, some ancient insect hatched out of its egg too soon and began groping around for food. It continued evolving until it could spend many weeks in this premature, caterpillar-like form, before finally metamorphosing into the long-belated nymph stage—which according to the authors had shortened and evolved into our modern pupa.

One fatal problem with this idea is that the underdeveloped 'pro-nymph,' as described in *Nature*, doesn't eat! It would need a fully formed digestive tract and the capability to bite, chew and swallow if it were to survive and grow into an adult. Truman and Riddiford argued that the 'pro-nymph' overcame these difficulties and gradually evolved until it could eat, move, and presumably defend itself. Defense is very important in a world of 'survival of the fittest,' and it boggles the mind how the first premature 'pro-nymph' was safer and better favored by evolution than a normal, fully developed nymph!

A better explanation is that four-stage metamorphosis was created independently and completely functional.

GOD DESIGNED UNMATCHED WEB SPINNERS

While Kevlar® is the 'gold medalist' of man-made fibers because of its bullet-stopping abilities, it's overshadowed in many ways by the humble spiderweb. "Spider silk is stronger and more elastic than Kevlar®, and Kevlar® is the strongest man-made fiber," according to Danish spider expert "Fritz Vollrath." Dragline silk, the main support for its web, is a hundred times stronger than steel—a cable of this silk a little thicker than a garden hose could support the weight of two full Boeing 737 aircraft. It can also stretch to %40 of its length, while the flagelliform silk in web spirals can stretch to over 200%.

The manufacture of Kevlar® requires harsh conditions, including the boiling of sulfuric acid and the leaving behind of dangerous chemicals that are expensive to dispose of.[1] But spiders need only ordinary temperatures, and they use a much milder acid bath, which is produced by special ducts.

Spiders can make silk at different speeds— up to 10 times faster when dropping to escape a predator—unlike most industrial chemical processes that would make 'gunk' if the speed was varied by that much. Spider silk is even environmentally friendly—spiders eat their own webs when they no longer need them, Figure 12.41.

Spider silk owes its amazing strength and elasticity to its 'complexity that makes synthetic fibers seem crude.' Man-made fibers are usually just simple strands of material, but a silk fiber has a core surrounded by concentric layers of *nanofibrils* (tiny threads). Some layers contain nanofibrils aligned parallel to the axis, while other layers contain nanofibrils coiling like a spiral staircase. The coiled ones allow the silk to be stretched, because they simply straighten up rather than break.

Fig.12.41: *Orb Web Spinner*

The nanofibrils themselves are very complicated, containing tiny protein crystals in an amorphous (shapeless) matrix of tangled protein chains. These *nanocrystals* contain electrical charges that stop the chains from slipping, so providing strength, while the amorphous material is rubbery and allows the fiber to stretch.

Some researchers have tried to make silk by forcing a solution of silk proteins, called *spidroin*, through tiny holes, but the fibers are less than half as strong as those produced by the spider. It seems that the spider produces the high complexity required by making the spidroin go through a *liquid crystal* phase, where rod-shaped molecules align parallel (Kevlar° manufacture also uses a liquid crystal phase).

Christopher Viney of Heriot–Watt University in Edinburgh believes that this enables them to flow more easily, thus saving energy. The liquid state also aligns the protein molecules so they can form the nanocrystals and coiled nanofibrils. This seems to occur in the spider's long *s-duct*, where water is both squeezed and pumped out. This brings *hydrophobic* (water-repelling) parts of the proteins to the outside and forms the nanocrystals and enables the fibres to form.

Spiders normally now use their webs for trapping insects and other prey, Figure 12.42. But some baby spiders catch pollen for food, providing a possible clue to a pre-Fall function for the spiderweb.

The Liquid crystalline spinning of spider silk, covers a number of important issues in detail, e.g., the high strength, stress *v* strain analysis, the composition of spidroins including some non-essential amino acids, liquid crystal spinning, the particular type of liquid crystal called the *nematic* phase where the rod-like molecules are aligned parallel to each other (the phase used in image display devices), the conventional external drawdown used in industrial spinning as well as the advanced internal spinning technology so far not duplicated in man-made processes, how we can learn much from the spider's design, the hyperbolic geometry of the s-duct so that the material elongates at a constant rate, preventing disclinations (a weakening defect analogous to dislocations in solid crystals), the structural complexity of silk — even greater than previously thought.

Alas, there is the usual homage to evolution as the designer without the slightest evidence. But Kevlar° expert Dr "Patrick Young" concurs with this *Creation* scientists' views that spider silk is evidence of a Designer.

LIQUID GOLD

The home of one of the biggest gold mines in the world, a Pacific volcanic island, is challenging the way geologists think about history's most coveted metal. Before now, geologists outside creation scientists' circles believed that gold deposits only form over long periods of time, perhaps

Fig.12.42: Spiders Normally Use Their Webs for Trapping Insects and other Prey

Fig.12.43: The Lihir gold deposit is located on an island to the north of Papua New Guinea (see arrow) to the north of Australia. Curtsy - iStockphoto

hundreds of thousands to millions of years. The "Lihir" gold deposit in Papua New Guinea is quickly changing that, Figure 12.43.

The island of "Lihir" is only a few miles long but sits on one of the most massive gold deposits known, the Ladolam deposit, Figure 12.43. This formed below-ground as magma (molten rock) heated mineral-rich water and forced the water under great pressure toward the surface. Inside the earth, these chemically aggressive solutions are able to dissolve all sorts of minerals. Approaching the surface, the water turned to steam, leaving behind its minerals, including the gold. Some geologists think many of the world's gold ore deposits formed in this way, Figure 12.44.

Fig.12.44: *Lihir's two open pits from the air, Minifie (foreground) and Lienetz (background). Curtesy - Lihir Gold Limited*

However, did they form quickly or slowly?

Quickly, say the authors of a paper published in the journal *Science* last year. Two New Zealand geologists, "Stuart Simmons" and "Kevin Brown," obtained a water sample from below Lihir mine by lowering a customized probe down a deep shaft.

The concentration of gold dissolved in the water was 15 parts per billion—seemingly small but 1,000 times greater than any concentration ever measured in hydrothermal surface water. Calculating the rate at which the steam was escaping, Simmons and Brown estimated that 24 kilograms (53 pounds) of gold are being *added* to the Ladolam deposit every year. They concluded the entire deposit could have been laid down in only 55,000 years. That's extremely quickly by secular, uniformitarian standards. But that's also assuming the rate has never changed.

In January 2007, *New Scientist* quoted two other researchers who think the Ladolam deposit was formed even more speedily, either because of higher concentrations of gold or as the result of a cataclysmic event.

Christoph Heinrich, a Swiss researcher who has studied gold deposits around the world, says he has found fluids locked in quartz crystals with concentrations 1,000 times higher than even what Simmons and Brown found. 'If you spin the same argument that they are using with a thousand times higher concentrations, then the time it takes might have been a thousand times shorter—50 or 60 years,' he said.

Fig.12.45: *Photo courtesy Lihir Gold Limited Gold pour at Lihir.*

"Greg Hall," a geologist who has worked for the mining company Placer Dome, told *New Scientist*, 'My gut feeling looking at Lihir is that it formed in the same time it took Mount St Helens to blow up—a month, a day, maybe as short as 5 hours, Figure 12.45.'

Mount St Helens erupted in 1980 and immediately became a working example of immense geologic change occurring in a short timeframe. Lihir, as a volcanic island, has been subject to similar processes.

THE FOUNTAINS OF THE GREAT DEEP

Creation scientists have a biblical explanation for these processes. The Bible clearly teaches that a global Flood covered the earth in Noah's day, a few thousand years ago, breaking up 'the fountains of the great deep.' Such a powerful event would have triggered immediate and ongoing volcanic activity throughout the earth. It would have been during this time, and afterwards, that many gold deposits formed, including the Ladolam deposit.

As scientists learn more about the earth, it becomes increasingly clear that long-age theories don't always hold 'flood water.' Instead, we're finding that the Creator's Book was accurate all along.

APPENDIX

i.   Some commentators differentiate between the term's 'image' and 'likenesses' of God. Others treat them as synonyms, meant to emphasize the idea of resemblance. Moses appears to do the latter: thus, he uses 'likeness' as an explanation of 'image' in Genesis 1:26 (there is no 'and' between the two phrases in the Hebrew); he uses 'image' twice with no mention of 'likeness' in Genesis 1:27; he uses 'likeness' (only) in Genesis 5:1, and 'image' (only) in Genesis 9:6.

ii. The making of Eve from Adam is an absolute stumbling block to theistic evolutionary interpretations. Some say 'dust to Adam' symbolizes evolution from lower forms, but what about 'rib to Eve'?

iii. In *Origin of Species*, Darwin actually devoted very little space to the issue of the origin of life. In an 1871 letter to the botanist Joseph Hooker, *Darwin* wrote: "It is mere rubbish, thinking at present of the origin of life; one might as well think of the origin of matter." *The Life and Letters of Charles Darwin*, edited by his son Francis Darwin, John Murray, London, 1887, Vol. 3, p. 18. Evolutionists still largely skirt around the origin-of-life issue (with some even disingenuously claiming that "evolution" has nothing to do with the origin of life despite the accepted term "chemical evolution"). Hardly surprising, given that there is no known non-intelligent cause that has ever been observed to generate even a small portion of the literally encyclopedic information required for life.

iv. See also John 1:3, 'All things were made by Him [i.e. the Word or Logos = Jesus Christ]; and without Him was not anything made that was made', and John 17:24, where Jesus speaks of God the Father's love for the Son, 'before the foundation of the world'.

v. Hume actually failed in his attempt to discredit teleology, and ended up affirming it. His 'multitude of creators', his 'stupid mechanic' and his 'creator of evil' are all intelligent agents working towards future purposes (however obscure, inadequate or distasteful they may appear to the naturalistic observer).

vi. Man, thus was given dominion over all things—animal, mineral and vegetable—not for personal exploitation or tyranny, but as God' representative. It is a delegated position of service and stewardship, and this implies accountability. In everything man does he is responsible to God (Revelation 20:12).

vii. Man is a three-fold being or tri-unity of body, soul, and spirit. Vegetation has body and unconscious life; animals have body and conscious life (Hebrew: *nephesh*, usually translated 'soul'); man has body, conscious life (or soul, comprising in man the intellect or mind, emotions or sensibility, and the power of choice or will), and spirit (God-consciousness). Man is the immortal and unique image of God in all three aspects. Some commentators compare this tri-unity of man to the Trinity within the Godhead of Father, Son, and Holy Spirit.

viii. The eyes are further shielded from excessive exposure to light and UVR by anatomical features: the normal horizontal orientation of the eyes when in the upright posture, the eyebrows (particularly for those with deep-set eyes), the nose and the cheeks. But any natural defense system can be overwhelmed and so it is sensible when necessary (just as it is to wear extra clothing in cold weather) to wear or use extra protection against UVR and blue light; all the more is this so with the depletion of our ozone layer

ix. The foveola subtends an angle of about 20 minutes of arc at the nodal point of the eye while the normal resolving power of the eye or the angle subtended by the minimum perceivable separation of two points is 1 minute of arc

x. Visual signals arising in the receptors are relayed in the retina first via the bipolar cells in the inner nuclear layer and then *via* the ganglion cells whose axons or nerve fibres form the nerve fiber layer of the retina.

xi. The reduction of overall visual field with the loss or occlusion of one eye amounts to 20–25% with the seeing eye looking straight ahead, mainly on account of the nose. The field of each eye is normally restricted by the facial contours mentioned in endnote 40. The field loss caused by the nose is largely recovered if the subject turns the head a little towards the blind side; this is often done unconsciously by a one-eyed person when looking intently.

xii. The plural form for 'God' (*Elohim*) and the plural pronouns 'us' and 'our' used with a singular verb in Genesis 1:26 foreshadow the doctrine of the Trinity. Compare *Genesis 1:1-3* and *Matthew 28:19*.

xiii. Animals communicate in several different ways, including voice, facial expression, posture and odors, but none is capable of speaking grammatical phrases or sentences.

xiv. Spectroscopic terms like 'singlet', 'doublet', 'triplet' etc. refer to the number of possible orientations of the total electronic spin of the molecule in a magnetic field. The ground (lowest energy) state of the O_2 molecule is a triplet state ($3\Sigma_g^-$) with two unpaired electrons. But when excited by a photon, it moves into a higher energy (thus more reactive) singlet state ($1\Delta_g$) with no unpaired electrons.

xv. The cerebral cortex is the grey cellular mantle (1–4 mm thick) forming the entire surface of the cerebral hemisphere of mammals. In man and the primates, part of the occipital cortex (at the posterior pole of each hemisphere) is specialized to receive signals from the two retinas and not until this level is reached by retinal signals is there conscious visual perception. The central 1.5 mm of the retina (the macula) has disproportionate representation in the visual cortex, amounting to about half of its area.

xvi. Identifying the precise location of xanthophyll, *i.e.,* the layers and structures in which it is present within the neurosensory retina has proved difficult for investigators but there is a consensus for what is given here.

REFERENCES

1. The contents of various editions of Darwin's *Origin*, (refs 3,4) along with much of his other writing (e.g. refs 8,9), can be viewed at: *The Complete Work of Charles Darwin Online*, darwin-online.org.uk.

2. Grigg, R., Darwin's arguments against God: How Darwin rejected the doctrines of Christianity, creation.com/darwinvgod, 13 June 2008.

3. Darwin, C., *On the origin of species by means of natural selection, or the preservation of favoured races in the struggle for life*, 1st edition, John Murray, London, 1859, p. 484.

4. Darwin, C., *On the origin of species by means of natural selection, or the preservation of favoured races in the struggle for life*, 2nd edition, John Murray, London, 1860, p. 484.

5. Roth, N., Review of Strick, J., *Darwin and the origin of life: A historical perspective*, American Association for the Advancement of Science—Origin of Life Workshop, 21–23 February, 2003; www.aaas.org.

6. In Roth's words (Ref. 5), "Darwin himself wanted not to alienate liberal Christians, noted Strick."

7. Grigg, R., *Darwin's bulldog—Thomas H. Huxley, Creation* 31(3):39–41, 2009.

8. Darwin, F., (ed.), *The Life and Letters of Charles Darwin*, John Murray, London, 1887, Vol. 2, p. 377.

9. Darwin erred in thinking that variety in offspring meant new features arising spontaneously, whereas we now know, following the pioneering work of the famous creationist scientist, Gregor Mendel, that the variety is due to the recombination of *existing* genes, not the creation of new ones. See Lester, L., *Genetics: no friend of evolution, Creation* 20(2):20–22, 1998.

10. For more on this topic, see Batten, D., Would Darwin be a Darwinist today? *Creation* **31**(4):48–51, 2009.

11. John Calvin, *Genesis*, translated and edited by John King, The Banner of Truth Trust, Edinburgh, p. 91, 1965.

12. Morris, H., *The Genesis Record*, Master Books, El Cajon, California, p. 74, 1976.

13. Adapted from Henry C. Thiessen, *Lectures in Systematic Theology*, Eerdmans, Michigan, pp. 156-7, 1977.

14. Denton, M., *Evolution: A Theory in Crisis,* Adler & Adler, Maryland, USA, pp. 329–330, 1985. See Sarfati, J., *By Design*, Creation Book Publishers, Brisbane, Australia, chapter 11, 2008.

15. Devine, D., Inexplicable insect metamorphosis, *Creation* 29(3):31–33, 2007; Return to text.

16. See Sarfati, J., *By Design,* chapter 13, Creation Book Publishers, 2008.

17. Gillen, A., and Sherwin, F., The origin of bubonic plague, *Journal of Creation* 20(1):7–8, 2006;

18. Wieland, C., Diseases on the Ark, *Journal of Creation* **8**(1):16–18, 1994; <creation.com/diseases>.

19. Meltzer, D., How Columbus sickened the New World, *New Scientist* 136(1842):38–41, 10 October 1992.

20. insect metamorphosis, *Nature* 401(6752):447–452, 30 September 1999.

21. Goor, R. and N., *Insect metamorphosis: from egg to adult*, Atheneum, New York, pp. 6–8, 1990.

22. Wigglesworth, V.B., *Insect hormones*, Scientific Publications Department, Carolina Biological Supply Company, Oxford University Press, 1983.

23. Chapman, R.F., *The insects: structure and function*, The English Universities Press Ltd, London, pp. 410–415, 1969.

24. Burnie, D. and Wilson, D., *Animal*, Smithsonian Institution, Dorling Kindersley, New York, p. 551, 2001.

25. Fox, D., The Spinners, *New Scientist* 162(2183):38–41, April 1999. See p. 1 for the quote on which our title here is based.

26. How spiders make their silk, *Discover* 19(10):34, October 1998.

27. Stokstad, E., Spider genes reveal flexible design, *Science* 287(5457):1378, February 2000 | PMID: 10722376.

28. *Nature Australia* 26(7):5, Summer 1999–2000.

29. Pollen-eating spiders, *Creation* 22(3):5, 2000.

30. McKenna, P., Gold, *New Scientist* 193(2587):32, 20 January 2007.

31. Gramling, C., Gold mine deposited rapidly, *Geotimes*, December 2006, <www.geotimes.org/dec06/NN_gold.html>, 7 May 2007.

32. Simmons, S. and Brown, K., Gold in magmatic hydrothermal solutions and the rapid formation of giant ore deposit, *Science* 314(5797):288–291, 13 October 2006.

33. Wieland, C., Darwin's real message: Have you missed it? *Creation* 14(4):16–19, 1992.

34. For refutations of some famous 'arguments from imperfection', see Gurney, P., Is our 'inverted' retina really 'bad design'? *Journal of Creation* 13(1):37–44, 1999; and Woodmorappe, J., The Panda thumbs its nose at the dysteleological arguments of the atheist Stephen Jay Gould, *Journal of Creation* 13(1):45–48, 1999.

35. Gould, S.J. and Lewontin, R.C., The Spandrels of San Marco and the Panglossian paradigm: a critique of the adaptationist programmed, *Proceedings of the Royal Society of London B* 205:581–598, 1979.

36. Wieland, C., CMI's views on the Intelligent Design Movement, 30 August 2002.

37. Ey, L. and Batten, D., Weasel, a flexible program for investigating deterministic computer 'demonstrations' of evolution, *Journal of Creation* 16(2):84–88, 2002.

38. Fabry, M., Now you know: which came first, the chicken or the egg? time.com, 21 September 2016.

39. Attenborough, D., *Attenborough's Wonder of Eggs*, screened on BBC 2, 31 March 2018.

40. Science Buddies, Porous science: How does a developing chick breathe inside its egg shell? scientificamerican.com, 3 May 2012.

41. Athanasiadou, D., and 14 others, Nanostructure, osteopontin, and mechanical properties of calcitic avian eggshell, *Science Advances* 4(3) eaar3219, 2018 | doi: 10.1126/sciadv.aar3219.

42. McGill University, Cracking eggshell nanostructure: New discovery could have important implications for food safety, phys.org, 30 March 2018.

43. Chien, Y.C., and 3 others, Ultrastructural matrix-mineral relationships in avian eggshell, and effects of osteopontin on calcite growth *in vitro*, *J. Structural Biology*, 163(1):84–99, 2008 | doi: 10.1016/j.jsb.2008.04.008.

44. Davis, N., Scientists solve eggshell mystery of how chicks hatch, theguardian.com, 30 March 2018.

45. Catchpoole, D., What's in an Egg? Unscrambling the mysteries, *Creation* 24(3):41–43, 2002; creation.com/egg.

46. Land, M., Eyes with mirror optics, *Journal of Optics A: Pure and Applied Optics* **2**(6):R44–R50, 2000.

47. Sarfati, J., Lobster eyes brilliant geometric design, *Creation* 23(3):12–13, 2001; creation.com/lobster.

48. Land, M., Superposition images are formed by reflection in the eyes of some oceanic decapod Crustacea, *Nature* 263(5580):764–765, 1976.

49. Gaten, E., *Optics and* phylogeny: is there an insight? The evolution of superposition eyes in the Decapoda (Crustacea), *Contributions to Zoology* 67(4):223–235, 1998.

50. University of Leicester, Lobster telescope has an eye for X-rays, sciencedaily.com, 5 April 2006.

51. U.S. Department of Homeland Security, Eye of the Lobster, *S&T Spotlight,* 1(7), November 2007.

52. Hall, M., Lobster serves as model for new X-ray device, *USA Today*, 20 December 2007.

53. Bergman, J., Did eyes evolve by Darwinian mechanisms?, *J. Creation* 22(2):67–74, 2008; creation.com/eyes-evolve.

54. Dawkins, R., *The Blind Watchmaker*, W. W. Norton & Company Inc., New York, p. 1, 1996.

55. Duke-Elder, S., *System of Ophthalmology*, Henry Kimpton, London, vol. 1, p. 147, 1958.

56. Dawkins, R., *The Blind Watchmaker: Why the evidence of evolution reveals a universe without design.* W.W. Norton and Company, New York, p. 1986 93.

57. Hogan, M.J., Alvarado, J.A., Weddell, J.E., The Retina. In *Histology of the Human Eye*, pp. ,522–393 1971. W.B. Saunders, Philadelphia. As cited in Tasman W., Jaeger E.A. (eds), *Foundations of Clinical Ophthalmology*, Lippincott–Raven, New York, vol. 1, ch. 21, 1998.

58. Zinn, K.M., Benjamin-Henkind, J., Anatomy of the human retinal pigment epithelium, 1979. In Zinn, K.M., Marmot, M.F. (eds.), *The Retinal Pigment Epithelium*, Harvard University Press, Cambridge, MA, pp. 3–31. As cited in Tasman W., Jaeger E.A. (eds.), Ref. 4.

59. LaVail, M.M., Outer segment disc shedding and phagocytosis in the outer retina, *Trans. Ophthalmol. Soc. UK* 103:397, 1983.

60. Steinberg, R.H., Research update: report from a workshop on cell biology of retinal detachment,*Exp. Eye Res.* 43:696–706, 1986.

61. Törnquist, P., Alm, A., Bill, A., Permeability of ocular vessels and transport across the blood-retinal barrier, *Eye* 4: 303–309, 1990.

62. Grierson, I., Hiscott, P., Hogg, P., Robey, H., Mazure, A., Larkin, G., Development, repair and regeneration of the retinal pigment epithelium, *Eye* 8: 255–262, 1994.

63. Kennon Guerry, R., Ham, W.T., Mueller, H.A. Light toxicity in the posterior segment, 1998. In Tasman W., Jaeger E.A. (eds.), *Clinical Ophthalmology*, Lippincott-Raven, New York, vol. 3, ch. 37.

64. Young, R.W., The Bowman Lecture: Biological renewal: Applications to the eye, *Trans. Ophthalmol. Soc. UK* 102:42–67, 1982.

65. Geeraets, W.J., Williams, R.C., Chan, G., Ham, W.T., Guerry, D., Schmidt, F.H., The loss of light energy in retina and choroid, *Arch. Ophthalmol.* 64:158, 1960. As cited by Parver, L.M. *et al.,* Ref. 18.

66. Photon energy (E) is inversely proportional to wavelength (λ): $E = {}^{hc}/_{\lambda}$, where h is Planck's Constant and c is the speed of light in a vacuum.

67. Noell, W.K., Walker, V.S., Kang B.S. *et al.,* Retinal damage by light in rats, *Invest. Ophthalmol.* **5**:450, 1966.

68. Friedman, E., Kuwubara, T., The retinal pigment epithelium: IV. The damaging effects of radiant energy, *Arch. Ophthalmol.* 80:265–279, 1968.

69. Parver, L.M., Auker, C., Carpenter, D.O., Choroidal blood flow as a heat dissipating mechanism in the macula, *Am. J. Ophthalmol.* 89:641–646, 1980.

70. Parver, L.M., Auker, C., Carpenter, D.O., Choroidal blood flow: III. Reflexive control in the human, *Arch. Ophthalmol.* 101:1604, 1983.

71. Parver, L.M., Temperature modulating action of choroidal blood flow, *Eye* 5:181, 1991.

72. Wieland, C., Seeing back to front: Are evolutionists right when they say our eyes are wired the wrong way? *Creation* 18(2):38–40, 1996.

73. Anon., An eye for creation: an interview with eye disease researcher Dr G. Marshall, University of Glasgow, Scotland , *Creation* 18(4):19–21, 1996.

74. Osterberg, G., Topography of the layer of rods and cones in the human retina, *Acta Ophthalmol. (suppl.)* 6:1, 1935. As cited in Tasman W., Jaeger E.A. (eds), Ref. 4, vol. 1, ch. 21.

75. Curio, C.A., Allen, K.A., Topography of ganglion cells in human retina, *J. Comp. Neurol.* 300:5, 1990. As cited in Tasman W., Jaeger E.A. (eds), Ref. 4, vol. 1, ch. 19.

76. Schein, S.J., Anatomy of macaque fovea and spatial densities of neurons in foveal representation, *J. Comp. Neurol.* 269:479, 1988. Cited in: Tasman W., Jaeger E.A. (eds), ref. 4, vol. 1, ch. 19.

77. Streeten, B.W., Development of the human retinal pigment epithelium and the posterior segment, *Arch. Ophthalmol.* 81:383–394, 1969.

78. Gass, J.D.M., Müller Cell Cone, an Overlooked Part of the Anatomy of the Fovea Centralis, *Arch Ophthalmol.* 117:821–823, 1999.

79. Sabates, F.N., Applied laser optics: Techniques for retinal laser surgery, 1997. In Tasman W., Jaeger E.A. (eds), 1997. *Clinical Ophthalmology,* Lippincott-Raven, New York, vol. 1, ch. 69A.

80. Nussbaum, J.J., Pruett, R.C., Delori, F.C., Historic perspectives. Macular yellow pigment. The first 200 years, *Retina* 1:296–310, 1981.

81. Ham, W.T. Jr., Mueller, H.A., Ruffolo, J.J. Jr. *et al.*, Action spectrum for retinal injury from near-ultraviolet radiation in the aphakic monkey, *Am. J. Ophthalmol.* 93:299, 1982. [The term *aphakia* means absence of the lens in the eye. The eyes of monkeys were subjected to the experiments after their lenses had been removed.]

82. Boettner, E.A., Wolter, J.R., Transmission of the ocular media, *Invest. Ophthalmol. Vis. Sci* 1:776, 1962. Cited in: Tasman W., Jaeger E.A. (eds), Ref. 11, vol. 5, ch. 55.

83. Williams, G.C., *Natural Selection: Domains, Levels and Challenges*, Oxford University Press, Oxford, pp. 72–73, 1992.

84. Traquair, H.M., *An Introduction to Clinical Perimetry,* The C V Mosby Co., St Louis, 1938. Return to text.

85. Wertheim, Z., *Psychol. Physiol. Sinnes.* 7:172, 1894. Cited in: Duke-Elder, S. (ed.), Ref. 1, vol. 4, p. 611, 1968. Return

86. Diamond, J., Voyage of the Overloaded Ark, *Discover*, June, pp. 82–92, 1985.

87. Mollusks, *Encyclopædia Britannica* 24:296-322, 15th ed., 1992; quote on p. 321.

88. Hanlon, R.T., and Messenger, J.B., *Cephalopod Behaviour*, Cambridge University Press, Cambridge, New York, p. 19, 1996.

89. Budelmann, B.U., Cephalopod sense organs, nerves and brain, 1994. In Pörtner, H.O., O'Dor, R.J. and Macmillan, D.L., ed., *Physiology of cephalopod molluscs: lifestyle and performance adaptations*, Gordon and Breach, Basel, Switzerland, p. 15, 1994.

90. Sensory Reception, *Encyclopædia Britannica* 27:114–221, 15th ed., 1992; quote on p. 147.

91. Kröger, R.H.H. and Biehlmaier, O., Space-saving advantage of an inverted retina, *Vision Research* 49(18): 2318–2321, 9 September 2009 | doi:10.1016/j.visres.2009.07.001.

92. Franze, K. *et al.*, Müller cells are living optical fibers in the vertebrate retina, *Proc. Nat. Acad. Sci. USA*, 10.1073/pnas.0611180104, 7 May 2007 | doi:10.1073/pnas.0611180104.

93. Labin, A.M. and Ribak, E.N., Retinal glial cells enhance human vision acuity, *Physical Review Letters* 104, 16 April 2010 | doi: 10.1103/PhysRevLett.104.158102.

94. Labin, A.M. *et al.*, Müller cells separate between wavelengths to improve day vision with minimal effect upon night vision, *Nature Communications* 5(4319), 8 July 2014 | doi:10.1038/ncomms5319. Return to text. Estimates within the literature vary from 50–60 km/h (30–37 mph).

95. Giraffe—the facts: Taxonomy, evolution and scientific classification, giraffeconservation.org, accessed July 2011.

96. Conger, C., If a giraffe's neck has only seven vertebrae, how is it so flexible? animals.howstuffworks.com, accessed July 2011.

97. Pedley, T., Giraffes' Necks & Fluid Mechanics, abc.net.au, (from transcript of broadcast 25 October 2003).

98. Fascia is a connective tissue that, in this instance, lies directly under the skin and serves as a means to flexibly bond the skin to the muscle's underneath it.

99. Hofland, L., Giraffes … animals that stand out in a crowd, *Creation* 18(4):10–13, 1996; creation.com/giraffe.

100. Lönnig, W.-E., The Evolution of the Long-Necked Giraffe (*Giraffa camelopardalis* L.) what Do We Really Know? (Part 1), weloennig.de, accessed March 2006.

101. Dagg, A and Foster J., *The Giraffe: Its Biology, Behavior, and Ecology*, Robert E. Krieger Publishing Company. Malabar, Florida, 1982.

102. African wildlife: Giraffe Facts, southafrican-wildlife.blogspot.com, accessed July 2011.

103. Holmes, B., How giraffes deal with sky-high blood pressure, *Knowable Magazine*, bbc.com, 4 Aug 2021.

104. Baccouche, B. and Natterson-Horowitz, B., Giraffe myocardial hypertrophy as an evolved adaptation and natural animal model of resistance to diastolic heart failure in humans, *International Society For Evolution, Medicine, and Public Health Conference*, 15 Aug 2019.

105. O'Brien. H. and Bourke, J. Focus: Vascular 'safety net' doesn't protect the brains of giraffes from dangerous pressure changes, anatomytoyou.com, 20 Jan 2019; Physical and computational fluid dynamics models for the hemodynamics of the artiodactyl carotid rete, *J. Theoretical Biology* 386:121–131, 7 Dec 2015.

106. Aalkjær, C and Wang, T., The remarkable cardiovascular system of giraffes, *Annual Review of Physiology* 83:1–15, Feb 2021.

107. Choi, D.S.J. and Ellis, R., Multiple hearts in animals other than *Barosaurus, Lancet* 352(9129):744, 29 Aug 1998.

108. Seymour, R.S., Cardiovascular physiology of dinosaurs, *Physiology* 31:430–441, 2016.